Simplified Business Statistics Using SPSS

Statistics are used throughout businesses to present and analyse data and decide on best practice. *Simplified Business Statistics Using SPSS* provides a practical approach to these concepts and their applications in business, economics and other areas of data analytics. This book guides the reader though these concepts without assuming prior knowledge and is an ideal reference for business analytics students and researchers in related fields.

Features

- Includes simplified statistical contents and a step-by-step guide on how to apply statistical concepts by perform analysis using Statistical Package for Social Sciences together with an interpretation of the statistical analysis output
- Provides a wide range of data sets to be used for examples and illustrations
- Designed to be accessible to readers with varied backgrounds

Gabriel Otieno Okello is a member of the statistics faculty at the School of Science and Technology, United States International University, Nairobi, Kenya.

Simplified Business Statistics Using SPSS

Gabriel Otieno Okello

CRC Press
Taylor & Francis Group
Boca Raton London New York

CRC Press is an imprint of the
Taylor & Francis Group, an **informa** business

A CHAPMAN & HALL BOOK

First edition published 2023
by CRC Press
6000 Broken Sound Parkway NW, Suite 300, Boca Raton, FL 33487-2742

and by CRC Press
4 Park Square, Milton Park, Abingdon, Oxon, OX14 4RN

CRC Press is an imprint of Taylor & Francis Group, LLC

© 2023 Gabriel Otieno Okello

Library of Congress Cataloging-in-Publication Data

Names: Otieno Okello, Gabriel, author.
Title: Simplified business statistics using SPSS / Gabriel Otieno Okello.
Description: First edition. | Boca Raton : CRC Press, [2023] | Includes
bibliographical references and index.
Identifiers: LCCN 2022013060 (print) | LCCN 2022013061 (ebook) | ISBN
9781032265179 (hbk) | ISBN 9781032274294 (pbk) | ISBN 9781003292654 (ebk)
Subjects: LCSH: Commercial statistics--Data processing. | Social
sciences--Statistical methods.
Classification: LCC HF1017 .O77 2023 (print) | LCC HF1017 (ebook) | DDC
519.5--dc23/eng/20220718
LC record available at https://lccn.loc.gov/2022013060
LC ebook record available at https://lccn.loc.gov/2022013061

ISBN: 978-1-032-27429-4 (hbk)
ISBN: 978-1-032-26517-9 (pbk)
ISBN: 978-1-003-29265-4 (ebk)

DOI: 10.1201/9781003292654

Typeset in Times
by KnowledgeWorks Global Ltd.

Contents

Part II   Probability Concepts

Part III   Introduction to Statistical Inference Concepts and Methods

Part IV   Special Topics in Statistical Analysis

Part I

Introduction to Statistical Analysis

Part 1

Introduction to Statistical Analysis

1

Introduction to Statistics, Data and Analysis

1.1 Basic Statistical Concepts

Statistics is the process of collecting, analyzing, presenting and interpreting data with an aim of deriving conclusions. Statistics is also the process of learning from data. With statistics one is able to generate meaningful information from the collected data.

From the broad definition of statistics, we can state that statistics has two branches namely: *descriptive statistics* and *inferential statistics*.

Descriptive statistics is part of statistics concerned with the description and summarization of data. It also refers to the presentation of a body of data in the form of tables, charts, graphs and other forms of graphic display together with the measures.

Inferential statistics is part of statistics concerned with drawing of conclusions about the properties of the whole population from sample.

Statistical inference is the process of using data from the sample to make estimates and test hypothesis about the characteristic of the population. Inferential statistics has two branches: *parametric statistics* and *nonparametric statistics*.

Parametric statistics requires certain assumptions about the distribution of the data. It requires interval or ratio data.

In *Nonparametric statistics* there is no assumptions about the distribution of the data. It requires data that are nominal and ordinal.

Element is an entity from which data are collected, e.g. patients.

Population is the total collection of elements of interest in a particular study, e.g. COVID-19 patients in Kenya. There are two types of population: finite and infinite. *Finite* is a population that is known. For example, the total number of students in a given class is known. *Infinite* is a population that is not known. For example, the total number of people in a given place as at today is not known. If the population is known, then it is denoted by N.

Sample is the subgroup of the population, e.g. COVID-19 patients from Nairobi. Sample is denoted by n. Usually sample is estimated from population using sample size formula, sample size tables and using general rules of thumb for estimating the sample size.

Parameter is a numerical measure describing characteristics of the population.

Statistic is a numerical measure describing characteristics of a sample.

A *statistic* is usually used to estimate the *parameter*. Parameter is the truth we desire to find out.

1.2 Application of Statistics

Statistics is used today practically in every profession and by everyone. Just to mention few professional who use statistics:

- Students use statistics to make choices of which product to buy online.
- A household member uses statistics to determine the number of household commodities that can be purchased in a supermarket.
- Businessperson may use statistics to tally the daily sales and comparisons of stock products.

DOI: 10.1201/9781003292654-2

- Politicians will use statistics to determine the voting patterns and preferences among their constituents.
- Meteorologists use statistics to forecast weather patterns.
- Employers use statistics to know the workforce distribution of employees, e.g. distribution of employees in terms of gender, age, etc.
- Political analysts use statistics to forecast voting patterns.
- Pharmaceutical researchers use statistics to test the effectiveness of a new drug.
- Finance analysts and investors use statistics to guide their investments recommendations for a given business product.

1.3 Data and Classification of Data

Data are information. Data are collected facts and figures. Data can also be defined as the values of a variable.
Elements are entities on which data are collected.

A *variable* is a characteristic of interest for the element. A variable can also be defined as a characteristic that varies from one person or thing to another. Variables can be continuous or discrete and in some cases variables can be dependent, independent, intervening or moderating.

Data can be classified using four different ways:

- Scales of measurement, i.e. nominal, ordinal and scale (ratio and interval)
- Either as numeric or non-numeric
- Either as cross-sectional or time series
- Either as primary or secondary

1.3.1 Scales of Measurement

These are the levels of data measurements. There are four measurement scales: nominal, ordinal, interval and ratio (NOIR). Since we will be using Statistical Package for Social Sciences (SPSS), then we will only have the three measurement scales that will be used: nominal, ordinal and scale.

Scale: Any quantitative data (data that is numerical in nature), e.g. age, salary, weight, volume, temperature, etc.

Nominal: Any categorical data that is not following any order, e.g. colors, type of cars, religion, etc.

Ordinal: Any categorical data that can be ranked or is following some order, e.g. level of education, job ranks, etc.

1.3.2 Numeric and Non-numeric Data

Data can be classified as either numeric or non-numeric.

Numeric data are those data sets that are numerical in nature. They are also known as quantitative data, e.g. salary, weight, age, etc. In SPSS, this will be the numeric type of data.

Non-numeric data are those data sets that are non-numerical in nature. Non-numeric data can either be qualitative or categorical. Non-numeric data can be grouped/classified as either qualitative or categorical. Qualitative data are those data sets that are inform texts, comments from open-ended surveys, videos, interviews, focused group discussions etc. Categorical data are those data sets that are in groups or categories, e.g. gender, grades, race, age groups, ethnicity and marital status. In SPSS, this will be the string type of data.

1.3.3 Cross-Sectional and Time Series Data

Data can be classified as either cross-sectional or time series especially when analyzing data.

Cross-sectional data are data collected at the same or approximately the same point in time, e.g. administering a survey questionnaire to students and the responses are given each and every day until sufficient responses are attained.

Time series data are data sets collected sequentially over time, e.g. weekly number of class attendees, elections, census, etc.

1.3.4 Primary and Secondary Data

Data can be classified as primary or secondary especially during data collection.

Primary data are data that are collected for the first time. These are data that have never existed, e.g. when one goes to the field and collects data using questionnaires for the first time.

Secondary data are data that are collected from already existing sources, e.g. data from published records like census data.

For the purpose of SPSS, data will be classified as either numeric or non-numeric using the three measurement scales (nominal, ordinal and scale).

The type of data determines the analytical procedure/techniques that can be done on the stated data. The type of data also determines the choice of statistical software for data analysis. Once the data have been collected, they should be prepared for analysis. Data analysis helps to generate meaningful information from the collected data using statistical software.

1.4 Data Sources

Data can be obtained from existing sources (secondary) or for the first time (primary) using various data collection methods.

List of Formulas

Name	Formula
Sturge's rule for estimating number of classes	$k = 1 + 3.322 \times logN$
Class width/range/interval	$interval = \dfrac{highest - lowest}{k}$
Relative frequency	$Relative\ frequency = \dfrac{Frequency\ of\ the\ class}{n}$
Arithmetic mean	$\bar{X} = \dfrac{\Sigma X_i}{n}$
Geometric mean	$\bar{X}_G = \sqrt[n]{X_1.X_2...X_n}$
Harmonic mean	$\bar{X}_H = \dfrac{N}{\Sigma\left(\dfrac{1}{X}\right)}$
Position of the percentile of interest	$i = \left(\dfrac{p \times n}{100}\right)$
Absolute deviation	$AD = \dfrac{\Sigma\left|X_i - \bar{X}\right|}{n}$
Variance	$s^2 = \dfrac{\Sigma\left(X_i - \bar{X}\right)^2}{n-1}$
Standard deviation	$s = \sqrt{\dfrac{\Sigma\left(X_i - \bar{X}\right)^2}{n-1}}$

Name	Formula
Interquartile range	$IQR = Q_3 - Q_1$
Quartile deviation	$Quartile\ Deviation = \dfrac{Q_3 - Q_1}{2}$
Coefficient of variation	$C.V. = \dfrac{s}{\overline{X}} \times 100$
Skewness (Pearson)	$S_k = \dfrac{Mean - Mode}{Standard\ Deviation}$
Skewness (Pearson)	$S_k = \dfrac{3(Mean - Median)}{Standard\ Deviation}$
Bowleys coefficient of skewness	$S_k = \dfrac{Q_3 + Q_1 - 2Q_2}{Q_3 - Q_1}$
Skewness based on moment	$\beta_1 = \dfrac{\mu_3^2}{\mu_2^3}$
Moment coefficient of skewness	$\gamma_1 = \sqrt{\beta_1} = \dfrac{\mu_3}{\mu_2^{3/2}}$
Second central moment	$\mu_2 = \dfrac{\sum(x_i - \overline{x})^2}{n}$
Third central moment	$\mu_3 = \dfrac{\sum(x_i - \overline{x})^3}{n}$
Fourth central moment	$\mu_4 = \dfrac{\sum(x_i - \overline{x})^4}{n}$
Kurtosis based on moment	$\beta_2 = \dfrac{\mu_4}{\mu_2^2}$
Moment coefficient of kurtosis	$\gamma_2 = \beta_2 - 3$
Pearson product moment coefficient of correlation	$r_{xy} = \dfrac{\dfrac{1}{n}\sum xy - \overline{x}\,\overline{y}}{\sqrt{\left(\dfrac{1}{n}\sum x^2 - \overline{x}^2\right)\left(\dfrac{1}{n}\sum y^2 - \overline{y}^2\right)}}$
Spearman coefficient of correlation	$\rho(x, y) = 1 - \left(\dfrac{6\sum d_i^2}{n(n^2 - 1)}\right)$
Additive law for mutually exclusive events	$P(A \cup B) = P(A) + P(B)$
Additive law for not mutually exclusive events	$P(A \cup B) = P(A) + P(B) - P(A \cap B)$
Conditional probability	$P(A \backslash B) = \dfrac{P(A \cap B)}{P(B)}$
Independent law	$P(B \mid A) = P(B)$
Independent events	$P(A \cap B) = P(A) \times P(B)$
Binomial distribution	$P(X = x) = C_x^n p^x (1 - p)^{n-x}, \ for\ x = 1, 2, \ldots, n$

Name	Formula
Mean for binomial distribution	$E[X] = \mu = np$
Variance for binomial distribution	$\sigma^2 = np(1-p)$
Poisson distribution	$P(X = x) = \dfrac{e^{-\mu}\mu^x}{x!}$
Uniform distribution	$f(x) = \dfrac{1}{\beta - \alpha}$, $for -\infty < \alpha \le x \le \beta < \infty$
Mean for uniform distribution	$E[X] = \dfrac{\alpha + \beta}{2}$
Variance for uniform distribution	$Var[X] = \dfrac{(\beta - \alpha)^2}{12}$
Normal distribution	$f(x) = \dfrac{1}{\sigma\sqrt{2\pi}} e^{\left\{ -\frac{1}{2}\left(\frac{x-\mu}{\sigma}\right)^2 \right\}}$, $for -\infty < x < \infty$
Z statistics	$Z = \dfrac{X - \mu}{\sigma}$
t distribution	$f(x:v) = \dfrac{1}{\sqrt{\pi v}} \dfrac{\Gamma\left(\dfrac{v+1}{2}\right)}{\Gamma\left(\dfrac{v}{2}\right)} \left(1 + \dfrac{x^2}{v}\right)^{-(v+1)/2}$
t statistics	$t = \dfrac{\bar{x} - \mu}{s/\sqrt{n}}$
F statistics	$F = \dfrac{s_1^2}{s_2^2}$
F distribution	$f(x) = \left[\dfrac{\Gamma\left(\dfrac{n+m}{2}\right)}{\Gamma\left(\dfrac{n}{2}\right)\Gamma\left(\dfrac{m}{2}\right)} \right] (n)^{\frac{n}{2}} (m)^{\frac{m}{2}} \dfrac{x^{\frac{n}{2}-1}}{(m+nx)^{\frac{n+m}{2}}}$
Chi-square distribution	$f(x;k) = \dfrac{1}{\Gamma\left(\dfrac{k}{2}\right)} \left(\dfrac{1}{2}\right)^{k/2} x^{\frac{k}{2}-1} e^{-\frac{1}{2}x}; x \ge 0$
Standard error of mean	$\sigma_{\bar{x}} = \sqrt{\dfrac{N-n}{N-1}} \left(\dfrac{\sigma}{\sqrt{n}}\right)$
Standard errors of mean	$\sigma_{\bar{x}} = \dfrac{\sigma}{\sqrt{n}}$
Z statistics (mean)	$Z = \dfrac{\bar{X} - \mu}{\sigma_{\bar{x}}}$
Proportion	$\bar{p} = \dfrac{x}{n}$
Standard error of proportion	$\sigma_{\bar{p}} = \sqrt{\dfrac{N-n}{N-1}} \sqrt{\dfrac{p(1-p)}{n}}$

Name	Formula
Standard error of proportion	$\sigma_{\bar{p}} = \sqrt{\dfrac{p(1-p)}{n}}$
Z statistics (proportion)	$Z = \dfrac{\bar{p} - E(p)}{\sigma_{\bar{p}}}$
Interval Estimate	$I.E. = P.E. \pm M.E.$
Margin of error	$M.E. = Z_{\alpha/2} \times S.E.$
Margin of error	$M.E. = t_{\alpha/2} \times S.E.$
Standard error of mean (one mean)	$S.E. = \dfrac{s}{\sqrt{n}}$
Standard error or mean (two means)	$S.E. = s_p \sqrt{\dfrac{1}{n_1} + \dfrac{1}{n_2}}$
Pooled standard deviation (two means)	$s_p = \sqrt{\dfrac{(n_1-1)\sigma_1^2 + (n_2-1)\sigma_2^2}{n_1 + n_2 - 2}}$
Standard error of mean (two means)	$S.E. = \sqrt{\left(\dfrac{\sigma_1^2}{n_1}\right) + \left(\dfrac{\sigma_2^2}{n_2}\right)}$
Pooled standard deviation (two means)	$s_p = \sqrt{\dfrac{(n_1-1)s_1^2 + (n_2-1)s_2^2}{n_1 + n_2 - 2}}$
Degree of freedom (two means)	$df = n_1 + n_2 - 2$
Standard error of mean (two means)	$S.E. = \sqrt{\left(\dfrac{s_1^2}{n_1}\right) + \left(\dfrac{s_2^2}{n_2}\right)}$
Degree of freedom (two means)	$df = \dfrac{\left(\dfrac{s_1^2}{n_1} + \dfrac{s_2^2}{n_2}\right)^2}{\left(\dfrac{1}{n_1-1}\right)\left(\dfrac{s_1^2}{n_1}\right)^2 + \left(\dfrac{1}{n_2-1}\right)\left(\dfrac{s_2^2}{n_2}\right)^2}$
Standard error of proportion (two proportions)	$S.E. = \sqrt{p(1-p) \times \left[\dfrac{1}{n_1} + \dfrac{1}{n_2}\right]}$
Pooled proportion (two proportions)	$p = \dfrac{p_1 \times n_1 + p_2 \times n_2}{n_1 + n_2}$
Standard error of proportion (two proportions)	$S.E. = \sqrt{\dfrac{p_1(1-p_1)}{n_1} + \dfrac{p_2(1-p_2)}{n_2}}$
Standard error of mean (paired samples)	$S.E. = \dfrac{s_d}{\sqrt{n}}$
Degree of freedom (one sample)	$df = n - 1$
Sample size formula (mean)	$n = \dfrac{Z_{\alpha/2}^2 \sigma^2}{E^2}$
Sample size formula (proportion)	$n = \dfrac{Z^2 pq}{E^2}$

Name	Formula
One sample Z-test (mean)	$z - test = \dfrac{\bar{x} - \mu_0}{S.E.}$
One sample t-test	$t - test = \dfrac{\bar{x} - \mu_0}{S.E.}$
One sample Z-test (proportion)	$Z - test = \dfrac{p - p_0}{S.E.}$
Standard error of proportion	$S.E. = \sqrt{\dfrac{p_0(1 - p_0)}{n}}$
Two sample Z-test (means)	$z - test = \dfrac{(\bar{x}_1 - \bar{x}_2) - \mu_0}{S.E.}$
Two sample t-test (means)	$t = \dfrac{(\bar{x}_1 - \bar{x}_2) - \mu_0}{S.E.}$
Two sample Z-test (proportions)	$Z - test = \dfrac{(p_1 - p_2) - p_0}{S.E.}$
Paired t-test	$t = \dfrac{\bar{d} - \mu_0}{S.E.}$
F-test (ANOVA)	$F = \dfrac{MSTR}{MSE}$
Mean square treatment (ANOVA)	$MSTR = \dfrac{SSTR}{k - 1}$
Sum of squares treatment (ANOVA)	$SSTR = n_1(\bar{x}_1 - \bar{x})^2 + n_2(\bar{x}_2 - \bar{x})^2 + \cdots + n_k(\bar{x}_k - \bar{x})^2$
Mean square error (ANOVA)	$MSE = \dfrac{SSE}{n - k}$
Sum of squares error (ANOVA)	$SSE = (n_1 - 1)s_1^2 + (n_2 - 1)s_2^2 + \cdots + (n_k - 1)s_k^2$
Expected frequency	$E_i = np_i$
Chi-square test statistics	$\chi^2 = \sum\limits_{i=1}^{k} \dfrac{(O_i - E_i)^2}{E_i}$
Degree of freedom (chi-square)	$df = k - 1$
Expected frequency	$E = \dfrac{\left(\begin{array}{c} \textit{Row Total of a given Obs Value} \times \\ \textit{Column Totals of a given Obs Value} \end{array} \right)}{\textit{Overall Total}}$
Degree of freedom (Chi-square)	$df = (r - 1) \times (c - 1)$
Regression equation (simple)	$y = \beta_0 + \beta_1 X + \varepsilon$
Regression equation (multiple)	$y = \beta_0 + \beta_1 X_1 + \beta_2 X_2 + \cdots + \beta_n X_n + \varepsilon$
Regression slope (simple)	$\beta_1 = \dfrac{\sum (x_i - \bar{x})(y_i - \bar{y})}{\sum (x_i - \bar{x})^2}$

Name	Formula
Regression constant (simple)	$\beta_0 = \bar{y} - b_1\bar{x}$
Regression slope (simple)	$\beta_1 = \dfrac{\sum xy - \dfrac{(\sum x \sum y)}{n}}{\sum x^2 - \dfrac{(\sum x)^2}{n}}$
Coefficient of determination (R Squared)	$r^2 = \dfrac{SSR}{SST}$
Total sum of squares (regression)	$SST = \sum(y_i - \bar{y})^2$
Regression sum of squares (regression)	$SSR = \sum(\hat{y}_i - \bar{y})^2$
Error sum of squares (regression)	$SSE = \sum(y_i - \hat{y}_i)^2$
t-test statistics (regression)	$t = \dfrac{\beta_1}{s_{b_1}}$
Standard error for regression slope	$s_{b_1} = \dfrac{s}{\sqrt{\sum(x_i - \bar{x})^2}}$
Total squared deviation from mean	$\sum(x_i - \bar{x})^2 = \sum x^2 - \dfrac{(\sum x)^2}{n}$
Standard deviation (regression)	$s = \sqrt{MSE} = \sqrt{\dfrac{SSE}{n-2}}$
Degree of freedom (regression t-test)	$df = n - 2$
Interval estimate for regression slope	$\beta_1 \pm t_{\alpha/2} \times s_{b_1}$
F-test statistics (regression)	$F = \dfrac{MSR}{MSE}$
Mean square error (regression)	$MSE = \dfrac{SSE}{n-2}$
Mean square regression (simple regression)	$MSR = \dfrac{SSR}{Number\ of\ independent\ variable = 1}$
Degree of freedom (regression F-test)	$df = \left\{ \begin{array}{l} Numerator = 1 \\ Denominator = n - 2 \end{array} \right\}$
Adjusted R Squared	$R_{Adj}^2 = 1 - \left[\dfrac{(1 - R^2)(n - 1)}{n - k - 1} \right]$
Z statistics (sign rank)	$z = \dfrac{(x + 0.5) - \dfrac{n}{2}}{\dfrac{\sqrt{n}}{2}}$
Mann Whitney U Test	$U_1 = n_1 n_2 + \dfrac{n_1(n_1 + 1)}{2} - R_1$ $U_2 = n_1 n_2 + \dfrac{n_2(n_2 + 1)}{2} - R_2$

Name	Formula
Kruskal Wallis test	$H = \dfrac{12}{N(N+1)} \sum \dfrac{T_i^2}{N_i} - 3(N+1)$
Control limits for $\bar{X}$ chart	$CL_{\bar{X}} = \bar{\bar{X}} \pm 3\dfrac{\hat{\sigma}}{\sqrt{n}} = \bar{\bar{X}} + A_2\bar{R}$
Lower control limit for R chart	$UCL_{\bar{R}} = D_4\bar{R}$
Upper control limit for R chart	$LCL_{\bar{R}} = D_3\bar{R}$
Lower control limit for S chart	$UCL_{\bar{S}} = B_4\bar{S}$
Upper control limit for S chart	$LCL_{\bar{S}} = B_4\bar{S}$
Control limits for p chart	$CL_p = \bar{p} \pm 3\sqrt{\dfrac{\bar{p}(1-\bar{p})}{n}}$
Control limits for np chart	$UCL_{np} = \bar{p} \pm 3\sqrt{n\bar{p}(1-\bar{p})}$
Control limits for c chart	$UCL_c = \bar{c} \pm 3\sqrt{\bar{c}}$
Control limits for u chart	$UCL_u = \bar{u} \pm 3\dfrac{\sqrt{\bar{u}}}{n_i}$

Practice Exercise

1. State the different data that you can collect in an organization. Thereafter, state the different measurement scales for those data sets that you will collect.
2. State other ways in which different people and professions use statistics.
3. Describe the different limitations of statistics.

2

Collecting and Preparing Data for Analysis

2.1 Data Collection Methods

Data collection is a process of gathering relevant information in a systematic way so as to answer the relevant research questions of interest. One must collect data so that collected data can be analyzed to generate meaningful information to answer the specific questions of interest. The commonly used methods for collecting data are: questionnaires, experiments, interviews, focused group discussions (FGDs) and observation check lists. There are several reasons that may lead to one choosing a particular data collection method.

Questionnaires are set of structured questions that are used to collect a given information. Questionnaires can either be open ended or closed ended and they can be administered physically or electronically. Questionnaires are quick and cheap method of data collection. However, when a questionnaire is used there may be high nonresponses since some questions maybe unanswered or misunderstood. Questionnaires can be used to collect quantitative/numerical data plus some qualitative data especially open ended.

An *Experiment* is an act that can generate information to understand the cause and effect relationships in a controlled environment. Experiments can be conducted in the field or in the laboratory. Data collected from experiment are very reliable but may be time consuming and expensive. Experiments can be used to collect quantitative/numerical, categorical and qualitative data.

An *Interview* is a set of structured or unstructured conversation aimed at collecting information by probing the respondents/asking questions from the respondents. Interviews may be one on one or face to face or even online or telephone interview where the interviewer obtains information from the interviewee. Interviews are flexible and comprehensive since it allows the researcher to collect as much information. However, interviews may be time consuming. Interviews are mostly used to collect qualitative data.

FGD is a moderated discussion of issues by a group of selected individuals who understand the issues being discussed. In FGDs we usually have 3–10 group members with a moderator. In FGDs individuals are asked opinions and their attitudes toward a given issue, products or services. FGDs may be single focus group or two-way focus group or online focus group. Focus groups helps to clarify and understand some issues. However, FGDs may be more expensive to administer and results may also be biased. FGDs are mainly used to collect qualitative data.

Observation guide is a checklist that items that can be collected by observing the activities of the respondents. Observations may be structured or unstructured. Using observation for data collection is cheap, flexible and fast to administer. However, it may be difficult to analyze data collected using observation. The collected data via observation may also be biased. Observations can be used to collect quantitative/numerical and qualitative data depending on how the data are collected.

The above data collection methods can be used to collect either qualitative or quantitative data.

The type of data (either quantitative or qualitative) determines the type of analysis and the choice of statistical software.

The collected data must be analyzed to be able to generate meaningful information form the data. This is usually done suing statistical software and the process is known as data analysis.

DOI: 10.1201/9781003292654-3

2.2 Preparing Data for Analysis

The collected data must be put into a format or structure that is analyzable. This will depend on the type of statistical software that will be used to conduct data analysis. Data preparation is the process of cleaning and transforming the raw collected data before conducting the data analysis. When data are properly prepared then analysis will be more efficient with limited errors and inaccuracies. Data preparation process can be a tedious process since it involves fixing errors and enriching the data so that it has minimal errors.

Steps for data preparation are as follows:

1. Data gathering – this is where you get all the right data form all the available sources.
2. Understanding data – at this stage, one tries to understand the data, what the data will do and for what purpose will the collected data be useful.
3. Cleaning the data – at this stage outliers in the data are deleted, fill the missing values and conforming the data into the standardized pattern. One may also go ahead and test for errors.
4. Transforming data – at this stage, data are updated into a format that is more easily understood by many people. For example, if data set had coded values then you can assign the values into the codes so that people can relate the meaning of the coded values.
5. Store the data – once the data has been prepared, it can be stored or transferred into other platforms or softwares for further use and analysis.

2.3 Data Analysis

The process of deriving meaningful information from the collected data using statistical software. The type of data (either quantitative or qualitative) determines the type of analysis and the choice of statistical software.

Statistical software is a specialized computer program designed to conduct statistical analyses for the collected data.

Statistical softwares can also be classified as follows:

- Open source statistical softwares
- Proprietary statistical softwares
- Public domain statistical softwares
- Freeware statistical softwares
- Microsoft Excel ad-ons

There are several statistical softwares. The common statistical softwares are:

- SPSS
- STATA
- SAS
- Epi Info
- Eviews
- Matlab
- Minitab
- R Statistical Computing
- Excel
- NVivo
- Atlas.ti and many more

The choice of statistical softwares depends on:

- The type of data whether qualitative or quantitative
- The type of analyses to be done
- Cost of statistical software
- Institutional preferences
- Learning curve
- Type of studies being conducted
- The technique of analysis to be used – click the drop down or use computing

Because of the *Business Statistics* aspect we will use *SPSS* since it's easier to learn how to use SPSS and it's a common statistical software with varied techniques for analysis. Most learning institutions also prefer using SPSS.

Practice Exercise

1. Looking at the different data collection methods, why would one choose a given data collection method against another data collection method?
2. Compare the different statistical softwares (SPSS, STATA, R, Nvivo, Atlas.ti). Why is one statistical software preferred to another?
3. Give examples of data in your organization that can be collected using the different types of questionnaires, experiments, interviews, focused group discussions (FGDs) and observation check lists.

3

Getting Started with SPSS

3.1 Starting SPSS

Statistical Package for Social Sciences (SPSS) current version is Version 27. Before starting SPSS, ensure that SPSS is installed in your computer.

Click Start, then click the IBM SPSS Statistics 27 folder in the list of programs, then click IBM SPSS Statistics 27 Icon and it will give a spreadsheet that has two SPSS windows at the bottom right: *Data View* and *Variable View*.

Alternatively: At the start button where there is a search window, type SPSS then click IBM SPSS Statistics 27 from the list of programs.

Once you have started your SPSS, a dialogue box may appear. Select Type in data then OK (Figures 3.1 and 3.2).

Data View: where you key in the data

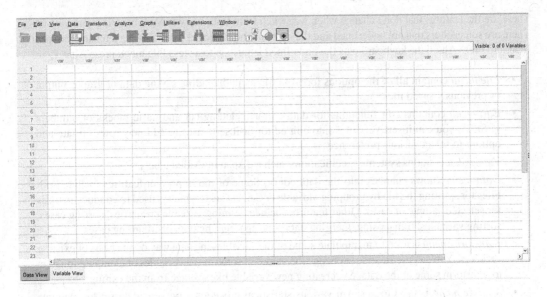

FIGURE 3.1 SPSS Data View Window

DOI: 10.1201/9781003292654-4

Variable View: where you create the variables

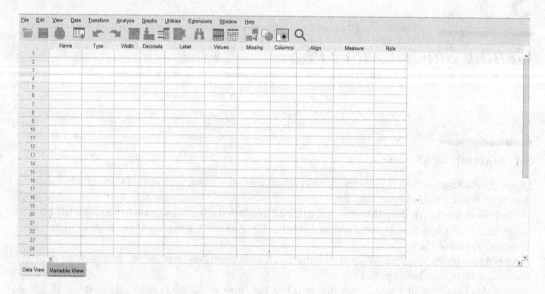

FIGURE 3.2   SPSS Variable View Window

3.1.1 Menus and Icons in SPSS

These are the features that don't change when we switch from Data View to Variable View and vice versa. There are some letters that are underlined and they present the keyboard short cut for exercising the specific function. Press Alt on the keyboard plus the underlined letter, e.g. to access File menu press Alt + F.

- *File menu* contain all of the options found in file menu including saving data or graphs output or previously saved files
- *Edit menu* contains edit functions for data editor including cut and paste blocks of numbers in SPSS, you can insert variables into edit editor or insert cases to add new rows of data and options to select various preferences
- *View menu* contains systems specifications such a displaying value labels
- *Data menu* allows changes into the data editor. Some of the important features include Spilt Files for splitting the file by grouping variable or Select Cases which is used to conduct analysis for selected sample of cases. Other features include defining variable properties, sorting cases, sorting variables, transposing, merging data sets, aggregating data or weighing cases
- *Transform menu* is used to manipulate some of the variables in some way on the basis of existing variables. You can use Recode to change the values of some of the variables or Compute function to transform some of the data, e.g. creating new variable like average from the exiting variables
- *Analyze menu* is used to perform various statistical procedures on your data set for example descriptive statistics, compare means, correlate, regression, etc.
- *Graphs menu* is used to access the chart builder and where you can create different types of graphs
- *Utilities* menu there is an option of data file comments for commenting on the data set and once can write notes from where the data came from and the date they were collected
- Extensions menu where we can install extension
- Windows menu allows you to switch from Window to Window, e.g. switching from Data View to Output
- Help is an invaluable menu since it offers online help for both the systems and the statistical tests

At the top of Data View there are sets of icons that are short cuts to frequently used facilities in the menu.

- Open Data Document 📁 open/view previously saved file, or data sets in other formats like Excel, Stata etc.
- Save this document 💾 allows you to save file in different formats
- Print 🖨 enables you to print what you are working on
- Recall recently used dialogs 🔲 activates the last dialog boxes
- Undo and redo user action ↺ ↻
- Go to case 🔳 takes you to case
- Go to variables 📊 takes you to the variable
- Variable icon ▤ shows the variables in the data set
- Run descriptive statistics 📊 shortcut for performing descriptive statistics
- Find 🔍 helps to search for words or numbers in the data set
- Split file ▦ used when performing analysis by groups
- Select cases ▦ helps to analyze some portion of data
- Value Labels 🔖 displays or hide the values of any coded variables
- Use Variables Sets 🔵 enables one to use variable sets to be used

3.2 Data Entry and Manipulation in SPSS

3.2.1 Data Entry

There are three different ways of reading the data into SPSS.

1. Importing the already existing SPSS data into SPSS.
   - Below File, there is a Folder *(Open Data Document), click the folder* and a window will populate. Look for the location of where you had saved *SPSS data* (.sav) *and select the file* (.sav) *then click Open* (Figure 3.3)

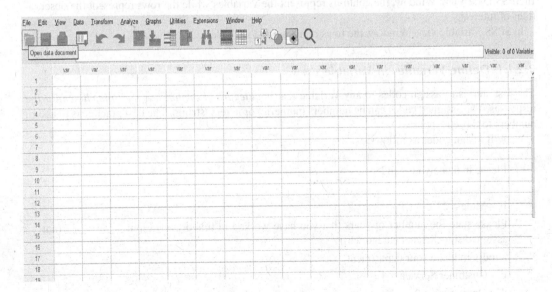

FIGURE 3.3 SPSS Open Data Document Menu Tab

2. Manually Enter Data into SPSS.

This is when you would like to use SPSS as a data entry platform and you would like to transfer data that you have collected using physical questionnaire/paper questionnaire. Let's use sample questionnaire to demonstrate how to create data entry screen into SPSS for manual data entry.

A questionnaire was administered to Employees of a Bank to determine their job satisfaction levels.

- Please indicate your gender
  - a. Female
  - b. Male
- Please state the number of years that you have worked in this organization: _____
- Kindly indicate your department
  - a. Customer Service
  - b. Operations
  - c. Card Support
  - d. Risk and Compliance
  - e. Finance and Investments
- Please indicate your level of agreement with the following statement about your work environment using a scale of 1–5, where 1 = Strongly Disagree, 2 = Disagree, 3 = Neutral, 4 = Agree and 5 = Strongly Agree.

Statement Describing Work Environment	Level of Agreement				
	1	2	3	4	5
a. Our office is cleaned on a daily basis					
b. Our supervisor always gets opinion from the employees					
c. Employees share tasks whenever they are given					
d. The working space in our office is sufficient					

- Please provide suggestions on how your organization can improve in terms of employee satisfaction: _____

In SPSS Data View Window, the columns represent the variables while the rows represent the observations of interest.

In SPSS Variable view Window, the rows represent the variables.

3.2.1.1 Define Variables in Variable View

The first step is to assign codes to any variable that is *categorical or that has multiple choice* for the responses. The assigned codes can be numbers (numeric) or letters (string). The best practice is to assign numbers to those responses.

We will assign codes as follows:

1. Please indicate your gender
   - a. Female = 1
   - b. Male = 2
2. Please state the number of years that you have worked in this organization: _____ (not assigned since it's not categorical)
3. Kindly indicate your department
   - c. Customer Service = 1
   - d. Operations = 2
   - e. Card Support = 3

    f. Risk and Compliance = 4

    g. Finance and Investments = 5

4. Please indicate your level of agreement with the following statement about your work environment using a scale of 1–5, where 1 = Strongly Disagree, 2 = Disagree, 3 = Neutral, 4 = Agree and 5 = Strongly Agree (the codes are already assigned)

5. Please provide suggestions on how your organization can improve in terms of employee satisfaction: _____ (not assigned since it's not categorical)

One we have assigned the codes to the questionnaire, we will go and create the five variables in the SPSS Variable View as follows:

Under column for Name: we give the abbreviation for the variable as follows:

    For question 1: gender

    For question 2: yrs

    For question 3: dept

    For question 4: qn4a, qn4b, qn4c and qn4d

    For question 5: suggest

Under column for Type: we select the type of data that we will key in (for categorical responses these are the codes) for the different variables as follows:

    For question 1: numeric (since we used numbers for the codes)

    For question 2: numeric (since we will be keying in number of years)

    For question 3: numeric (since we used numbers for the codes)

    For question 4: qn4a, qn4b, qn4c and qn4d – numeric (since we used numbers for the codes)

    For question 5: string (since we will be typing statements)

Under column for Width: we select the number of characters that we will key in (for categorical responses these are the codes) for the different variables as follows. Usually 8 is used as the default except for cases where we type statements

    For question 1: 8

    For question 2: 8

    For question 3: 8

    For question 4: qn4a, qn4b, qn4c and qn4d – 8

    For question 5: 1000

Under column for Decimals: we select the number of decimals for the data that we will key for the different variables as follows. For this questionnaire, the collected data will not have decimals. If you are keying in data with decimals, then select the number of decimals that your data set has.

    For question 1: 0

    For question 2: 0

    For question 3: 0

    For question 4: qn4a, qn4b, qn4c and qn4d – 0

    For question 5: 0

Under column for Labels: we type the complete description of the variables in full as follows.

    For question 1: Please indicate your gender

    For question 2: Please state the number of years that you have worked in this organization

    For question 3: Kindly indicate your department

    For question 4: qn4a – Our office is cleaned on a daily basis

    For question 4: qn4b – Our supervisor always gets opinion from the employees

    For question 4: qn4c – Employees share tasks whenever they are given

    For question 4: qn4d – The working space in our office is sufficient

For question 5: Please provide suggestions on how your organization can improve in terms of employee satisfaction (Figure 3.4)

	Name	Type	Width	Decimals	Label	Values	Missing	Columns	Align	Measure
1	gender	Numeric	8	0	Please indicate your gender	None	None	8	Right	Unknown
2	yrs	Numeric	8	0	Please state the number of years that you have worked in this organization	None	None	8	Right	Unknown
3	dept	Numeric	8	0	Kindly indicate your department	None	None	8	Right	Unknown
4	qn4a	Numeric	8	0	Our office is cleaned on a daily basis	None	None	8	Right	Unknown
5	qn4b	Numeric	8	0	Our supervisor always gets opinion from the employees	None	None	8	Right	Unknown
6	qn4c	Numeric	8	0	Employees share tasks whenever they are given	None	None	8	Right	Unknown
7	qn4d	Numeric	8	0	The working space in our office is sufficient	None	None	8	Right	Unknown
8	suggest	String	1000	0	Please provide suggestions on how your organization can improve in terms of employee satisfaction	None	None	8	Left	Nominal

FIGURE 3.4 SPSS Variable View Window with all the Variables from the Sample Questionnaire

Under column for Values: we type the codes for the categorical variables/variables with multiple choice responses (Figures 3.5 and 3.6).

For question 1:
Put Value 1 for Label Female
Put Value 2 for Label Male

For question 2: Select None since there are no codes

For question 3:
Put Value 1 for Label Customer Service
Put Value 2 for Label Operations
Put Value 3 for Card Support
Put Value 4 for Label Risk and Compliance
Put Value 5 for Label Finance and Investments

For question 4: qn4a –
Put Value 1 for Label Strongly Disagree
Put Value 2 for Label Disagree
Put Value 3 for Label Neutral
Put Value 4 for Label Agree
Put Value 5 for Label Strongly Agree

For question 4: qn4b –
Put Value 1 for Label Strongly Disagree
Put Value 2 for Label Disagree
Put Value 3 for Label Neutral
Put Value 4 for Label Agree
Put Value 5 for Label Strongly Agree

For question 4: qn4c –
Put Value 1 for Label Strongly Disagree

Put Value 2 for Label Disagree

Put Value 3 for Label Neutral

Put Value 4 for Label Agree

Put Value 5 for Label Strongly Agree

For question 4: qn4d –

Put Value 1 for Label Strongly Disagree

Put Value 2 for Label Disagree

Put Value 3 for Label Neutral

Put Value 4 for Label Agree

Put Value 5 for Label Strongly Agree

For question 5: Select None since there are no codes

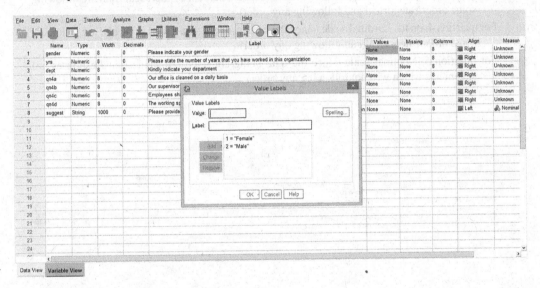

FIGURE 3.5 SPSS Variable View Window Illustrating Value Labels

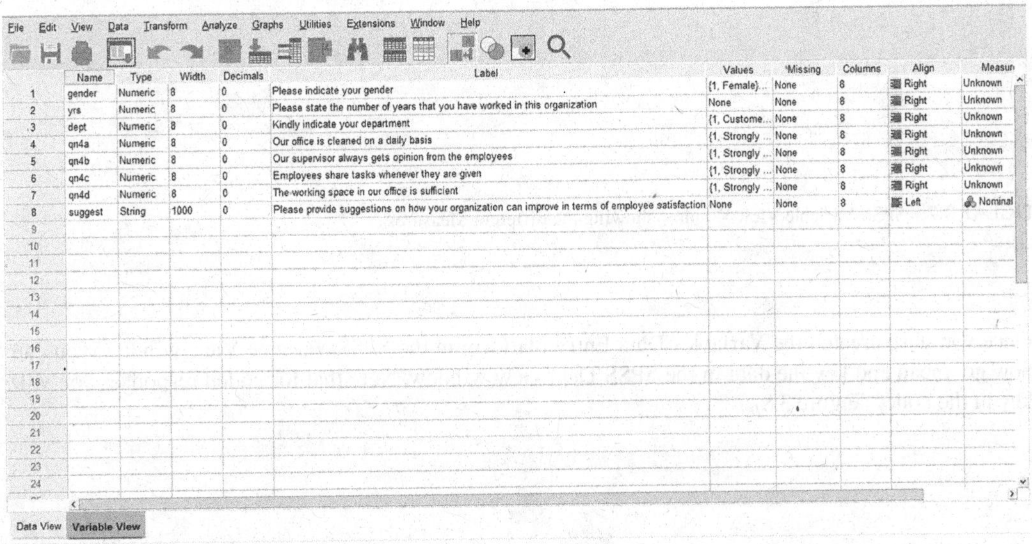

FIGURE 3.6 SPSS Variable View Window with Coded Variables

Under column for Measure: we select the appropriate measurement scale depending on the type of data being collected (Figure 3.7).

For question 1: Nominal scale (since we are collecting data on gender/categorical and not following any order)

For question 2: Scale (since we are collecting data on years which are quantitative/numerical)

For question 3: Nominal scale (since we are collecting data on departments/categorical and not following any order)

For question 4: qn4a – Ordinal scale (since the responses are on a Likert scale and are following some order/categorical and following any order)

For question 4: qn4b – Ordinal scale (since the responses are on a Likert scale and are following some order/categorical and following any order)

For question 4: qn4c – Ordinal scale (since the responses are on a Likert scale and are following some order/categorical and following any order)

For question 4: qn4d – Ordinal scale (since the responses are on a Likert scale and are following some order/categorical and following any order)

For question 5: Nominal scale (since the data being collected are suggestions/responses are statements which will not following any order)

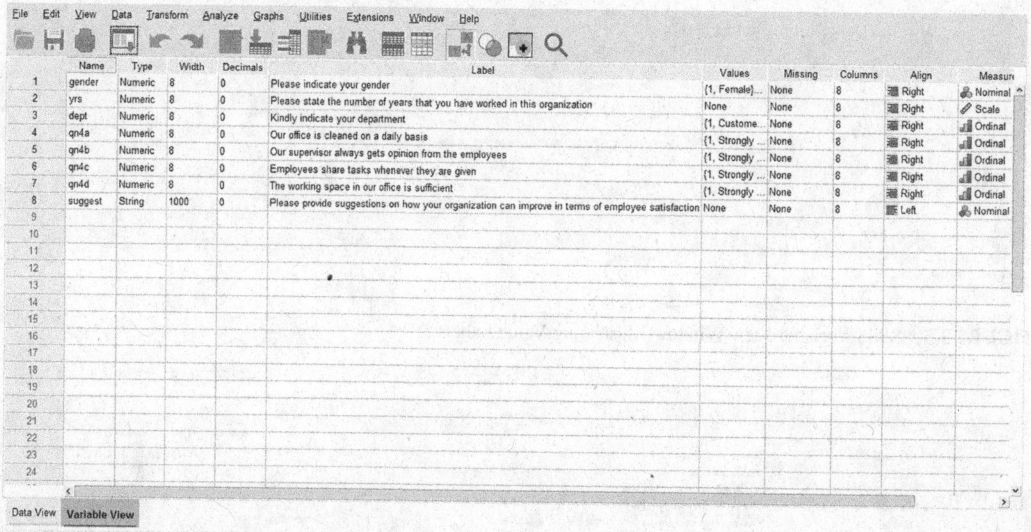

FIGURE 3.7   SPSS Variable View Window showing the different Measures

Once you have created the Variables/Data Entry platform in the SPSS Variable View window, you can now go ahead and key the data in the SPSS Data view window. Note that for coded responses you will key in the codes (Figure 3.8).

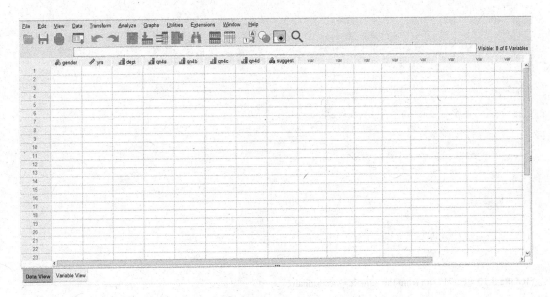

FIGURE 3.8   SPSS Data View Window with all the Variables from the Sample Questionnaire

3. Read Data from Excel, Text Data, SAS or Stata

Create the data entry screen in Excel by putting the variables in the column as shown in Figure 3.9.

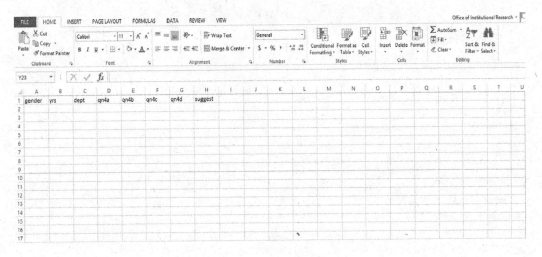

FIGURE 3.9   Excel with all the Variables from the Sample Questionnaire

Then key in the coded values for categorical/multiple response in the Excel data entry screen as shown in Figure 3.10.

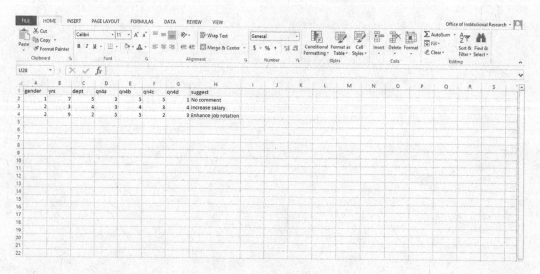

FIGURE 3.10 Excel Data from the Sample Questionnaire

Click File > Import Data > select Excel > there will be a pop up window for Open Data > (Figure 3.11).

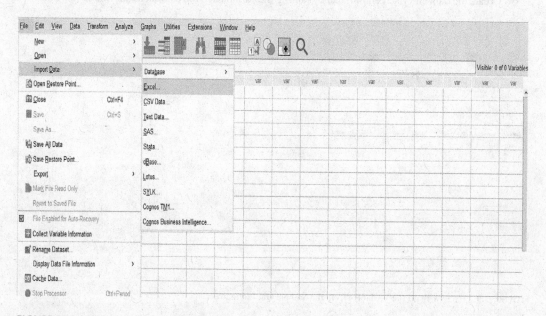

FIGURE 3.11 SPSS Menu for Importing Excel Data

Look for the location of where you want to import the appropriate data set in Excel and select the data to be imported > there will be a pop up window for Read Excel File (Figure 3.12).

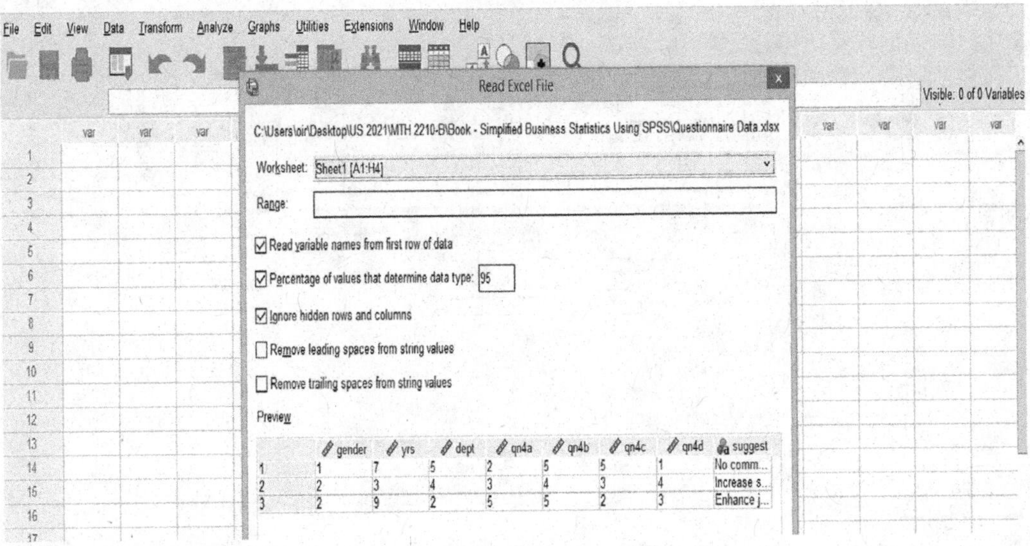

FIGURE 3.12   SPSS Pop Up Window for Read Excel File

Click OK and the Excel data will be imported into SPSS Data View Window (Figure 3.13).

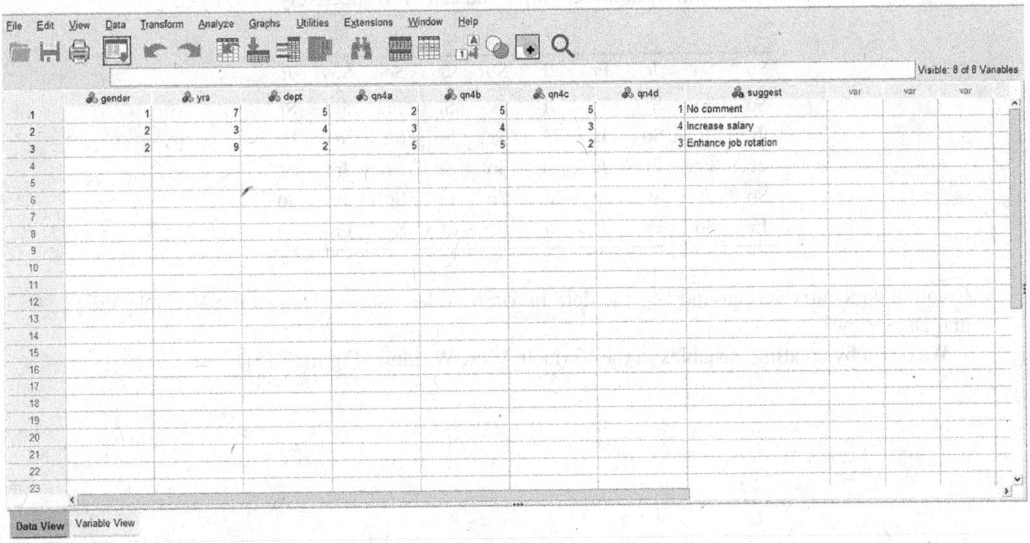

FIGURE 3.13   SPSS Data View for the Imported Excel Data

Go to SPSS Variable View insert the complete description of the variable under the column for Label and create the labels represented by the imported codes (Figure 3.14).

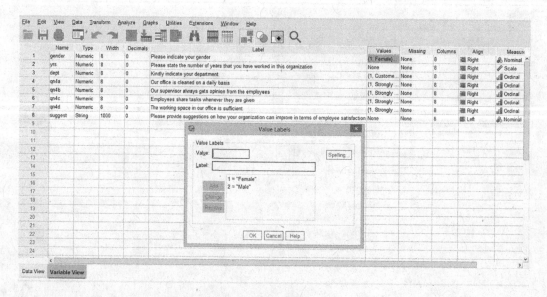

FIGURE 3.14   SPSS Variable View Window showing Value Labels

Example: University Class Levels Data

Consider the following data for the class levels of undergraduate university students, where Fr, So, Jr and Sr denote freshman, sophomore, junior and senior, respectively.

So	So	Jr	Fr	Jr	So	Jr	So	So	Jr
So	So	Sr	So	Jr	Jr	Sr	Fr	Sr	Sr
Jr	Jr	So	Jr	Fr	Sr	Jr	So	Sr	Fr
Jr	Fr	Fr	Jr	Sr	So	Sr	Sr	Jr	Fr
So	Jr	So	Sr	So	So	Fr	So	Jr	So
Fr	So	Fr	Jr	Jr	Fr	Jr	So	Fr	So

Create a data entry screen and key the data in SPSS. Take a screen shot of the Variable View and Data View.

We begin by creating variables in the Variable View Window (Figure 3.15).

Variable View

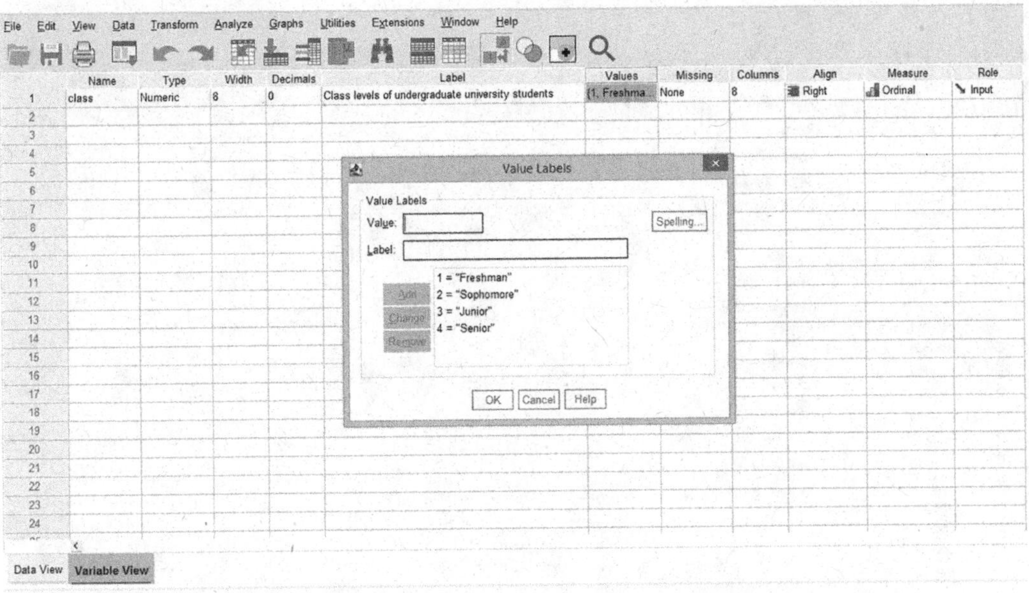

FIGURE 3.15 SPSS Variable View Window for Class Levels of Undergraduate University Students

Data View

We then key the data in the Data View Window (Figure 3.16).

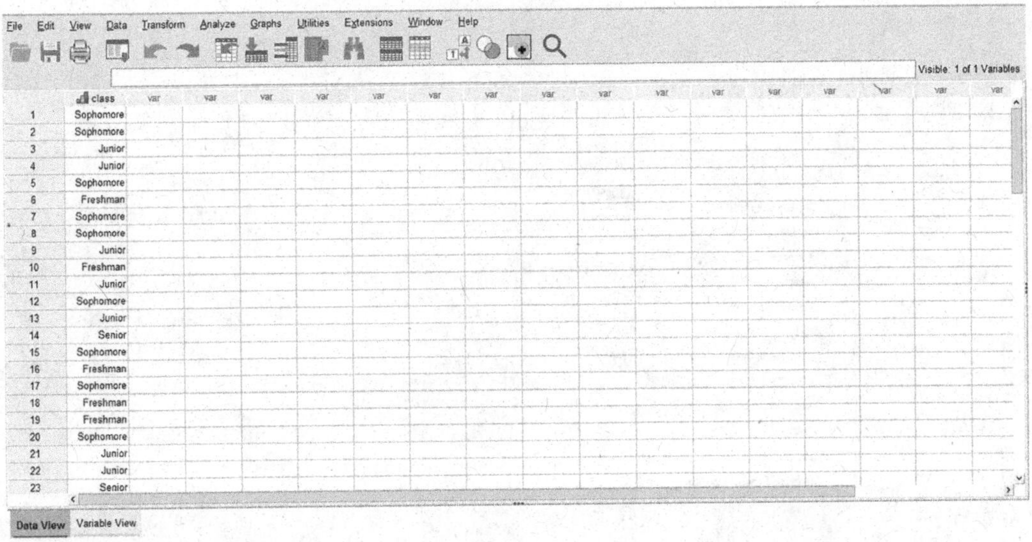

FIGURE 3.16 SPSS Data View Window for Class Levels of Undergraduate University Students

Use the Analyze Tab to conduct the analysis where we click and select the variables (Figure 3.17).

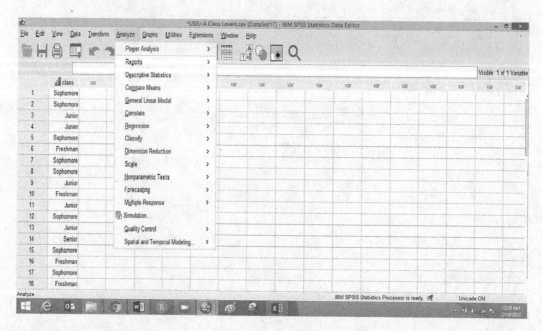

FIGURE 3.17   SPSS Menu for Performing Analysis

Conducting Basic Descriptive Statistics Analysis: Let us generate a *frequency table* for class levels of undergraduate university students (Figure 3.18).
We click *Analyze > Descriptive Statistics > Frequencies*

FIGURE 3.18   SPSS Menu for Performing Descriptive Statistics Analysis

Frequencies dialog box will pop up > select Class levels of undergraduate university students variable > select the variable for class levels of undergraduate university students and click the right arrow to move class levels of undergraduate university students into the Variable(s) Window > Ensure that Display Frequency Table and Create American Psychological Association (APA) Style Table is checked then click OK (Figure 3.19).

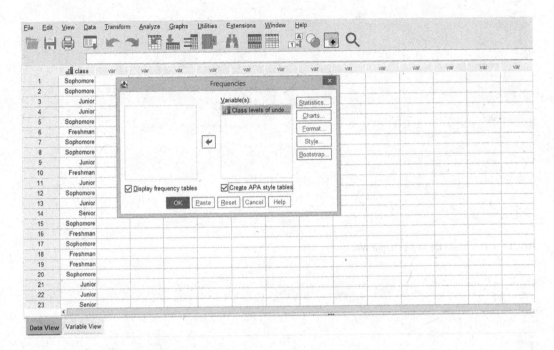

FIGURE 3.19 SPSS Pop Up Window for Frequencies

The SPSS output window will have the frequency table for undergraduate class levels (Figure 3.20).

Class Levels of Undergraduate University Students

	N	*%*
Freshman	12	20.0%
Sophomore	20	33.3%
Junior	18	30.0%
Senior	10	16.7%

FIGURE 3.20 Frequency Table for Class Levels of Undergraduate University Students

Interpretation: There were more sophomores (33%) in the undergraduate class levels.

Examining SPSS Output: The output of analyses is in the output window. Click Window then a drop down of active windows will pop up and then select Output (Figures 3.21 and 3.22).

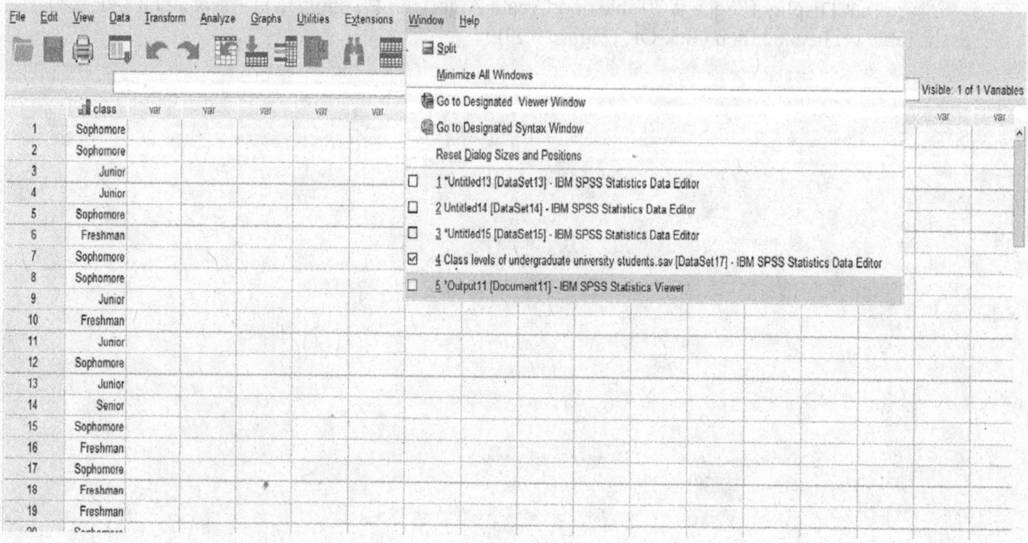

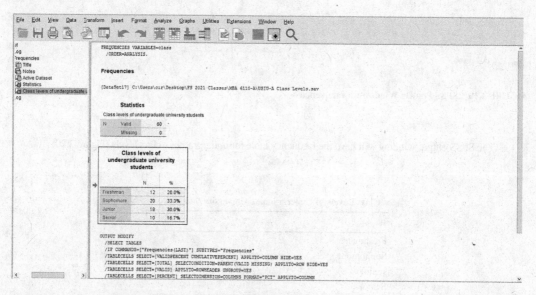

FIGURE 3.21 Menu for SPSS Output Window

FIGURE 3.22 SPSS Output Window displaying the Frequency Table

3.2.2 Menu for Data Management and Analysis

The SPSS Data menu tab for performing basic data manipulations. Click Data and all data management procedures will appear on the drop down (Figure 3.23).

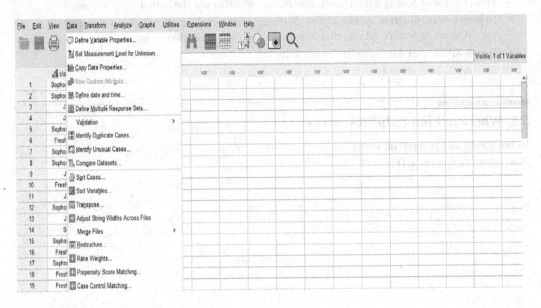

FIGURE 3.23 SPSS Data Management Menu

There is also Transform menu tab where once can still perform other data manipulations. Click Transform and there will be several data manipulation techniques on the drop down (Figure 3.24).

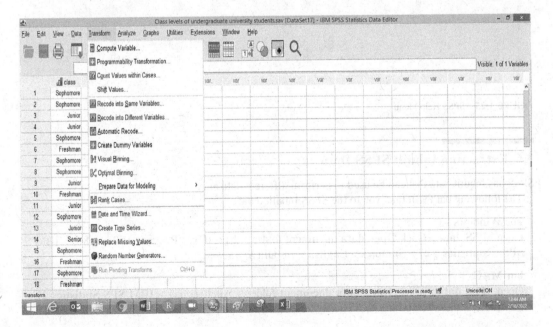

FIGURE 3.24 SPSS Transform Menu

3.4 Saving, Printing and Exiting SPSS

- To save data files, program syntax, output or charts, first go to the window of interest (Data View, Output or Syntax Editor) click *File* menu and either use the option *Save or Save As*
- To print output, program syntax, charts, or the data file, first go to the window of interest (Data View, Output or Syntax Editor) click File menu and then the option *Print*
- To exit SPSS, click File menu then click Exit

3.5 Where to Find Help/Resources

In-built help and printed manuals
- SPSS Help System (Figure 3.25).

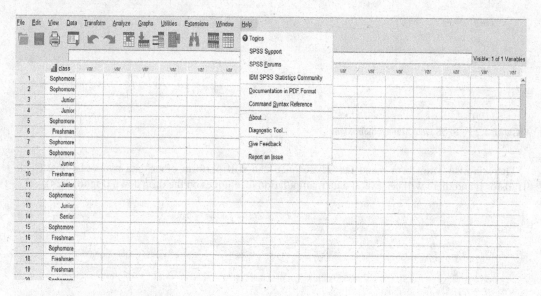

FIGURE 3.25   SPSS Help Menu

3.6 Extracting Inbuilt SPSS Data

SPSS has inbuilt data that can be used to perform different data analysis.
    The following path will take you to the SPSS inbuilt data:

For Windows
Open **SPSS data** > Click **Open data document** folder > drive **(C):** > **Program Files (x86)** > **IBM** > **SPS** > **Statistics** > **27** > **Samples** > **English** > In-built *SPSS data files in. sav extension.*

For MAC
Open **SPSS data** > Click **Open data document** folder > Applications > **IBM SPS Statistics 27** > **Samples** > **English** > In-built *SPSS data files in. sav extension.*

Practice Exercise

1. Design a questionnaire with five survey/research questions. The questionnaire should ensure that you are able to collect categorical (qualitative) and quantitative data from your organization. These questions should also have the three different measurement scales – nominal, ordinal and scale

2. Using the self-designed questionnaire in (1) above create a data entry screen in SPSS and enter ten responses for your questionnaire (take a screen shot of the variable view and data view)

3. You have been provided with "ATM withdrawal Data" in Excel. Transfer the data set into SPSS and name all the variables as shown, according to the description provided. Take a screen shot of the Variable View and Data View

44	27	31	36	40	38	32	39	22	34
61	36	24	45	38	43	32	28	31	55
37	34	20	23	34	47	25	31	57	61
34	30	43	22	37	26	31	40	25	33
61	54	59	55	53	44	46	29	42	29
42	31	24	35	125	29	34	21	29	45
60	58	52	58	59	58	51	14	36	44
31	30	134	26	45	24	40	43	52	30
35	35	34	20	42	34	27	40	59	31
47	42	56	57	49	41	43	58	47	25
30	41	28	14	40	36	38	24	21	26

Data description: Number of ATM withdrawals from one of the ATMs located in region A.

4. Using bankloan.sav data, select the appropriate variable of your choice and perform the data management procedure using Data and Transform menu

4

Descriptive Statistics Methods

4.1 Descriptive Statistics

Summaries of data which may be tabular, graphical or numerical are referred to as descriptive statistics. The analyzed data can be presented using tables (frequency tables), graphs/charts and numerical measures.

4.1.1 Tabular Presentation of Data Using SPSS

The analyzed data can be presented in form of a frequency table or frequency distribution. *Frequency distribution* is a tabular summary of data showing the items in groups or classes and the number of times those items are appearing (frequency of items).

There are generally two types of frequency tables:

- Frequency table with class boundaries/intervals/width
- Frequency table without class boundaries/intervals/width

While constructing a frequency table, we try to group/classify data. There are two methods of data classifications:

- *Exclusive method* – excludes the last value in the interval/class
- *Inclusive method* – includes the last value in the interval/class

4.1.2 Frequency Table without Intervals

This is a frequency table that groups similar items together. The groups or classes don't have ranges or intervals as shown below. This can be done for both qualitative (categorical) and quantitative (numeric) data.

Example of frequency table without intervals:

Age (x)	Frequency (f)
20	1
34	6
42	4
50	2

Frequency table without intervals can be easily done in SPSS using the path: Analyze > Descriptive Statistics > Frequencies > Select the variable of interest (either categorical or numerical) > ensure that Display Frequency Table is checked/ticked then click OK. There will be a frequency table in the SPSS Output

DOI: 10.1201/9781003292654-5

Example

Using *customer_subste.sav* data (inbuilt SPSS data), generate a frequency table for Gender.

For Windows
To locate *customer_subset.sav* data we use the path: Click **Open data document** folder > drive
(C): > **Program Files (x86)** > **IBM** > **SPS** > **Statistics** > **27** > **Samples** > **English** > customer_
subset.sav data (Figure 4.1) (In-built *SPSS data files*).

For MAC
Open **SPSS data** > Click **Open data document** folder > Applications > **IBM SPS Statistics 27** >
Samples > **English** > customer_subset.sav data (Figure 4.1) (In-built *SPSS data files*).

Customer_subset.sav data

	custid	region	townsize	gender	age	agecat	birthmonth	ed	edcat	jobcat	union	employ	empcat	retire	
1	4557-ABRNJI-5YB	Zone 2	> 250,000	Female	39	35-49	December	17	College d...	Managerial...	No	4	2 to 5	No	
2	3675-MPXAID-868	Zone 4	10,000-49,999	Male	43	35-49	May	16	Some col...	Sales and ...	No	2	2 to 5	No	
3	8970-ZQVBIX-Z0U	Zone 5	10,000-49,999	Male	49	35-49	August	14	High sch...	Sales and ...	No	4	2 to 5	No	
4	0300-MXYQVV-WKF	Zone 5	> 250,000	Male	35	35-49	January	9	Did not c...	Managerial...	No	2	2 to 5	No	
5	9827-QPOKBY-LKO	Zone 5	50,000-249,999	Male	32	25-34	September	16	Some col...	Service	No	5	2 to 5	No	
6	5739-BYRLOH-4E9	Zone 4	10,000-49,999	Female	21	18-24	July	14	High sch...	Precision ...	No	1	Less than 2	No	
7	4030-WSTXBV-BTC	Zone 3	50,000-249,999	Female	79	>65	June	12	High sch...	Sales and ...	No	18	More than 15	Yes	
8	9484-XPXYRM-EKK	Zone 1	> 250,000	Male	64	50-64	March	11	Did not c...	Managerial...	No	16	More than 15	No	
9	2681-FGWTIM-PR4	Zone 3	< 2,500	Male	62	50-64	January	16	Some col...	Agricultural...	No	21	More than 15	Yes	
10	9381-GBLQVX-22A	Zone 1	50,000-249,999	Male	61	50-64	August	9	Did not c...	Managerial...	Yes	22	More than 15	No	
11	6677-HFMVVT-AM5	Zone 5	10,000-49,999	Female	48	35-49	June	17	College d...	Agricultural...	No	7	6 to 10	No	
12	1196-LXFRGW-4PI	Zone 5	< 2,500	Female	35	35-49	October	16	Some col...	Sales and ...	No	0	Less than 2	No	
13	2873-UAHQBO-KUX	Zone 5	< 2,500	Female	25	25-34	March	18	College d...	Sales and ...	No	0	Less than 2	No	
14	0194-YAJMQD-D7C	Zone 4	< 2,500	Female	61	50-64	June	16	Some col...	Sales and ...	No	4	2 to 5	No	
15	0491-YMZGVA-RWZ	Zone 4	50,000-249,999	Male	57	50-64	July	16	Some col...	Precision ...	No	19	More than 15	No	
16	4314-SOPKIL-OLR	Zone 4	10,000-49,999	Male	52	50-64	April	18	College d...	Operation,...	No	13	11 to 15	No	
17	6875-OCAMBG-ZAB	Zone 5	50,000-249,999	Female	19	18-24	November	14	High sch...	Operation,...	Yes	0	Less than 2	No	
18	7042-XUPZNS-LII	Zone 1	2,500-9,999	Male	65	>65	February	14	High sch...	Sales and ...	No	13	11 to 15	No	
19	2038-OVSVLN-8NO	Zone 2	50,000-249,999	Female	22	18-24	April	17	College d...	Managerial...	No	0	Less than 2	No	
20	4762-CPVKNE-QZ0	Zone 1	2,500-9,999	Female	31	25-34	May	13	High sch...	Sales and ...	No	3	2 to 5	No	
21	4254-GIHFGQ-694	Zone 3	> 250,000	Female	43	35-49	March	9	Did not c...	Sales and ...	No	4	2 to 5	No	
22	2886-TPWOUY-VXI	Zone 1	> 250,000	Female	54	50-64	February	18	College d...	Operation,...	No	22	More than 15	No	
23	9716-WKBIJBO-YPB	Zone 1	10,000-49,999	Female	43	35-49	August	14	High sch...	Precision ...	No	9	6 to 10	No	

FIGURE 4.1   Customer-subset.sav data

To generate a frequency table without intervals for gender we use the following SPSS path:
Analyze > Descriptive Statistics > Frequencies (Figure 4.2) > There will be a pop-up window
for Frequencies (Figure 4.3) > Select Gender under Variable(s) > ensure that Display Frequency
Table and Create APA Style Table is checked/ticked then click OK

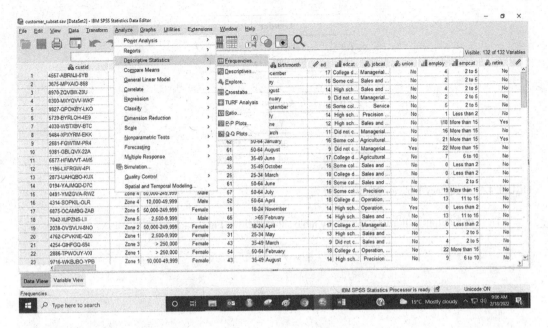

FIGURE 4.2 SPSS Path for Frequency Table

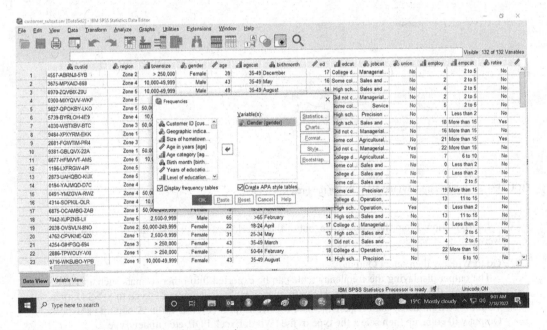

FIGURE 4.3 SPSS Pop Up Window for Frequencies

	Gender	
N	Valid	80
	Missing	0

Gender		
	N	%
Male	36	45.0%
Female	44	55.0%

FIGURE 4.4   Frequency Table for Gender

The SPSS output will give a frequency table shown below (Figure 4.5).

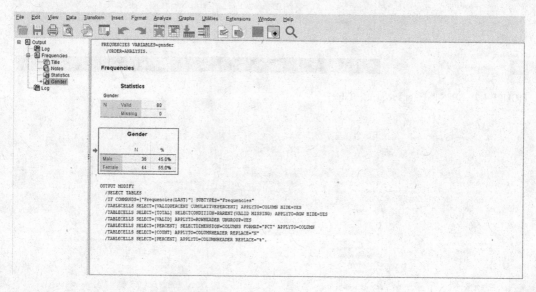

FIGURE 4.5   SPSS Output Window displaying the Frequency Table

Interpretation: From the above table, out of the 80 selected customers, most of the customers were female at 55%.

Go to Window then select Output

The output displaying the frequency table can be exported into other formats such as Word, Excel, PDF, Power Pont, etc. To export the frequency table in the output, we go to SPSS Output Window > Right Click the Frequency Table > Click Export (Figure 4.6) > A pop up of Export Output will come up then select the type of file (Word, Excel, PDF, etc.) under Type

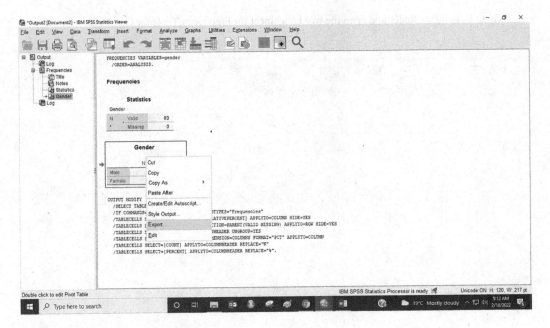

FIGURE 4.6 SPSS Pop Up Window for Export Output

Click Browse to choose the folder where you would like to save the SPSS output and give it a name (e.g. Frequency Table for Gender) (Figure 4.7). Click Save the OK and it will export the output

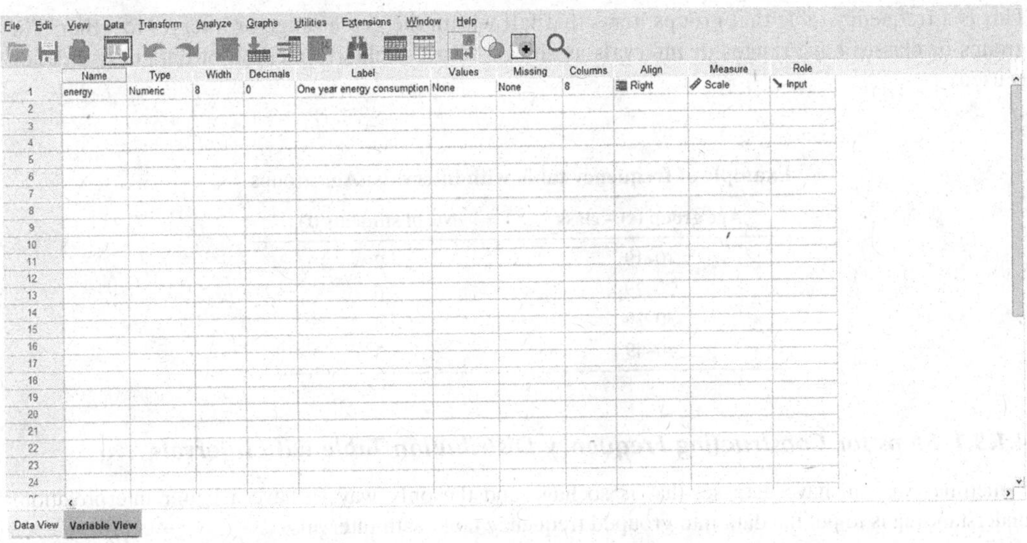

FIGURE 4.7 SPSS Variable View for One Year Energy Consumption Data

Go to Documents folder and open the word document for Frequency Table for Gender

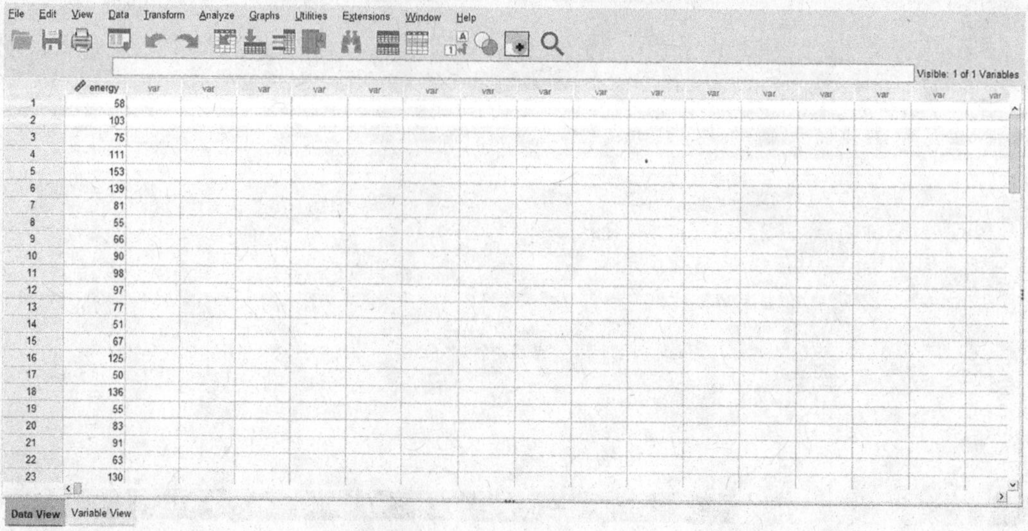

FIGURE 4.8   SPSS Data View for One Year Energy Consumption Data

Alternatively, one may export the frequency table from the SPSS output by copying the frequency table from the SPSS output and then pasting it into the Word or Excel document (Figure 4.8).

4.1.3 Frequency Table with Intervals

This is a frequency table that groups items that fall within the given range or interval together. The groups or classes have ranges or intervals as shown below. Inclusive method of data classification is used.

Example of frequency table with intervals: Age groups

Age group (x) – class	No. of students (f)
10–19	2
20–29	6
30–39	4
40–49	3

4.1.3.1 Steps for Constructing Frequency Distribution Table with Intervals

Sometimes we can have data set that is so huge and the only way to have a better interpretation/ understanding is to put the data into grouped frequency table with intervals.

1. Obtain the lowest value, highest value and the total number of observations.

   We go to Data View Window > Right click the variable of interest > select Sort Descending and Sort Ascending. This arranges the data in ascending or descending order.

   Or

Go to Analyze > Descriptive Statistics > Frequencies > there will be a pop-up window for Frequencies > Select the variable of interest under Variable(s) > Under Statistics select Minimum and Maximum > then click OK

The SPSS output will give the total number of observations (N), Minimum and Maximum values

2. Determine the number of classes using the Sturge's Rule

$$k = 1 + 3.322 \log_{10} N$$

where k is the number of classes and N is the total number of observations

In SPSS, go to Transform > Compute Variable > Put classes under Target Variable > Click All in the Function group > Under Functions and Special Variables select lg10 > in the Numeric Expression widow type the formula 1+(3.322*LG10(N)) and substitute with the value of N then click OK > In the SPSS Data View Window look for the values appearing in the "classes" variable.

3. Determine the class width/interval

$$width = \frac{highest\ value - lowest\ value}{k}$$

In SPSS, go to Transform > Compute Variable > Put **interval** under Target Variable > in the Numeric Expression window type the formula (Maximum − Minimum)/k and substitute the values then click OK > In the SPSS Data View Window look for the values appearing in the "interval" variable.

4. Group the data using the estimated width/interval starting with the lowest value

Groups are created manually and codes are assigned to each group. We then create these codes and groups in SPSS.

In SPSS, go to Transform > Recode into Different Variables > there is a pop up SPSS window for Recode into Different Variables > Select the numerical variable of interest **Input Variable** > Put **grp_data** as the Name in the **Output Variable** and **Grouped Data** as the **Label** then click Change

Click **Old and New Values** > Under **Old Values** select **Range** and type in the intervals/groups that were created manually (e.g., 43 through 57) and under the **New Values** put the code for the range where there is value (i.e., 1 for 43–57) then Add. Repeat the same for the other intervals/groups and their respective codes and once done click Continue and OK

Go to SPSS Variable View and look up for the last variable (newly created variable)

Under the column for Values for the newly created variable (grp_data), put the codes as the values and label as the different ranges/groups then click OK.

To generate frequency table with Intervals, go to Analyze > Descriptive Statistics > Frequencies > Reset > Selected the Grouped Data variable and put it in Variable(s) > Ensure that Display frequency table and Create APA style tables is checked > OK.

Example

The following table gives one year's energy consumption for sampled households in the Nairobi.

58	103	75	111	153	139	81	55	66	90	98
97	77	51	67	125	50	136	55	83	91	63
130	55	45	64	147	66	60	80	102	62	106
54	86	100	78	93	113	111	104	96	113	119
95	74	57	69	105	43	48	67	115	123	87
96	87	129	109	69	94	99	97	83	97	49
57	62	78								

Key the data into SPSS then generate grouped frequency distribution (with intervals) for the data. Interpret the output:

Steps:
1. Create one year's energy consumption variable in the Variable View in SPSS:
   a. Name: energy
   b. Type: Numeric
   c. Width: 8
   d. Decimal: zero
   e. Label: One-year energy consumption
   f. Values: None
   g. Measure: Scale

Variable View
2. Key data in the Data View

Data View

Step 1:

Obtain the highest and the lowest value

We go to Analyze > Descriptive Statistics > Frequencies (Figure 4.9)

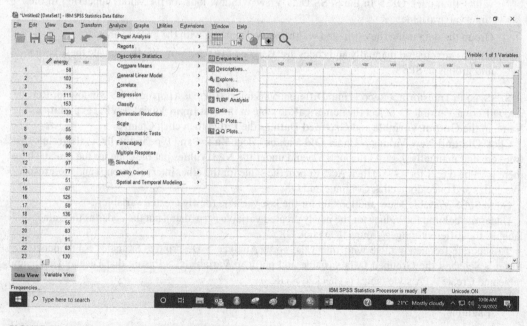

FIGURE 4.9 SPSS Menu for Descriptive Statistics Analysis

There will be a pop-up window for Frequencies > Select Variable of interest (One Year Energy consumption) > Under Statistics select Minimum and Maximum > click Continue then OK (Figure 4.10)

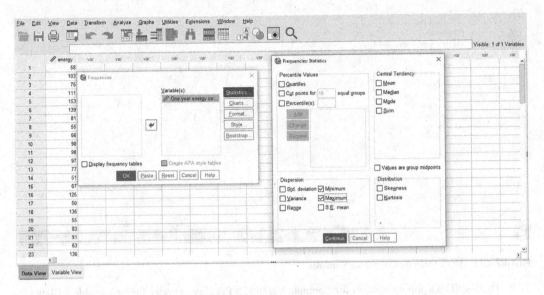

FIGURE 4.10 SPSS Pop Up Window for Frequencies displaying Statistics

SPSS Output

Statistics		
One-Year Energy Consumption		
N	Valid	69
	Missing	0
Minimum		43
Maximum		153

FIGURE 4.11 Minimum and Maximum Values for One Year Energy Consumption Data

From the SPSS output Lowest (Minimum) = 43, Highest (Maximum) = 153 and number of observations (N) = 69

Step 2: Estimate the number of classes using Sturge's rule

$$k = 1 + 3.322 \times logN$$

$$k = 1 + 3.322 \times log69 = 7.108657$$

This will be approximately 7–8 classes

In SPSS, go to Transform > Compute Variable (Figure 4.11)

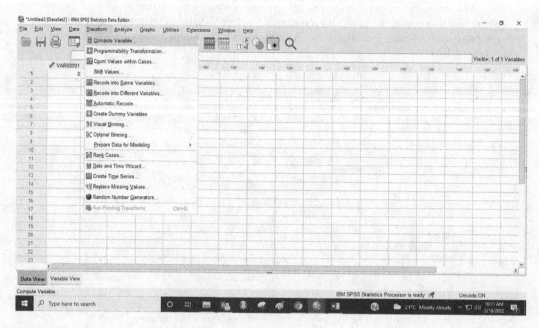

FIGURE 4.12 SPSS Menu for Computing Variables

There will be a pop-up window for Compute Variable > Put classes under Target Variable > Click All in the Function group > Under Functions and Special Variables select lg10 > in the Numeric Expression widow type the formula 1+(3.322*LG10(69)) then click OK (Figure 4.13) > In the SPSS Data View Window look for the values appearing in the "classes" variable

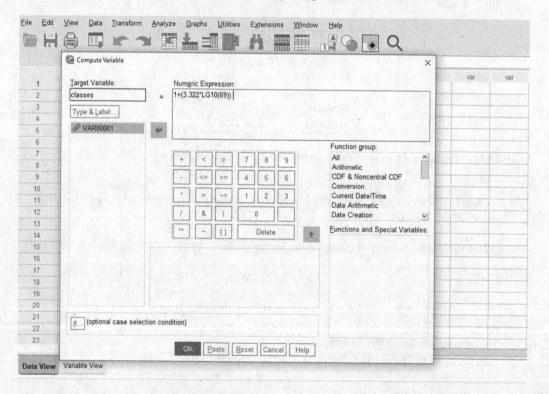

FIGURE 4.13 SPSS Pop Up Window for Computing Number of Class

The value of k will be in the SPSS Data View Window (Figure 4.14)

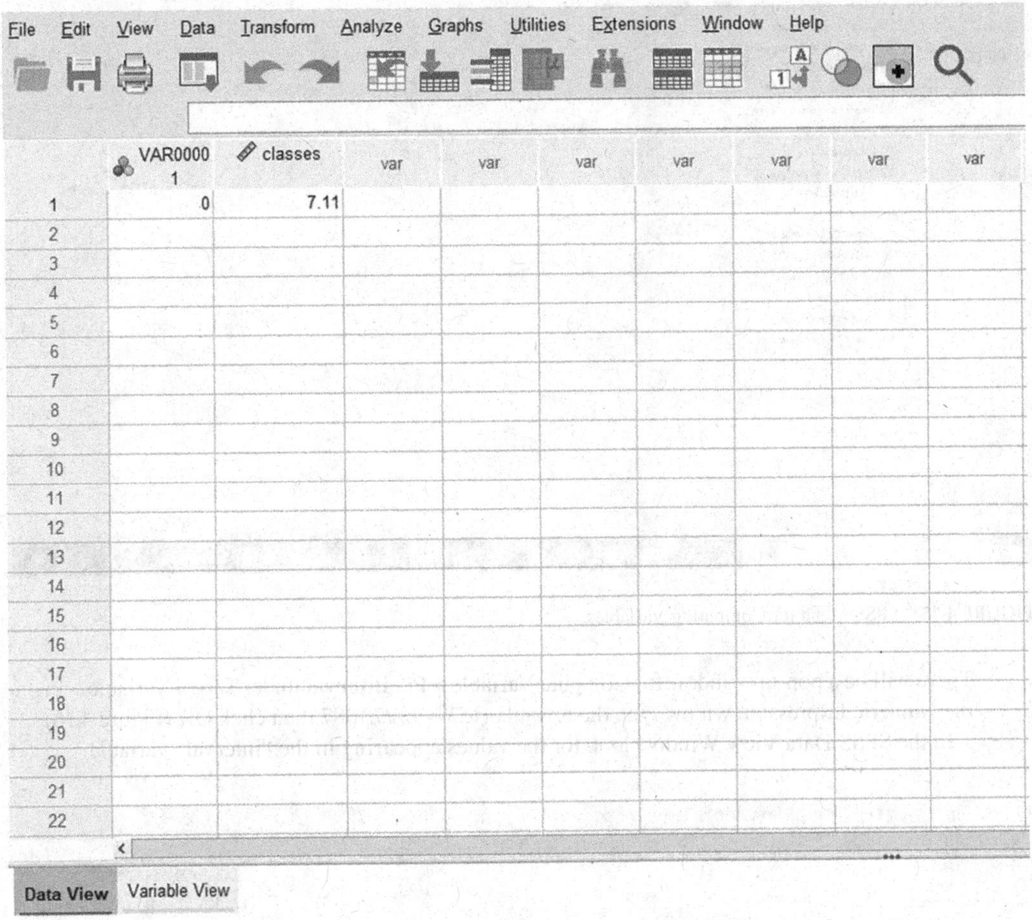

FIGURE 4.14 SPSS Data View Displaying Number of Classes

$$k = 7.1087$$

Step 3: Estimate the class interval/width

$$interval = \frac{highest - lowest}{k} = \frac{153 - 43}{7.108657} = 15.474$$

In SPSS, go to Transform > Compute Variable (Figure 4.15)

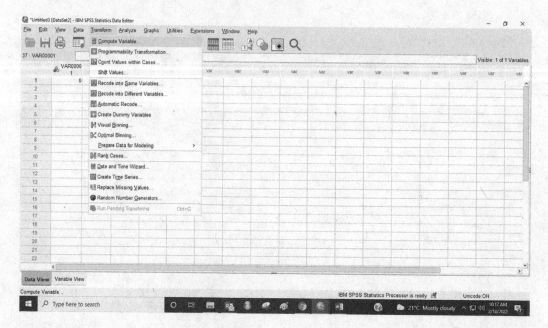

FIGURE 4.15   SPSS Menu for Computing Variables

There will be a pop-up window for Compute Variable > Put **interval** under Target Variable > in the Numeric Expression widow type the formula (153 – 43)/7.1087 then click OK (Figure 4.16) > In the SPSS Data View Window look for the values appearing in the "interval" variable

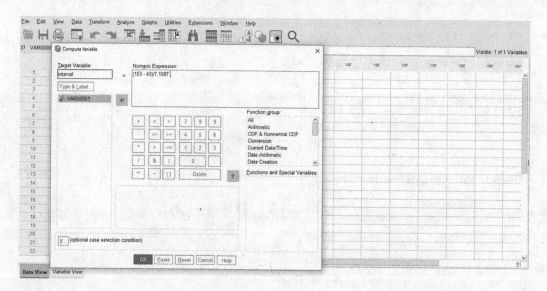

FIGURE 4.16   SPSS Pop Up Window for Computing Class Interval

The value of interval will be in the SPSS Data View Window (Figure 4.17)

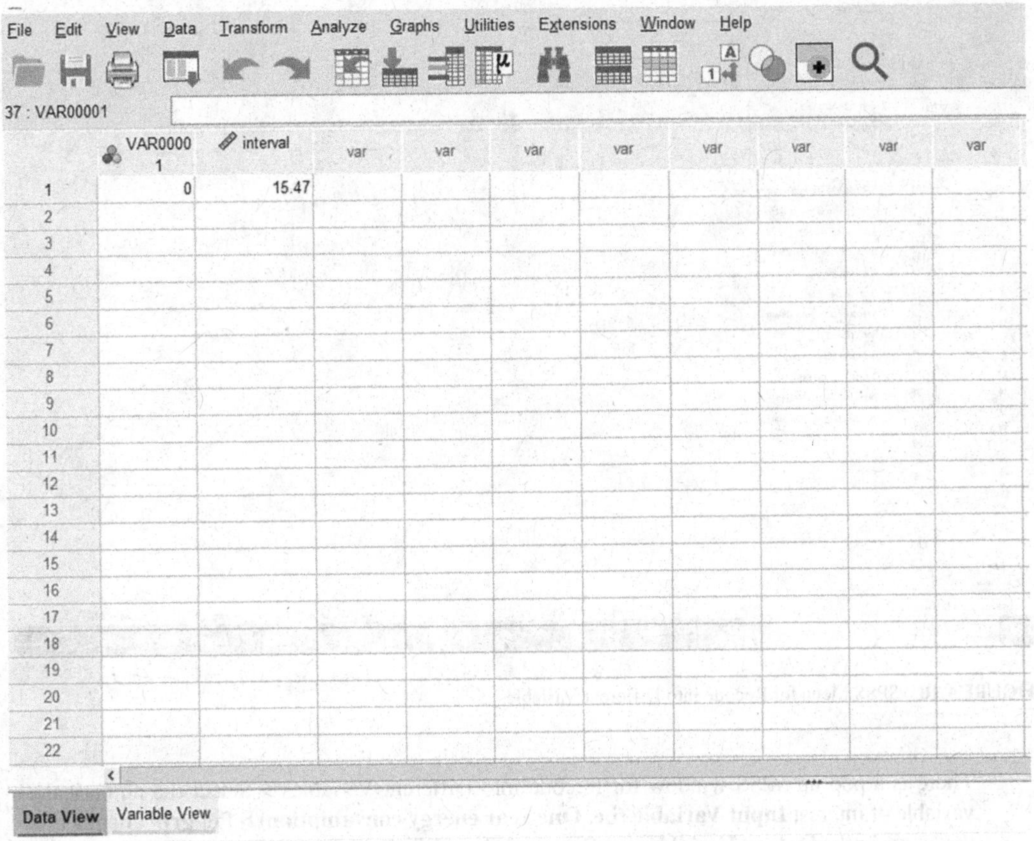

FIGURE 4.17   SPSS Data View Displaying Class Interval

Width = 15.474 which is approximately 15

Step 4: Manually create the groups and give them codes starting with the lowest value and taking into account the estimated interval/width. The interval must be uniform.

Groups	Codes
43–57	1
58–72	2
73–87	3
88–102	4
103–117	5
118–132	6
133–147	7
148–162	8

Step 5: Go to SPSSS and create the groups plus the codes

Transform > Recode into Different Variables (Figure 4.18)

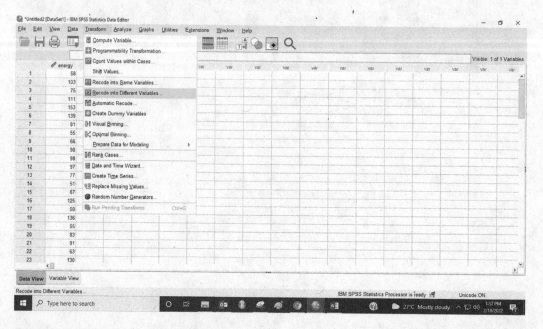

FIGURE 4.18   SPSS Menu for Recode into Different Variables

There is a pop up SPSS window for Recode into Different Variables > Select the numerical variable of interest **Input Variable (i.e. One year energy consumption)** > Put **grp_enegry** as the Name in the **Output Variable** and **Grouped One Year Energy Consumption** as the **Label** then click Change (Figure 4.19)

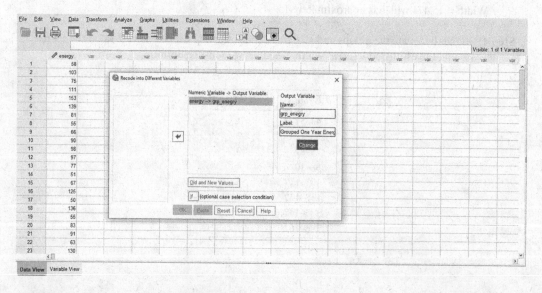

FIGURE 4.19   SPSS Pop up Window for Recode into Different Variables

Click **Old and New Values** > Under **Old Values** select **Range** and type in the intervals/groups that were created manually (e.g. 43 through 57) and under the **New Values** put the code for the range where there is value (i.e. 1 for 43–57) then Add. Repeat the same for the other intervals/groups and their respective codes and once done click Continue and OK (Figure 4.20)

Step 1: Under Recode into Different Variables

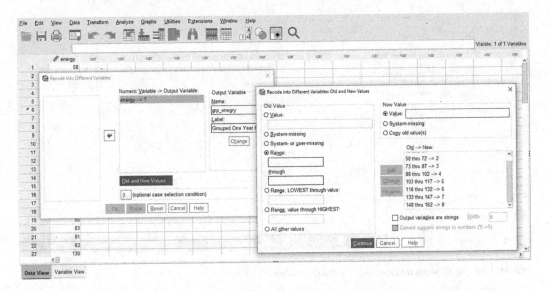

FIGURE 4.20   SPSS Pop Up Window for Recode into Different Variables displaying Old and New Values

Step 2: Go to SPSS Variable View and look up for the last variable (newly created variable)

Under the column for Values for the newly created variable (grp_energy), put the codes as the values and label as the different ranges/groups then click OK (Figure 4.21).

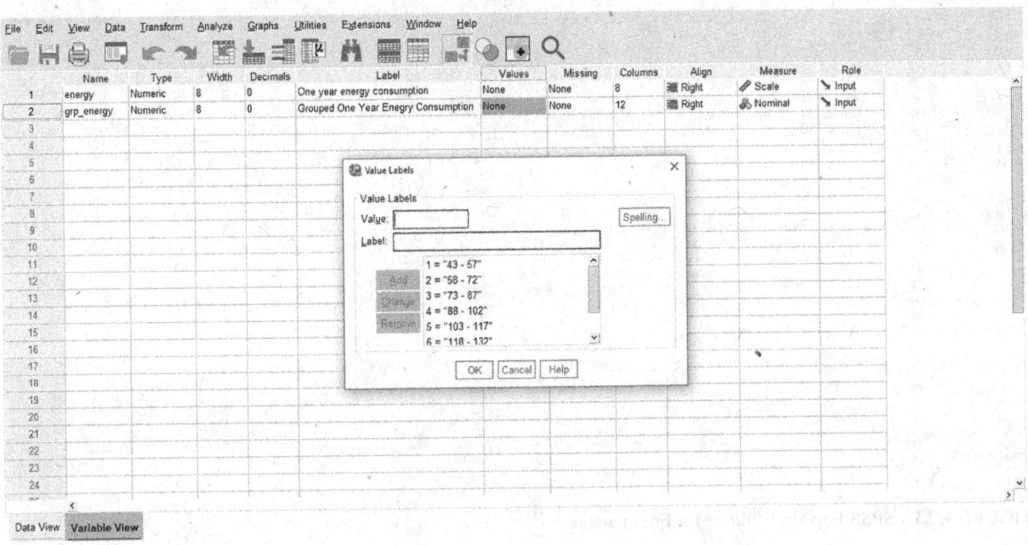

FIGURE 4.21   SPSS Variable View displaying Value Labels

Step 3: Generate Frequency Table with Intervals

SPSS path: Analyze > Descriptive Statistics > Frequencies (Figure 4.22)

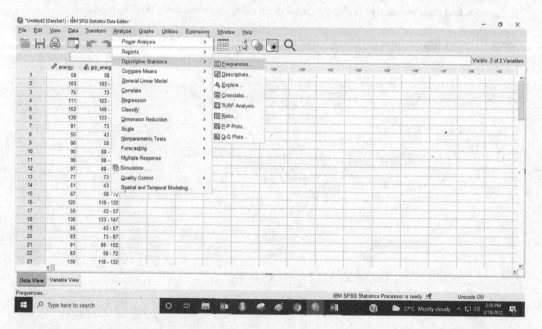

FIGURE 4.22   SPSS Menu for Descriptive Statistics Analysis

There will be SPSS pop-up window for Frequencies > Selected the Grouped Energy Consumption variable and put it in Variable(s) > Ensure that Display frequency table and Create APA style tables is checked > OK (Figure 4.23).

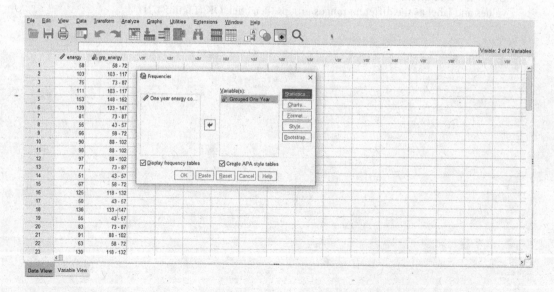

FIGURE 4.23   SPSS Pop Up Window for Frequencies

The SPSS output will give the frequency table with intervals for one year energy consumption (Figure 4.24).

Grouped One-Year Energy Consumption		
	N	%
43–57	12	17.4%
58–72	12	17.4%
73–87	12	17.4%
88–102	14	20.3%
103–117	10	14.5%
118–132	5	7.2%
133–147	3	4.3%
148–162	1	1.4%

FIGURE 4.24 Frequency Table with Intervals for the One Year Energy Consumption Data

Interpretation. Majority of the 69 respondents had consumed between 88 and 102 at 20.3%

We can obtain some measures and graphs from the frequency table including Cumulative frequency, Relative frequency, Percent relative frequency, Cumulative frequency curve, Histogram, Frequency polygon, Line graph, Bar chart and Pie chart.

Cumulative Frequency

A cumulative frequency distribution shows the total number of observations in all class up to and including that class. Cumulative frequency is abbreviated as c.f. In SPSS, we have cumulative Percent as shown below:

		Grouped One-Year Energy Consumption			
		Frequency	Percent	Valid Percent	Cumulative Percent
Valid	43–57	12	17.4	17.4	17.4
	58–72	12	17.4	17.4	34.8
	73–87	12	17.4	17.4	52.2
	88–102	14	20.3	20.3	72.5
	103–117	10	14.5	14.5	87.0
	118–132	5	7.2	7.2	94.2
	133–147	3	4.3	4.3	98.6
	148–162	1	1.4	1.4	100.0
	Total	69	100.0	100.0	

There are two types of cumulative frequencies: - *"Less than" cumulative frequency* and *"More than" cumulative frequency*

- For "less than c.f." we add frequencies downwards
- For "more than c.f." we add frequencies upwards

Note: Cumulative frequencies (c.f.) can be used to generate cumulative frequency curve and also compute the median and partition values.

Relative Frequency

$$Relative \ frequency = \frac{Frequency \ of \ the \ class}{n}$$

where n is the total number of observation

Percent Relative Frequency

Percent relative frequency of the class is the relative frequency multiplied by 100.

Note: R.F and % RF can be used to generate pie charts and probabilities.

In SPSS, % relative frequency is represented by Valid Percent.

4.2 Graphical Presentation of Data

The type of data determines the type of graph/chart

Categorical/Qualitative Data

- Bar graph
- Pie chart

Quantitative/Numeric Data

- Histogram
- Line graph
- Stem and leaf plot
- Frequency polygon
- Scatter plot
- Box plot
- Cumulative frequency curve/ogive

There are usually three different ways of generating graphs and charts in SPSS.

SPSS path: Analyze > Descriptive Statistics > Frequencies > there will be a popup window for Frequencies > Select the variable of interest > Click Charts > Select the appropriate chart for the variable of interest

Alternatively: Graphs > Legacy Dialogs > Choose the chart/graph of interest for the given variable

Alternatively: Analyze > Descriptive Statistics > Explore > there will be a popup window for Explore > Choose Numerical Variable of Interest under Dependent List – Statistics

4.2.1 Bar Graph

A bar graph is a plot of frequency (y-axis) and categories/groups (x-axis). To generate a bar graph in SPSS we use the path:

Analyze > Descriptive Statistics > Frequencies > There will be a pop-up window for Frequencies > From the list of variables select the categorical data of interest and use the right arrow to move the selected variables to Variable(s) window > Click Charts > There will be a pop-up window for Charts > Select Bar charts > Under Chart Values select Percentages > Click Continue then OK.

Alternatively, one may use:

Graphs > Legacy Dialogs > Bar > There will be a pop-up window for Bar > Click Simple > Under Data in Chart are select Summaries for groups of cases then click Define > There will be a pop-up window for Summaries for Groups of Cases > Select the categorical variable of interest and put it under Categorical Axis then click OK.

Example

Using *customer_subset.sav* data (inbuilt SPSS data), generate a bar graph for Gender

For Windows
To locate *customer_subset.sav* data we use the path: Click **Open data document** folder > drive **(C): > Program Files (x86) > IBM > SPS > Statistics > 27 > Samples > English** > customer_subset.sav data (In-built *SPSS data files*).

For MAC
Open **SPSS data** > Click **Open data document** folder > Applications > **IBM SPS Statistics 27 > Samples > English** > customer_subset.sav data (In-built *SPSS data files*).

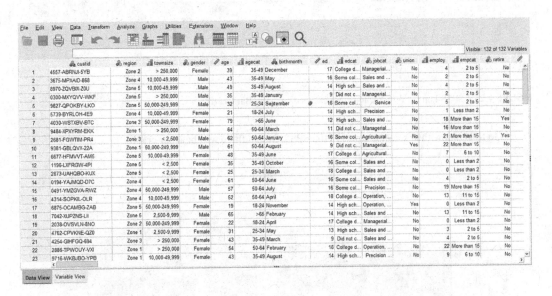

Customer-subset.sav data

To generate a bar graph for gender: go to Analyze > Descriptive Statistics > Frequencies > There will be a pop-up window for Frequencies > From the list of variables select Gender and use the right arrow to move Gender to Variable(s) window > Click Charts > There will be a pop-up window for Charts > Select Bar charts > Under Chart Values select Percentages > Click Continue then OK (Figure 4.25).

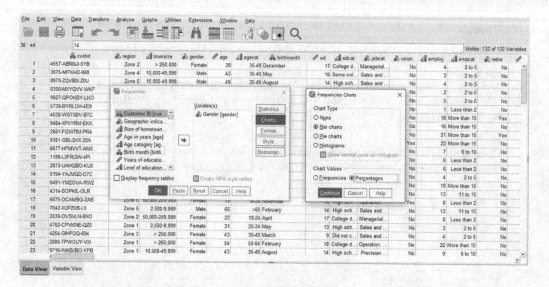

FIGURE 4.25   SPSS Pop Up Window for Frequencies displaying Bar Charts

Go to Window then select Output. In the SPSS Output there will be a bar graph

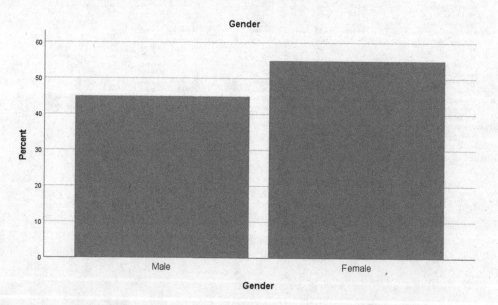

FIGURE 4.26   Bar Chart for Gender

In the SPSS Output Window, we can edit the bar graph to include percentage values, make it a 3-D and even change colors as follows:

Double click the bar graph > There will be a pop-up window for Chart Editor (Figure 4.27)

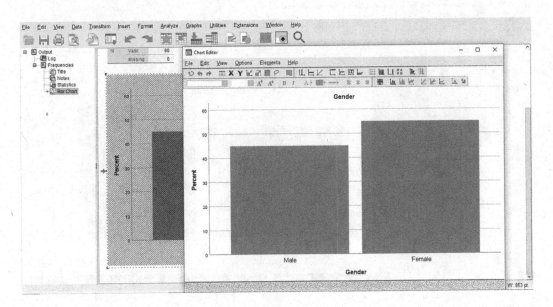

FIGURE 4.27 Chart Editor for the Bar Chart

THEN double click the bar chart in the Chart Editor Window and a pop-up window for Properties will populate (Figure 4.28)

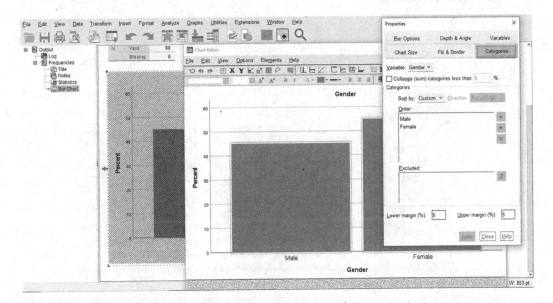

FIGURE 4.28 SPSS Pop Up Window for Chart Editor displaying Properties

Close the Properties window > then look for Data Label Mode (Figure 4.29)

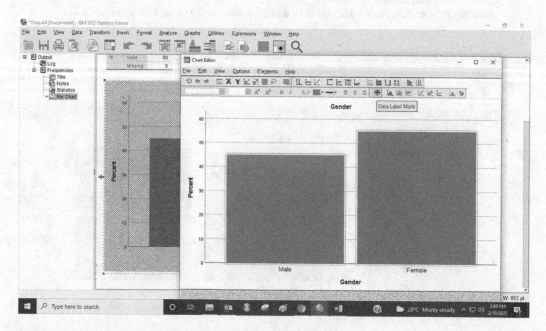

FIGURE 4.29   SPSS Pop Up Window for Chart Editor displaying Data Label Mode

Then click Data Label Mode once, then click each of the bars in the bar chart and percentages will appear. Once the percentages have appeared on the charts, click Data Label Mode again (Figure 4.30)

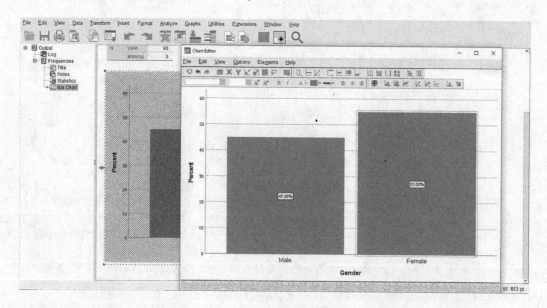

FIGURE 4.30   SPSS Chart Editor displaying Percentages on the Bar Chart

You can change the bar graphs to 3-D as follows:

In the Char Editor Window, double click any of the bars and there will be a pop-up window for Properties> Select Depth & Angle and under Effects Select 3-D and Apply. There will be a 3-D bar graph (Figure 4.31)

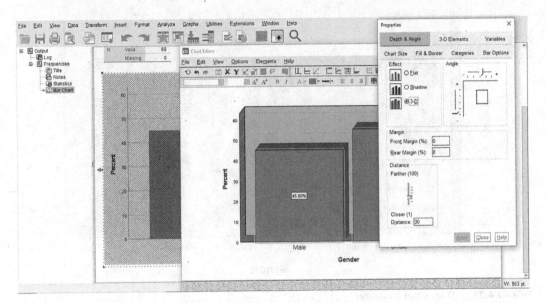

FIGURE 4.31   SPSS Pop Up Window for Chart Editor displaying Properties for Depth & Angle

You can change the colors for the graph as follows:

In the Chart Editor Window, double click any of the bars and there will be a pop-up window for Properties > Select Fill & Border and select the color of choice then click Apply (Figure 4.32)

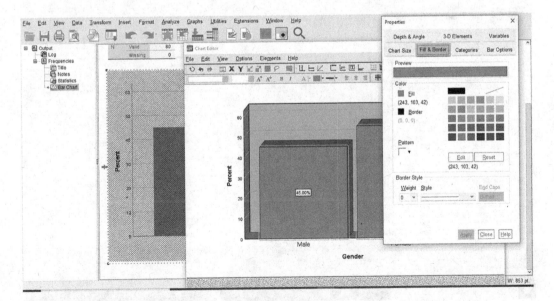

FIGURE 4.32   SPSS Pop Up Window for Chart Editor displaying Properties for Fill & Border

Once all the edits have been done in the Chart Editor, close the Chart Editor and it will save all the changes in the bar graph as shown in SPSS Output Window (Figure 4.33)

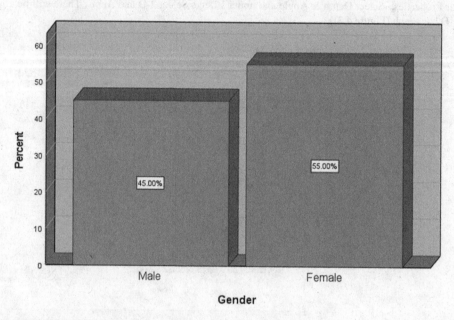

FIGURE 4.33   3-D Bar Chart for Gender

Interpretation: from the graph above, it is evident that most of the respondents were female at 55%

We can also generate Bar graph using the other SPSS path:
  Graphs > Legacy Dialogs > Bar (Figure 4.34)

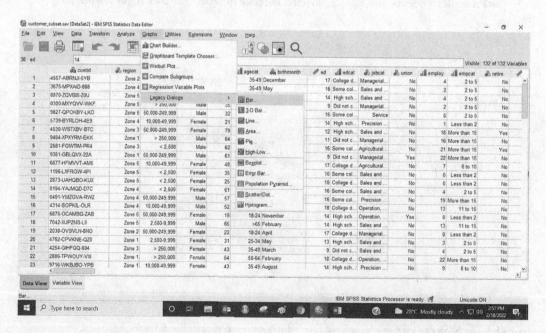

FIGURE 4.34   SPSS Graph Menu for Bar Chart

There will be a pop-up window for Bar > Click Simple > Under Data in Chart are select Summaries for groups of cases (Figure 4.35)

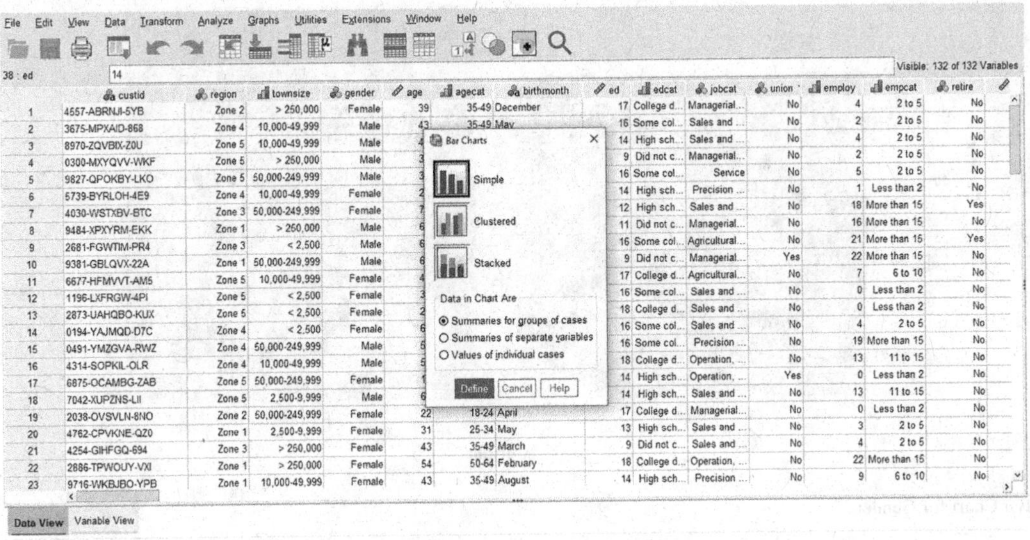

FIGURE 4.35 SPSS Pop Up Window for Bar Chart

Then click Define > There will be a pop-up window for Summaries for Groups of Cases > Select Gender and put it under Categorical Axis then click OK (Figure 4.36).

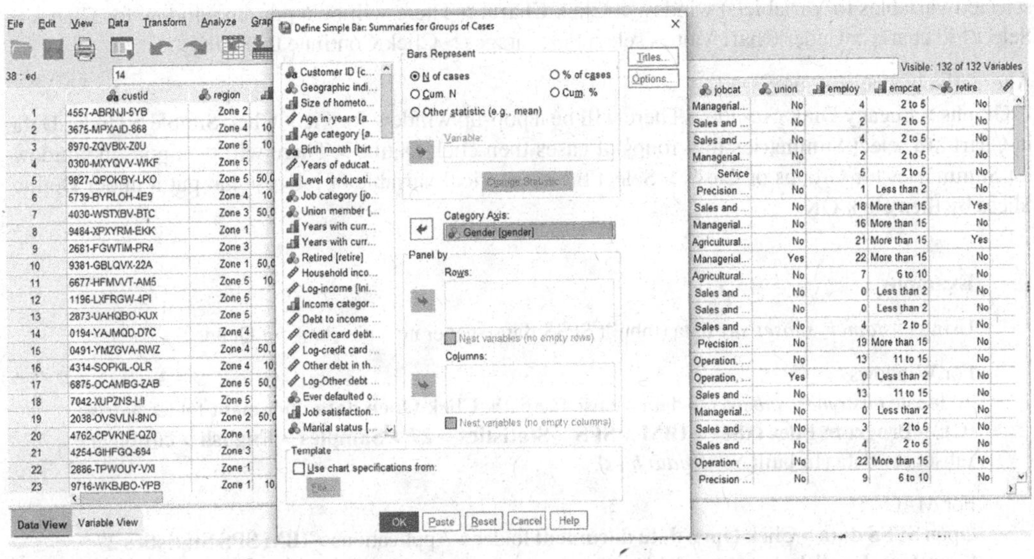

FIGURE 4.36 SPSS Pop Up Window for Simple Bar Chart displaying Summaries for Group of Cases

In the SPSS Output Window, a bar graph will be generated.

Bar Chart for Gender

To edit the graph, we follow the same process described above in the Chart Editor.

4.2.2 Pie Chart

A pie chart is a circular graph for presenting data. The slices represent the different categories and shows the relative size of the data. To generate a bar graph in SPSS we use the path:

Analyze > Descriptive Statistics > Frequencies > There will be a pop-up window for Frequencies > From the list of variables select the categorical data of interest and use the right arrow to move the selected variables to Variable(s) window > Click Charts > There will be a pop-up window for Charts > Select Pie charts > Under Chart Values select Percentages > Click Continue then OK.

Alternatively, one may use:

Graphs > Legacy Dialogs > Pie > There will be a pop-up window for Pie > Click Simple > Under Data in Chart are select Summaries for groups of cases then click Define > There will be a pop-up window for Summaries for Groups of Cases > Select the categorical variable of interest and put it under Define Slices by then click OK.

Example

Using *customer_subset.sav* data (inbuilt SPSS data), generate a pie chart for Gender

For Windows
To locate *customer_subset.sav* data we use the path: Click **Open data document** folder > drive **(C):** > **Program Files (x86)** > **IBM** > **SPS** > **Statistics** > **27** > **Samples** > **English** > customer_subset.sav data (In-built *SPSS data files*).

For MAC
Open **SPSS data** > Click **Open data document** folder > Applications > **IBM SPS Statistics 27** > **Samples** > **English** > customer_subset.sav data (In-built *SPSS data files*).

In SPSS Data View Window, go to Analyze > Descriptive Statistics > Frequencies > there will be a pop-up window for Frequencies > From the list of variables select the categorical data of interest and use the right arrow to move the selected variables to Variable(s) window > Click Charts > there will be a pop-up window for Charts > Select Pie charts > Under Chart Values select Percentages > Click Continue then OK (Figure 4.37).

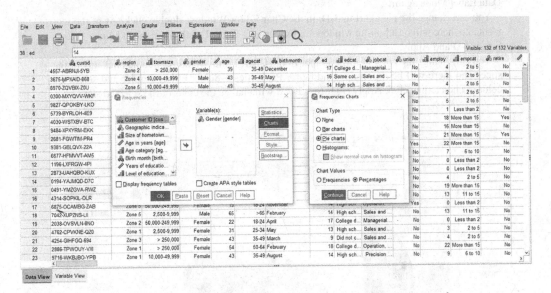

FIGURE 4.37 SPSS Pop Up Window for Frequencies displaying Pie Chart

In SPSS Data View Window, go to Window then select Output and obtain the pie chart (Figure 4.38)

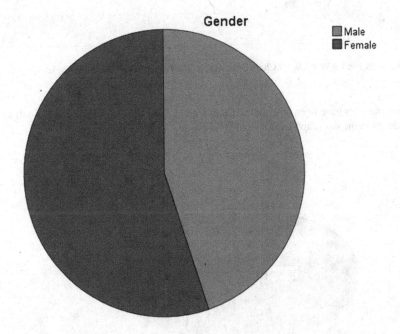

FIGURE 4.38 Pie Chart for Gender

In the SPSS Output Window, we edit the pie chart the same way we edited bar graphs as follows:

Double click the pie chart > There will be a pop-up window for Chart Editor and click Data Editor Mode. Look for Data Label Mode and click it once, then click each of the segment in the pie chart and percentages will appear. Once the percentages have appeared on the pie chart, click Data Label Mode again.

You can change the pie chart to 3-D: in the Char Editor Window, double click any of the segment and there will be a pop-up window for Properties> Select Depth & Angle and under Effects Select 3-D and Apply.

You can change the colors for the graph: In the Chart Editor Window, double click any of the slices and there will be a pop-up window for Properties > Select Fill & Border and select the color of choice then click Apply (Figure 4.39). It will change the color of the selected slice.

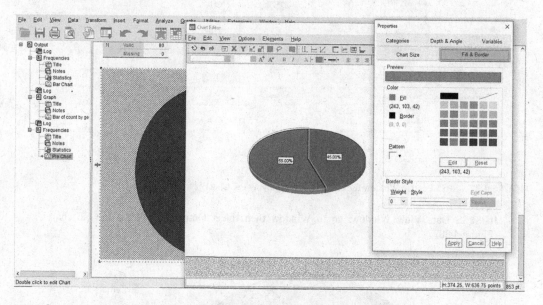

FIGURE 4.39   SPSS Pop Up Window for Chart Editor displaying Properties for Depth & Angle

Once all the edits have been done in the Chart Editor, close the Chart Editor and it will save all the changes in the bar graph as shown in SPSS Output Window (Figure 4.40).

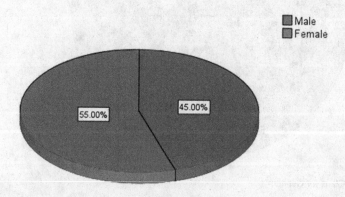

FIGURE 4.40   3-D Pie Chart for Gender

Interpretation: from the graph above, it is evident that most of the respondents were female at 55%.

Alternatively, one may also generate a pie chart using:
Graphs > Legacy Dialogs > Pie (Figure 4.41)

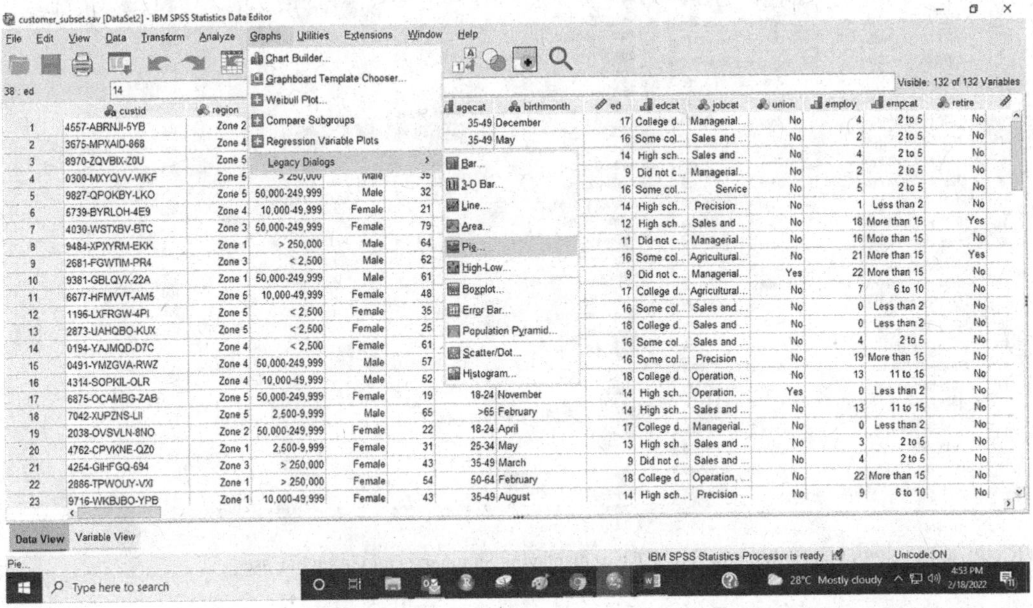

FIGURE 4.41 SPSS Graph Menu for Pie Chart

There will be a pop-up window for Bar > Click Simple > Under Data in Chart are select Summaries for groups of cases then click Define (Figure 4.42)

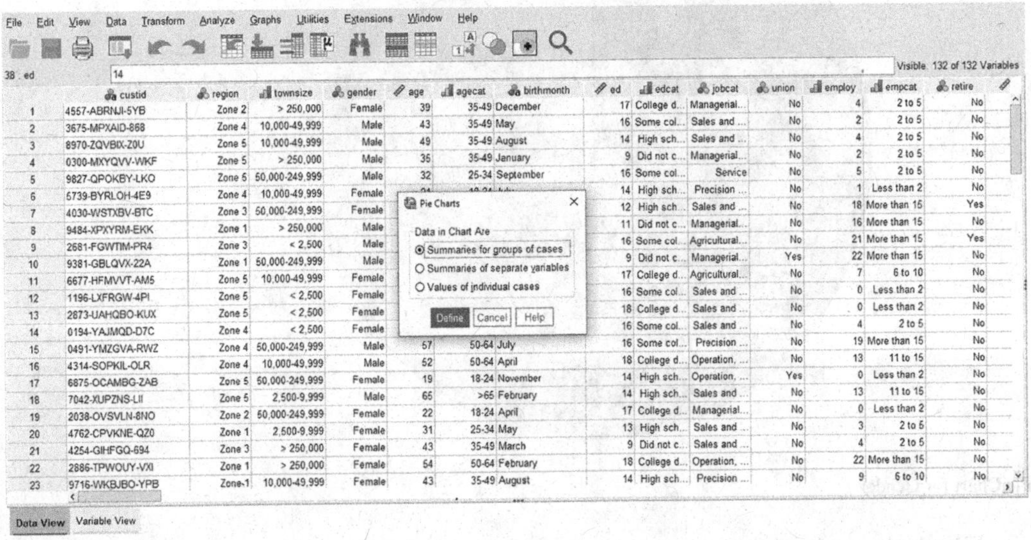

FIGURE 4.42 SPSS Pop Up Window for Pie Chart

There will be a pop-up window for Summaries for Groups of Cases > Select the categorical variable of interest and put it under Define Slices by then click OK (Figure 4.43).

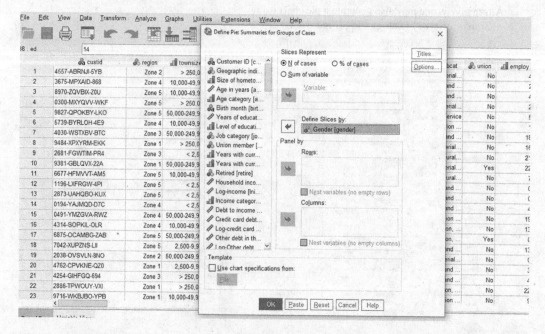

FIGURE 4.43 SPSS Pop Up Window for Pie Chart displaying Summaries for Group of Cases

The SPSS output window will give the pie chart

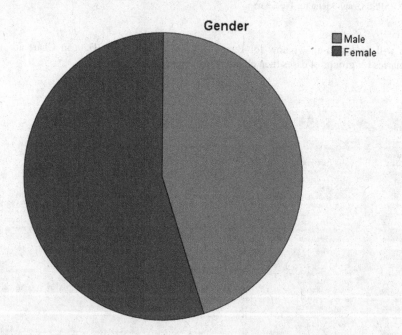

Pie Chart for Gender

To edit the graph to have percentages, into 3-D and even add more colors, we follow the same process described above in the Chart Editor.

4.2.3 Histogram

A Histogram displays data using continuous bars. Histogram displays the shape and spread of continuous/numerical data. To generate a bar graph in SPSS we use the path:

Analyze > Descriptive Statistics > Frequencies > There will be a pop-up window for Frequencies > From the list of variables select the numerical data of interest and use the right arrow to move the selected variables to Variable(s) window > Click Charts > There will be a pop-up window for Charts > Select Histograms > Click Continue then OK.

Alternatively, one may use:

Graphs > Legacy Dialogs > Histogram > There will be a pop-up window for Histogram > Select the numerical variable of interest and put it under Variable then click OK.

Alternatively, one may use:

Analyze > Descriptive Statistics > Explore > there will be a pop-up window for Explore > from the variables list select the numerical variable of interest and put it under Dependent List > Under Plots select Histogram then Continue > Under Display in the Explore popup window select Plots then click OK.

Example

Using *customer_subset.sav* data (inbuilt SPSS data), generate a histogram for Age in Years.

For Windows
To locate *customer_subset.sav* data we use the path: Click **Open data document** folder > drive (C): > **Program Files (x86) > IBM > SPS > Statistics > 27 > Samples > English** > customer_subset.sav data (In-built *SPSS data files*).

For MAC
Open **SPSS data** > Click **Open data document** folder > Applications > **IBM SPS Statistics 27 > Samples > English** > customer_subset.sav data (In-built *SPSS data files*).

Analyze > Descriptive Statistics > Frequencies > there will be a pop-up window for Frequencies > From the list of variables select Age in Years and use the right arrow to move the selected variables to Variable(s) window > Click Charts > there will be a pop-up window for Charts > Select Histograms > Click Continue then OK (Figure 4.44).

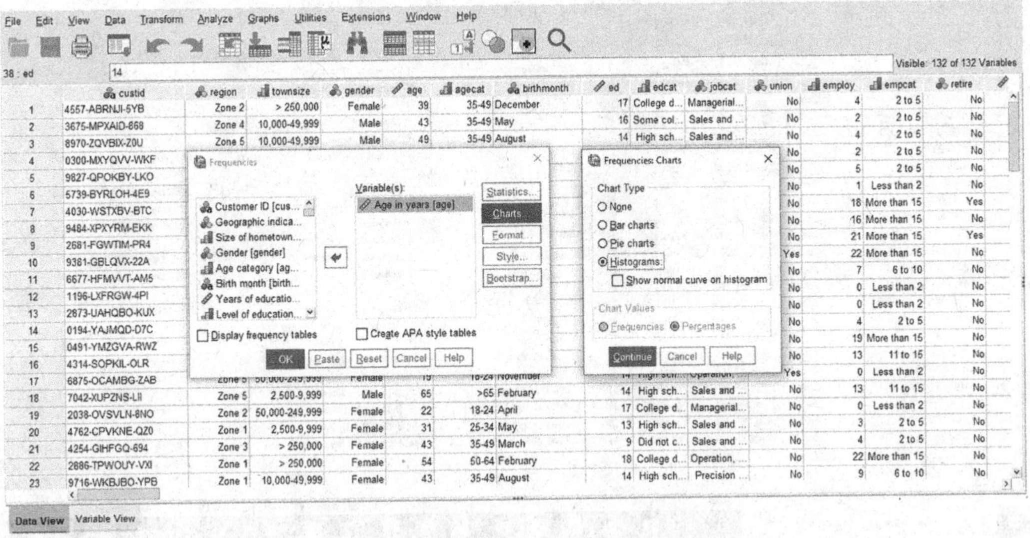

FIGURE 4.44 SPSS Pop Up Window for Frequencies displaying Histogram

In the SPSS Data View Window, go to Window then select Output to retrieve the Histogram (Figure 4.45)

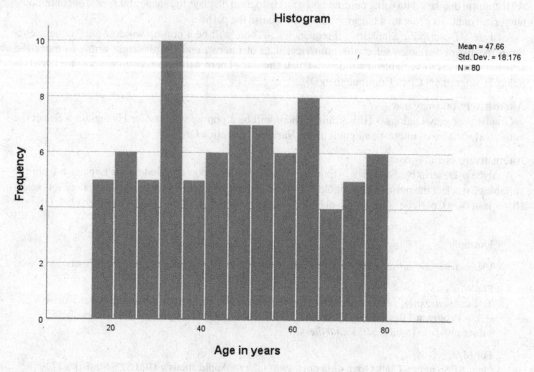

FIGURE 4.45 Histogram for Age in Years

Interpretation: The age in years is approximately normally distributed

Using the second SPSS path:
Graphs > Legacy Dialogs > Histogram (Figure 4.46)

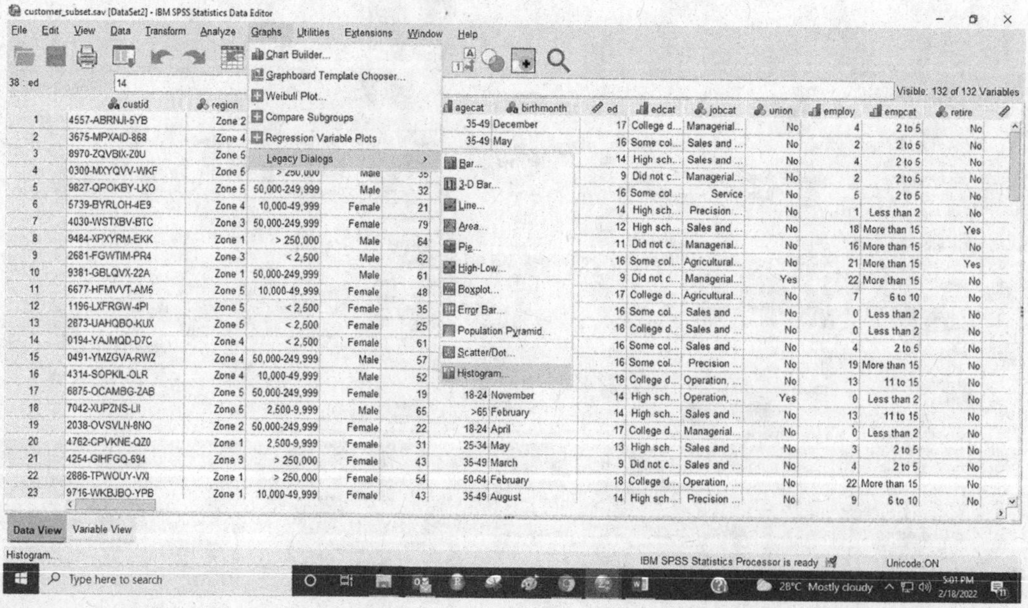

FIGURE 4.46 SPSS Graph Menu for Histogram

There will be a pop-up window for Histogram > Select Age in Years and put it under Variable then click OK (Figure 4.47).

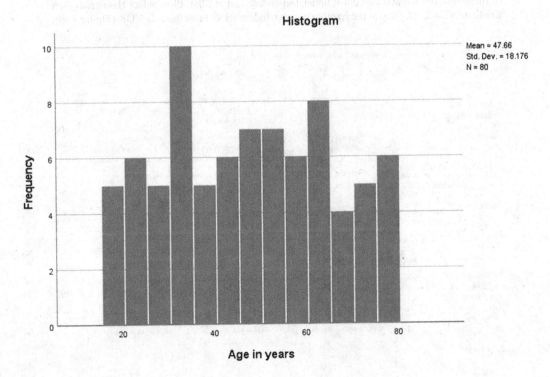

FIGURE 4.47 SPSS Pop Up Window for Histogram

In the SPSS Data View Window, go to Window then select Output to retrieve the Histogram.

Histogram

Mean = 47.66
Std. Dev. = 18.176
N = 80

Histogram for Age in Years

Using the third SPSS path:
Analyze > Descriptive Statistics > Explore (Figure 4.48)

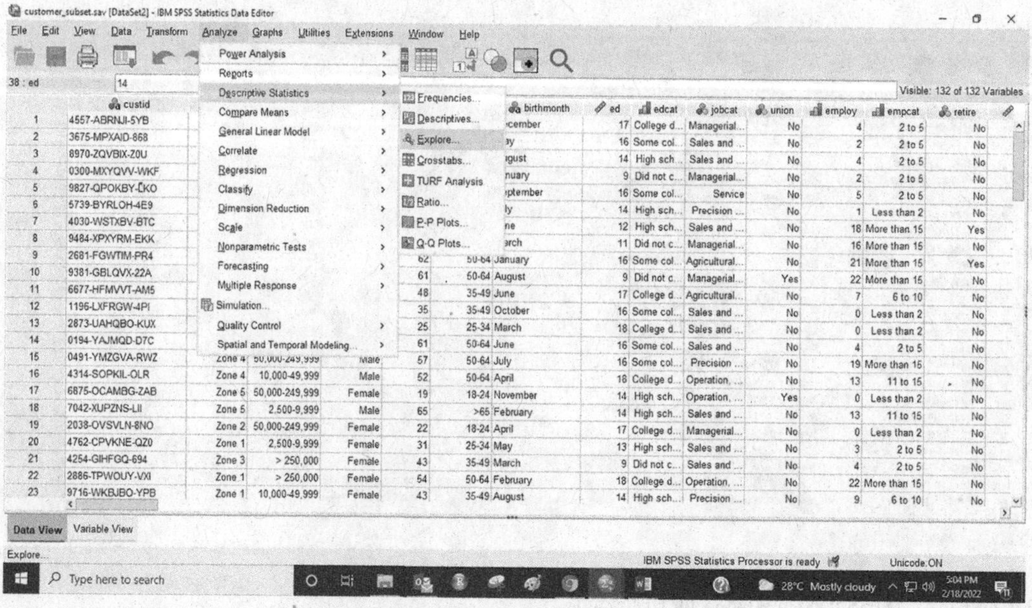

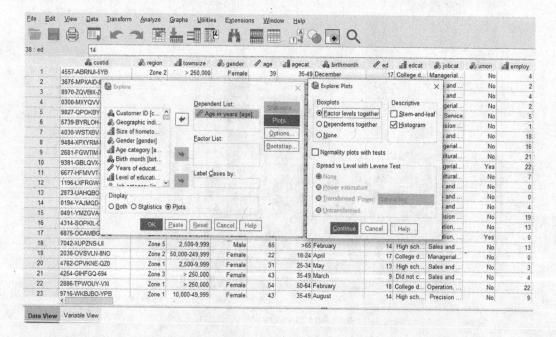

FIGURE 4.48   SPSS Menu for Explore

There will be a pop-up window for Explore > from the variables list select the numerical variable of interest (Age in years) and put it under Dependent List > Click Plots select Histogram then Continue > Under Display in the Explore popup window select Plots then click OK (Figure 4.49).

FIGURE 4.49   SPSS Pop Up Window for Explore displaying Histogram Plot

In the SPSS Data View Window, go to Window then select Output to retrieve the Histogram.

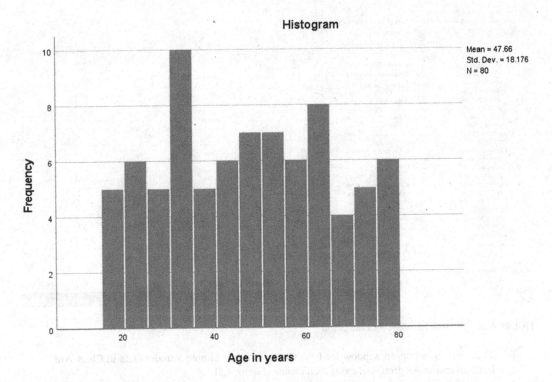

Histogram for Age in Years

4.2.4 Line Graph

A line graph is a plot of distinct data values on the horizontal axis (x-axis) and frequencies (y-axis). To generate a bar graph in SPSS we use the path:

Graphs > Legacy Dialogs > Line > There will be a pop-up window for Line Charts > Select Simple > under Data in Chart Are select Summaries for groups of cases then Define > There will be a pop-up window for Summaries for Groups of Cases > select the numerical variable of interest and put it under Category Axis then click OK.

Example

Using *customer_subset.sav* data (inbuilt SPSS data), generate a line graph for Age in Years.

For Windows
To locate *customer_subset.sav* data we use the path: Click **Open data document** folder > drive **(C): > Program Files (x86) > IBM > SPS > Statistics > 27 > Samples > English** > customer_subset.sav data (In-built *SPSS data files*).

For MAC
Open **SPSS data** > Click **Open data document** folder > Applications > **IBM SPS Statistics 27 > Samples > English** > customer_subset.sav data (In-built *SPSS data files*).

In SPSS go to Graphs > Legacy Dialogs > Line (Figure 4.50)

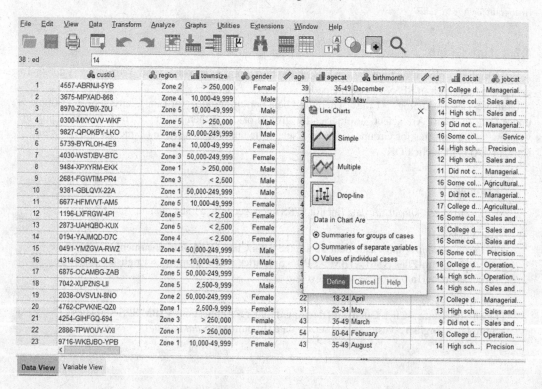

FIGURE 4.50   SPSS Graph Menu for Line Graph

There will be a pop-up window for Line Charts > Select Simple > under Data in Chart Are select Summaries for groups of cases then Define (Figure 4.51).

FIGURE 4.51   SPSS Pop Up Window for Line Charts

There will be a pop-up window for Summaries for Groups of Cases > select Age in years and put it under Category Axis then click OK (Figure 4.52).

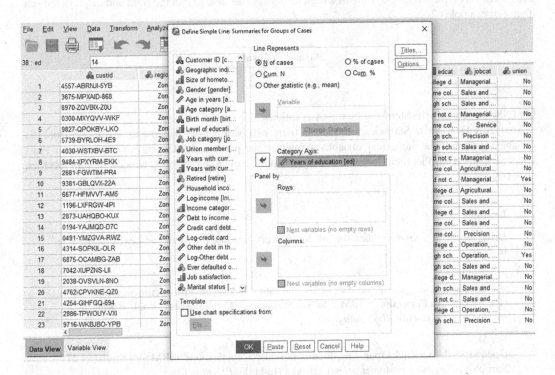

FIGURE 4.52 SPSS Pop Up Window for Line Charts displaying Summaries for Group of Cases

In the SPSS Data View Window, go to Window then select Output to retrieve the line graph (Figure 4.53).

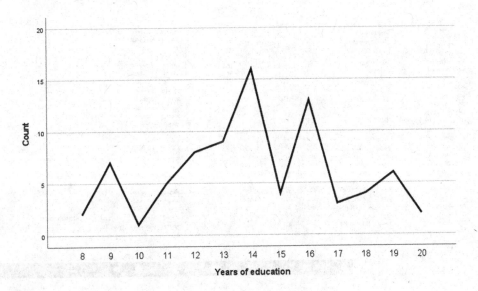

FIGURE 4.53 Line Graph for Years of Education

4.2.5 Stem and Leaf Plot

A stem and leaf plot is obtained by first dividing the data values into two parts – its stem and its leaf. The "**stem**" is on the left displays the first digit or digits. The "**leaf**" is on the right and displays the last digit. For example 62

Stem	Leaf
6	2

In SPSS, we generate the Stem and Leaf plot as follows:

Analyze > Descriptive Statistics > Explore > there will be a pop-up window for Explore > from the variables list select the numerical variable of interest and put it under Dependent List > Under Plots select Stem and Leaf Plot then Continue > Under Display in the Explore popup window select Plots then click OK.

Example

Using *customer_subset.sav* data (inbuilt SPSS data), generate the Stem and Leaf Plot for Age in Years.

For Windows
> To locate *customer_subset.sav* data we use the path: Click **Open data document** folder > drive **(C):** > **Program Files (x86)** > **IBM** > **SPS** > **Statistics** > **27** > **Samples** > **English** > customer_subset.sav data (In-built *SPSS data files*).

For MAC
Open **SPSS data** > Click **Open data document** folder > Applications > **IBM SPS Statistics 27** > **Samples** > **English** > customer_subset.sav data (In-built *SPSS data files*).

In SPSS, go to Analyze > Descriptive Statistics > Explore

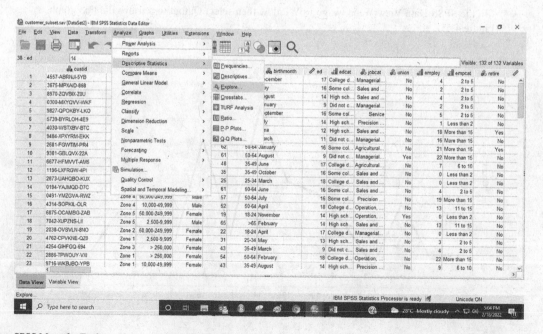

SPSS Menu for Explore

There will be a pop-up window for Explore > from the variables list select Age in years and put it under Dependent List > Under Plots select Stem and Leaf Plot then Continue > Under Display in the Explore popup window select Plots then click OK (Figure 4.54).

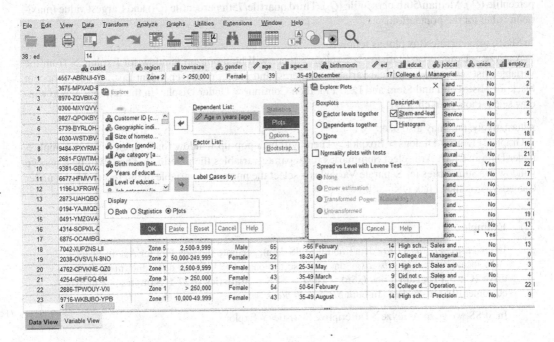

FIGURE 4.54 SPSS Pop Up Window for Explore displaying Stem-and-Leaf Plot

In the SPSS Data View Window, go to Window then select Output to retrieve the Stem and Leaf Plot (Figure 4.55).

Age in Years		
Age in Years Stem-and-Leaf Plot		
Frequency	Stem &	Leaf
5.00	1.	88899
11.00	2.	11124567899
14.00	3.	12222235557889
11.00	4.	01233357789
16.00	5.	0002234455677899
11.00	6.	11123445789
12.00	7.	012245778999

FIGURE 4.55 Stem-and-Leaf Plot for Age in Years

Interpretation: Just like the shape of the histogram, age in years is approximately normally distributed.

4.2.6 Box Plot

A box plot is a graphical summary that is based on the five-number summary. **The five-number summary includes five items:** Smallest value (minimum for the population of interest), First quartile/25th percentile (Q_1), Median/50th percentile (Q_2), Third quartile/75th percentile (Q_3) and Largest value (maximum value for the population).

In SPSS, we generate Box plot as follows:

Analyze > Descriptive Statistics > Explore > there will be a pop-up window for Explore > from the variables list select the numerical variable of interest and put it under Dependent List > Under Plots put black for Histogram and Stem and Leaf Plot then Continue > Under Display in the Explore popup window select Plots then click OK.

Alternatively, one may also use:

Graphs > Legacy Dialogs > Boxplot > There will be a pop-up window for Boxplot > Select Simple > under Data in Chart Are select Summaries for separate variables then Define > There will be a pop-up window for Summaries for Separate Variables > select the numerical variable of interest and put it under Category Axis then click OK.

Example

Using *customer_subset.sav* data (inbuilt SPSS data), generate the Box Plot for Age in Years.

To locate *customer_subset.sav* data we use the path: Click **Open data document** folder > drive **(C): > Program Files (x86) > IBM > SPS > Statistics > 27 > Samples > English >** customer_subset.sav data (In-built *SPSS data files*).

In SPSS, we go to Analyze > Descriptive Statistics > Explore

SPSS Menu for Explore

There will be a pop-up window for Explore > from the variables list select Age in years and put it under Dependent List > Under Plots don't select for Histogram and Stem and Leaf Plot then Continue > Under Display in the Explore popup window select Plots then click OK (Figure 4.56).

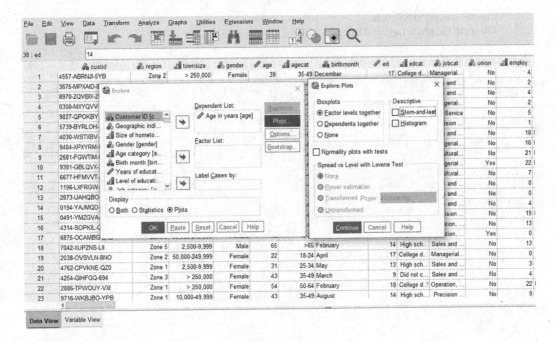

FIGURE 4.56 SPSS Pop Up Window for Explore displaying Plots

In the SPSS Data View Window, go to Window then select Output to retrieve the Boxplot (Figure 4.57).

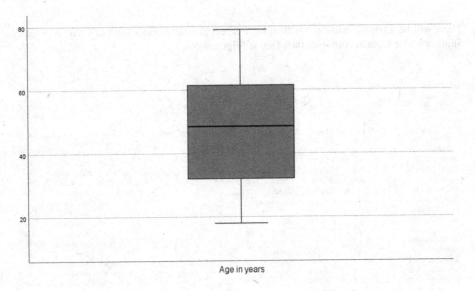

Age in years

FIGURE 4.57 Box Plot for Age in Years

Interpretation: to interpret the boxplot we can look at the outliers if present. In this case there were no outliers for variable age in years.

Usually values above and below the whisker are outliers. We can also interpret the 5 points (i.e., the minimum, maximum, 25th percentile, 50th percentile and 75th percentile). In this case, from the Boxplot, the maximum/ highest age for this population of interest is about 80 years. 75% or three-quarter of the population have about 60 years. 50% or half of the population have about 50 years. 25% or a quarter of the population have about 30 years and the lowest age for this population was about 19 years.

Using the second SPSS path:
  We go to Graphs > Legacy Dialogs > Boxplot (Figure 4.58)

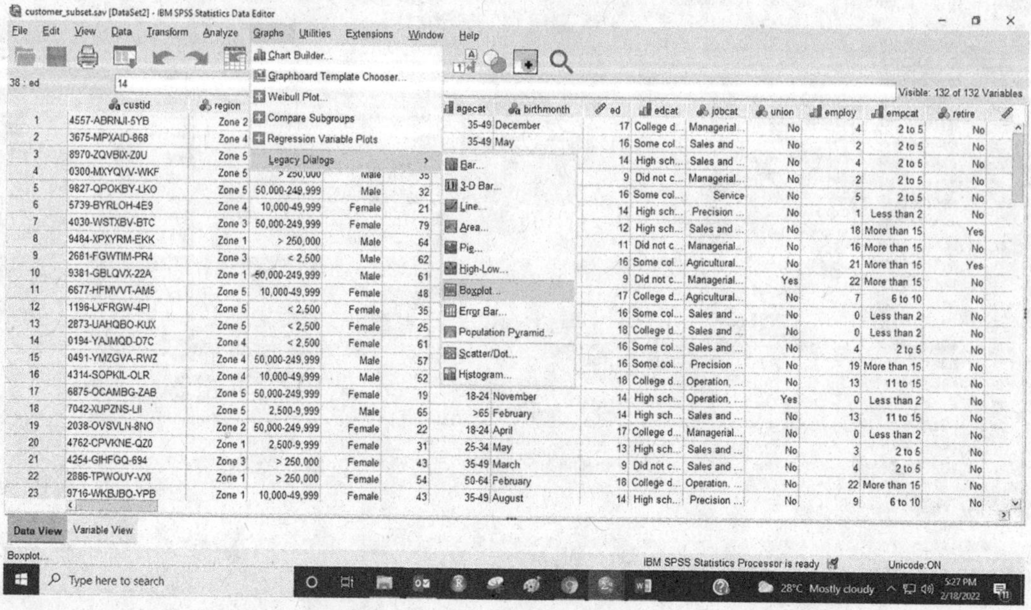

FIGURE 4.58 SPSS Graph Menu for Box Plot

There will be a pop-up window for Boxplot > Select Simple > under Data in Chart Are select Summaries for separate variables then Define (Figure 4.59).

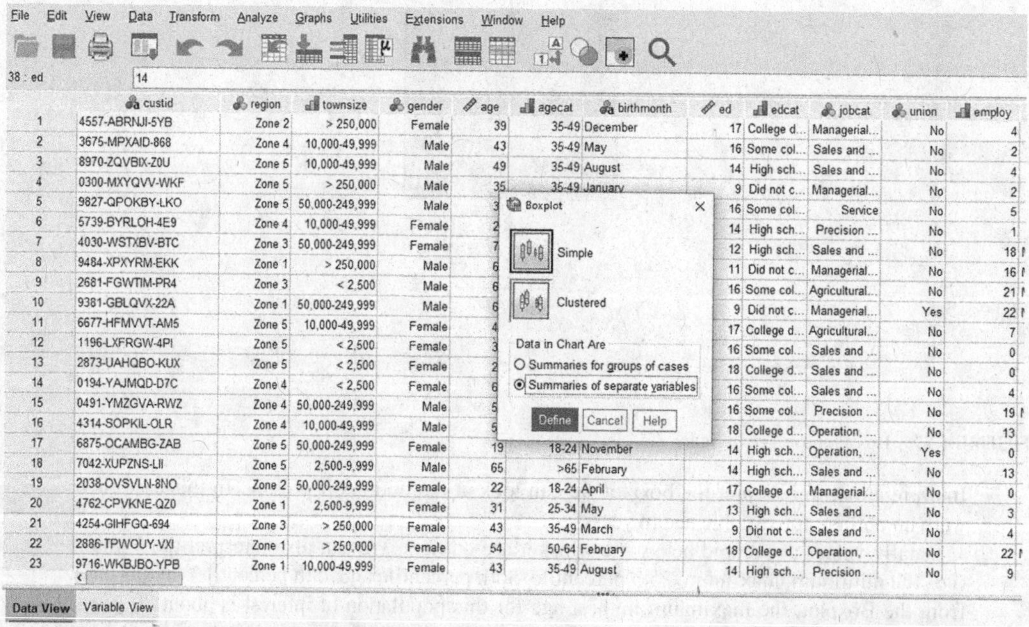

FIGURE 4.59 SPSS Pop Up Window for defining Box Plot

There will be a pop-up window for Summaries for Separate Variables > select the numerical variable of interest and put it under Category Axis then click OK (Figure 4.60).

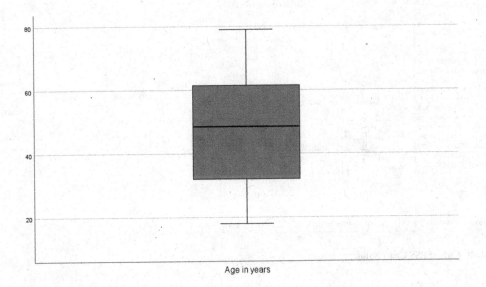

FIGURE 4.60   SPSS Pop Up Window for Simple Box Plot displaying Summaries of Separate Variables

In the SPSS Data View Window, go to Window then select Output to retrieve the Boxplot.

Box Plot for Age in Years

4.2.7 Frequency Polygon

A frequency polygon is a plot of frequencies of different data values on the vertical axis (y-axis) and distinct points/mid points (x-axis). To generate a frequency polygon in SPSS.

Graphs > Chart Builder > OK > There will be a pop up window for Chart Builder > Under the Gallery click Histogram and it will bring four charts from which you will select the third graph (Frequency Polygon) > Drag the frequency polygon to SPSS window that is on the right of the variables list > Drag the numerical variable of interest and put it on the – Axis then Click OK.

Example

Using *customer_subset.sav* data (inbuilt SPSS data), generate a Frequency Polygon for Age in Years.

For Windows
To locate *customer_subset.sav* data we use the path: Click **Open data document** folder > drive **(C):** > **Program Files (x86)** > **IBM** > **SPS** > **Statistics** > **27** > **Samples** > **English** > customer_subset.sav data (In-built *SPSS data files*).

For MAC
Open **SPSS data** > Click **Open data document** folder > Applications > **IBM SPS Statistics 27** > **Samples** > **English** > customer_subset.sav data (In-built *SPSS data files*).

Graphs > Chart Builder > Select OK > There will be a pop-up window for Chart Builder > Under the Gallery click Histogram (Figure 4.61) and it will bring four charts from which you will select the third graph (Frequency Polygon)

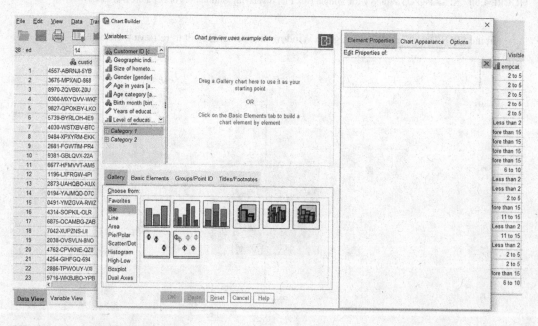

FIGURE 4.61 SPSS Chart Builder

Drag the frequency polygon (the third graph) to SPSS window that is on the right of the variables list (Figure 4.62).

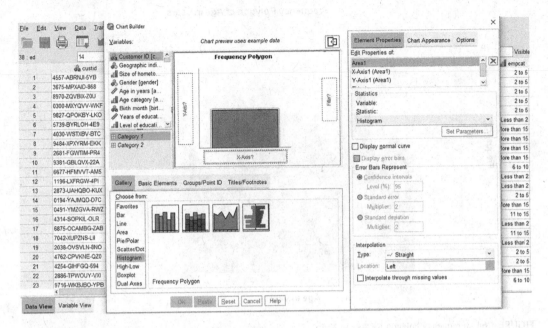

FIGURE 4.62   SPSS Char Builder for Frequency Polygon

Drag Age in years from the list of Variables and put it on the X–Axis then Click OK (Figure 4.63).

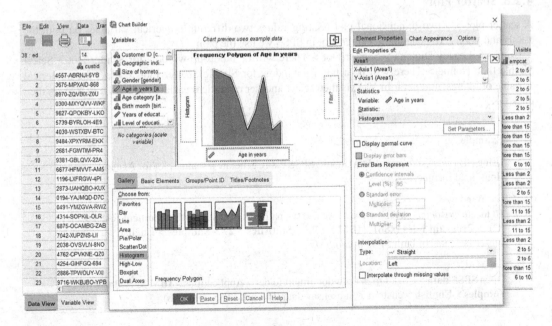

FIGURE 4.63   SPSS Chart Builder for Frequency Polygon for Age in Years

In SPSS Data View, click Window then Output (Figure 4.64).

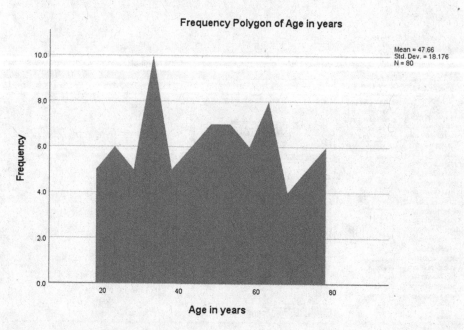

FIGURE 4.64   Frequency Polygon for Age in Years

Interpretation: Majority of the respondents had about 30 years, followed closely by 60 years.

4.2.8 Scatter Plot

A scatter plot is a two-dimensional plot relating values two different numerical variables on both the x-axis and y-axis. To generate a Scatter Plot in SPSS we use the path:

Graphs > Legacy Dialogs > Scatter/Dot > There will be a pop-up window for Scatter/Dot > Select Simple then Define > There will be a pop-up window for Simple Scatter Plot > select the numerical variable of interest and put it in Y Axis then select another numerical variable and put it in X Axis then click OK.

Example

Using *customer_subset.sav* data (inbuilt SPSS data), generate a scatter plot for Age in Years and Household Income.

For Windows
To locate *customer_subset.sav* data we use the path: Click **Open data document** folder > drive **(C): > Program Files (x86) > IBM > SPS > Statistics > 27 > Samples > English** > customer_subset.sav data (In-built *SPSS data files*).

For MAC
Open **SPSS data** > Click **Open data document** folder > Applications > **IBM SPS Statistics 27 > Samples > English** > customer_subset.sav data (In-built *SPSS data files*).

In SPSS go to Graphs > Legacy Dialogs > Scatter (Figure 4.65)

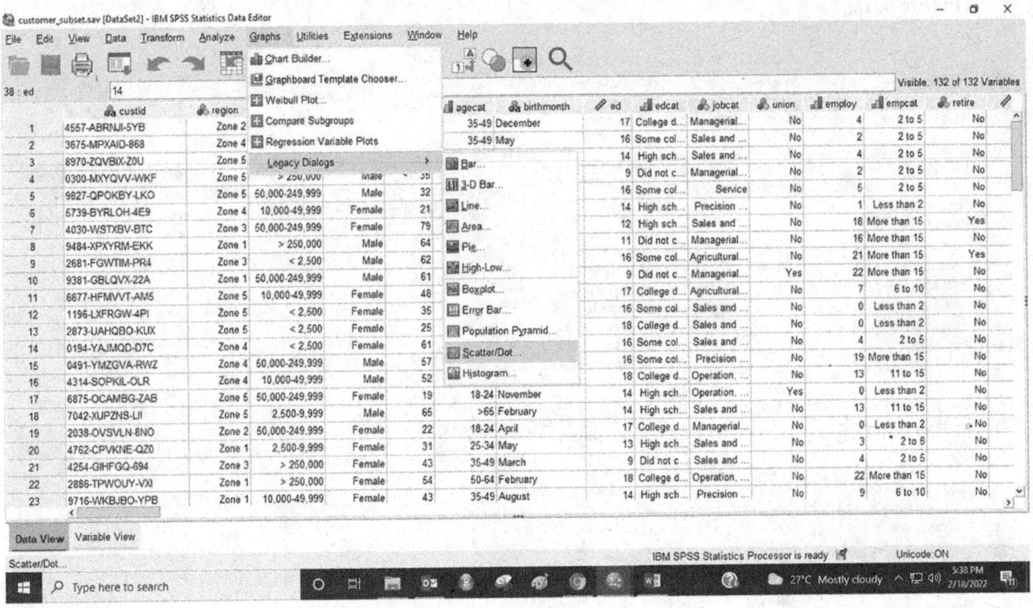

FIGURE 4.65 SPSS Graph Menu for Scatterplot

There will be a pop-up window for Scatter/Dot > Select Simple then Define (Figure 4.66)

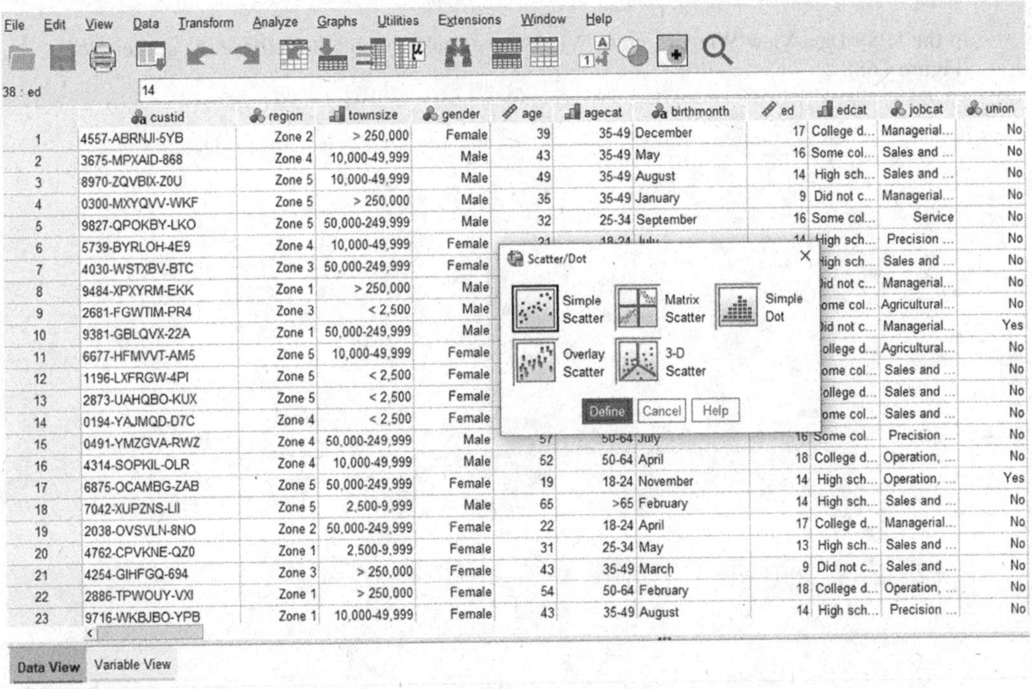

FIGURE 4.66 SPSS Pop Up Window for defining Scatterplot

There will be a pop-up window for Simple Scatter Plot > select Age in years and put it under Y Axis then select Household income in thousands and put it under X Axis then click OK (Figure 4.67).

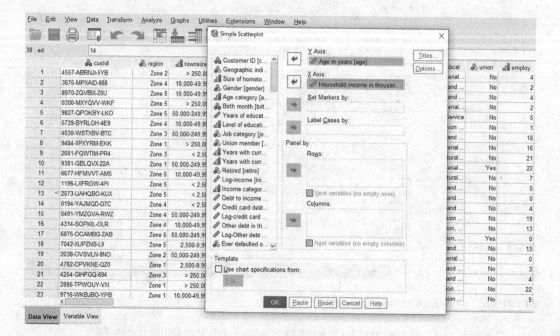

FIGURE 4.67   SPSS Pop Up Window for selecting variables for Simple Scatterplot

In the SPSS Data View Window, go to Window then select Output to retrieve the scatter plot (Figure 4.68).

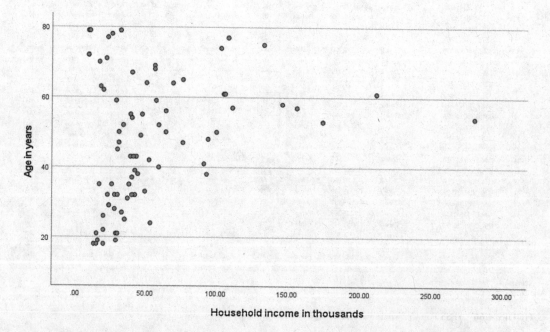

FIGURE 4.68   Scatter Plot for Age in Years and Household Income in Thousands

Interpretation: The above graph shows that there is a positive relationship between household income in thousands and age in years.

4.2.9 Cumulative Frequency Curve/Ogive

A cumulative frequency curve is a graph of cumulative frequencies (on y-axis) against the class limits on (x-axis). Also known as an Ogive. There are usually two types of ogives

- "Less than ogive" is when less than c.f. is plotted against the upper limit of the corresponding classes
- "More than ogive" is when the more than c.f. is plotted against the upper limit of the corresponding classes

In SPSS we can only generate a cumulative frequency curve from a given frequency table, i.e. relative ogive using the following path:

We first generate a frequency table with cumulative percentages using Analyze > Descriptive Statistics > Frequencies > A pop up window for Frequencies will appear > Select numeric variable of interest > ensure that Display Frequency Table is checked/ticked then click OK

Once you have the frequency table double click on the table > select the cumulative percentages > right click in the selection > create graph > select line

Example:

Using *customer_subset.sav* data (inbuilt SPSS data), generate a cumulative frequency curve for Age in years

For Windows
To locate *customer_subset.sav* data we use the path: Click **Open data document** folder > drive (C): > **Program Files (x86) > IBM > SPS > Statistics > 27 > Samples > English** > customer_subset.sav data (In-built *SPSS data files*).

For MAC
Open **SPSS data** > Click **Open data document** folder > Applications > **IBM SPS Statistics 27 > Samples > English** > customer_subset.sav data (In-built *SPSS data files*).

To generate a frequency table without intervals for gender we use the following SPSS path:
Analyze > Descriptive Statistics > Frequencies > A pop up window for Frequencies will appear > Select Age in Years > ensure that Display Frequency Table is checked/ticked then click OK (Figure 4.69)

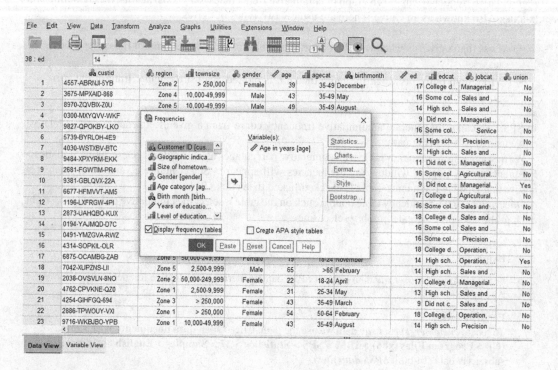

FIGURE 4.69   SPSS Pop Up Window for generating Frequency Tables

In the SPSS Data View Window, select Window the Output to view the frequency table

Age in Years

		Frequency	Percent	Valid Percent	Cumulative Percent
Valid	18	3	3.8	3.8	3.8
	19	2	2.5	2.5	6.3
	21	3	3.8	3.8	10.0
	22	1	1.3	1.3	11.3
	24	1	1.3	1.3	12.5
	25	1	1.3	1.3	13.8
	26	1	1.3	1.3	15.0
	27	1	1.3	1.3	16.3
	28	1	1.3	1.3	17.5
	29	2	2.5	2.5	20.0
	31	1	1.3	1.3	21.3
	32	5	6.3	6.3	27.5
	33	1	1.3	1.3	28.7
	35	3	3.8	3.8	32.5
	37	1	1.3	1.3	33.8
	38	2	2.5	2.5	36.3
	39	1	1.3	1.3	37.5
	40	1	1.3	1.3	38.8
	41	1	1.3	1.3	40.0
	42	1	1.3	1.3	41.3
	43	3	3.8	3.8	45.0
	45	1	1.3	1.3	46.3
	47	2	2.5	2.5	48.8
	48	1	1.3	1.3	50.0
	49	1	1.3	1.3	51.2
	50	3	3.8	3.8	55.0
	52	2	2.5	2.5	57.5
	53	1	1.3	1.3	58.8
	54	2	2.5	2.5	61.3
	55	2	2.5	2.5	63.7
	56	1	1.3	1.3	65.0
	57	2	2.5	2.5	67.5
	58	1	1.3	1.3	68.8
	59	2	2.5	2.5	71.3
	61	3	3.8	3.8	75.0
	62	1	1.3	1.3	76.3
	63	1	1.3	1.3	77.5
	64	2	2.5	2.5	80.0
	65	1	1.3	1.3	81.3
	67	1	1.3	1.3	82.5
	68	1	1.3	1.3	83.8
	69	1	1.3	1.3	85.0
	70	1	1.3	1.3	86.3
	71	1	1.3	1.3	87.5
	72	2	2.5	2.5	90.0
	74	1	1.3	1.3	91.3
	75	1	1.3	1.3	92.5
	77	2	2.5	2.5	95.0
	78	1	1.3	1.3	96.3
	79	3	3.8	3.8	100.0
	Total	80	100.0	100.0	

FIGURE 4.70 Frequency Table without Intervals for Age in Years

Go to SPSS output window where there is a frequency table for Age in years and double click on the table there will be a pop-up window for Pivot Table Age in Years > select the cumulative percentages > right click in the selection > create graph > select line then close the window for Pivot Table Age in Years (Figure 4.71)

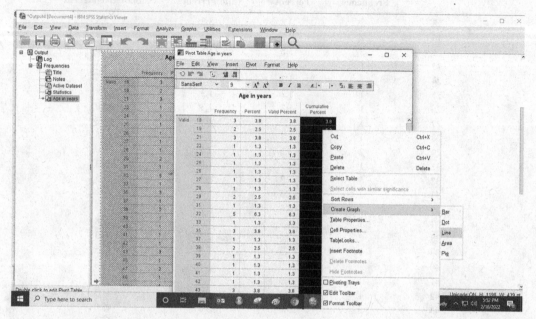

FIGURE 4.71   Generating Line Graph from the Frequency Table Chart Editor

In the SPSS output, we will have the cumulative frequency curve (Figure 4.72).

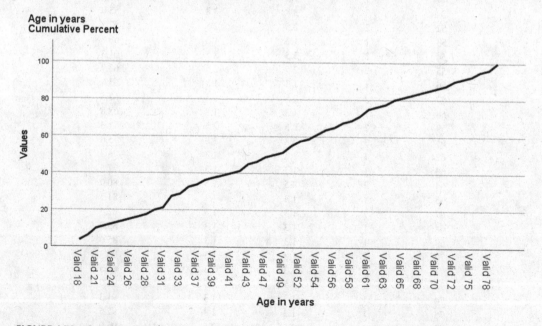

FIGURE 4.72   Cumulative Percent Graph for Age in Years

Interpretation: The total cumulative proportion of ages for individuals have been increasing gradually to 100%. The ogive can also be used to estimate the percentage of individuals with a given age in years.

4.3 Numerical Measures

There are four different numerical measures of presenting data

- Measures of central tendency
  - Mean, Median, Mode, Partition (Quartiles, Decile, Percentile)
- Measures of variability/dispersion/spread
  - Range, Interquartile range, Variance, Standard deviation, Coefficient of variation
- Measures of shape – shape of distribution
  - Skewness and Kurtosis
- Measures of association
  - Correlation

Most of these numerical measures are applicable to quantitative data. Only mode is applicable to categorical data.

4.3.1 Measures of Central Tendencies

i. Mean

Arithmetic Mean

The arithmetic mean or average, denoted by $\bar{X}$, is calculated as follows

$$\bar{X} = \frac{\sum X_i}{n}$$

$$\bar{X} = \frac{X_1 + X_2 + \cdots + X_n}{total \ number \ of \ observations} = \frac{\sum X_i}{n}$$

Example: Compute the arithmetic mean value for the one-year energy consumption data below

| 66 | 103 | 75 | 111 | 153 | 139 | 81 | 55 | 66 | 90 | 98 | 97 |

$$\bar{X} = \frac{\sum X_i}{n} = \frac{66 + 103 + 75 + 111 + 153 + 139 + \cdots + 90 + 98 + 97}{12} = 94.50$$

The arithmetic mean is the most commonly used mean and can be easily generated using SPSS.
Other types of mean

Geometric Mean

Used when working with percentages. Geometric mean, denoted by $\bar{X}_G$ for n positive number, is the nth root of their product.

$$\bar{X}_G = \sqrt[n]{X_1.X_2 \ldots X_n}$$

Harmonic Mean

Harmonic mean is used to average ratios/rates. The Harmonic mean, denoted by $\bar{X}_H$, is computed as

$$\bar{X}_H = \frac{N}{\sum \left(\frac{1}{X} \right)}$$

In SPSS, we can compute the Harmonic and Geometric mean using the path:

Analyze > Compare Means > Means > There will be a pop-up SPSS window for Mean > Select the Numerical Variable of interest and put it under Dependent List Window > Click Options > There will be a pop up SPSS window for Means: Options > Under Statistics select Harmonic Mean and Geometric Mean and move them to Cell Statistics Window > Click Continue then OK.

Example: Generate the geometric and harmonic mean for the following sampled data for one-year energy consumption by different households

66   103   75   111   153   139   81   55   66   90   98   97

Geometric mean is computed using

$$\bar{X}_G = \sqrt[n]{X_1.X_2...X_n}$$

$$\bar{X}_G = \sqrt[12]{66 \times 103 \times 75 \times 111 \times 153 \times 139 \times ... \times 90 \times 98 \times 97} = 90.5229$$

Harmonic mean is computed as

$$\bar{X}_H = \frac{N}{\Sigma\left(\frac{1}{X}\right)}$$

$$\bar{X}_H = \frac{12}{\left(\frac{1}{66} + \frac{1}{103} + \frac{1}{75} + \frac{1}{111} + \frac{1}{153} + \frac{1}{139} \cdots + \frac{1}{90} + \frac{1}{98} + \frac{1}{97}\right)} = \frac{12}{0.138236} = 86.8079$$

Using SPSS, we first create the variable in Variable View, (Figure 4.73) then key the data into data view.

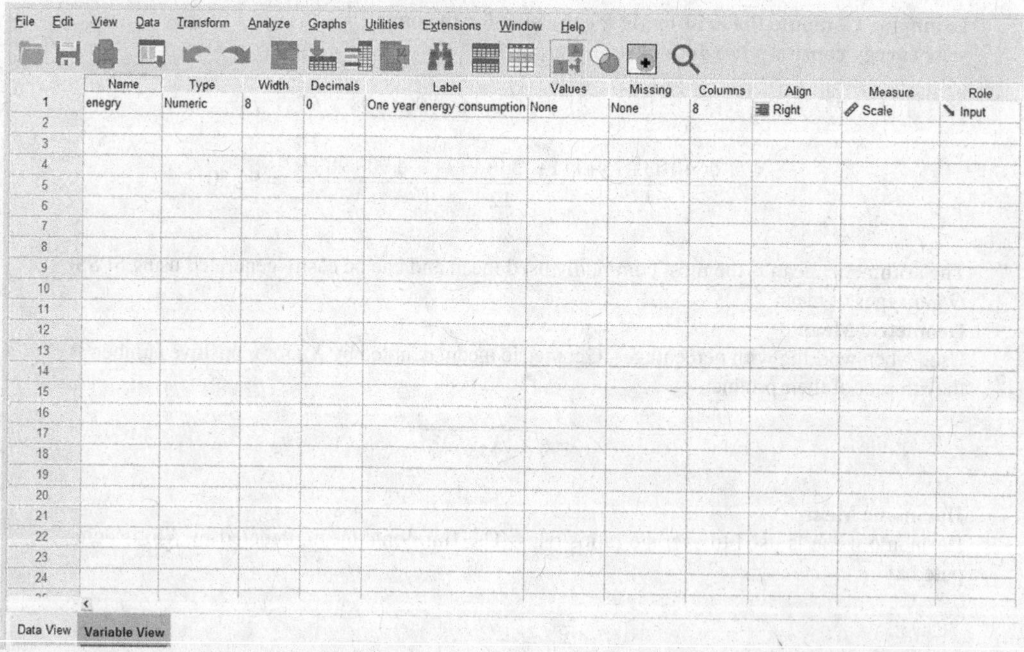

FIGURE 4.73   SPSS Variable View Window for One Year Energy Consumption Data

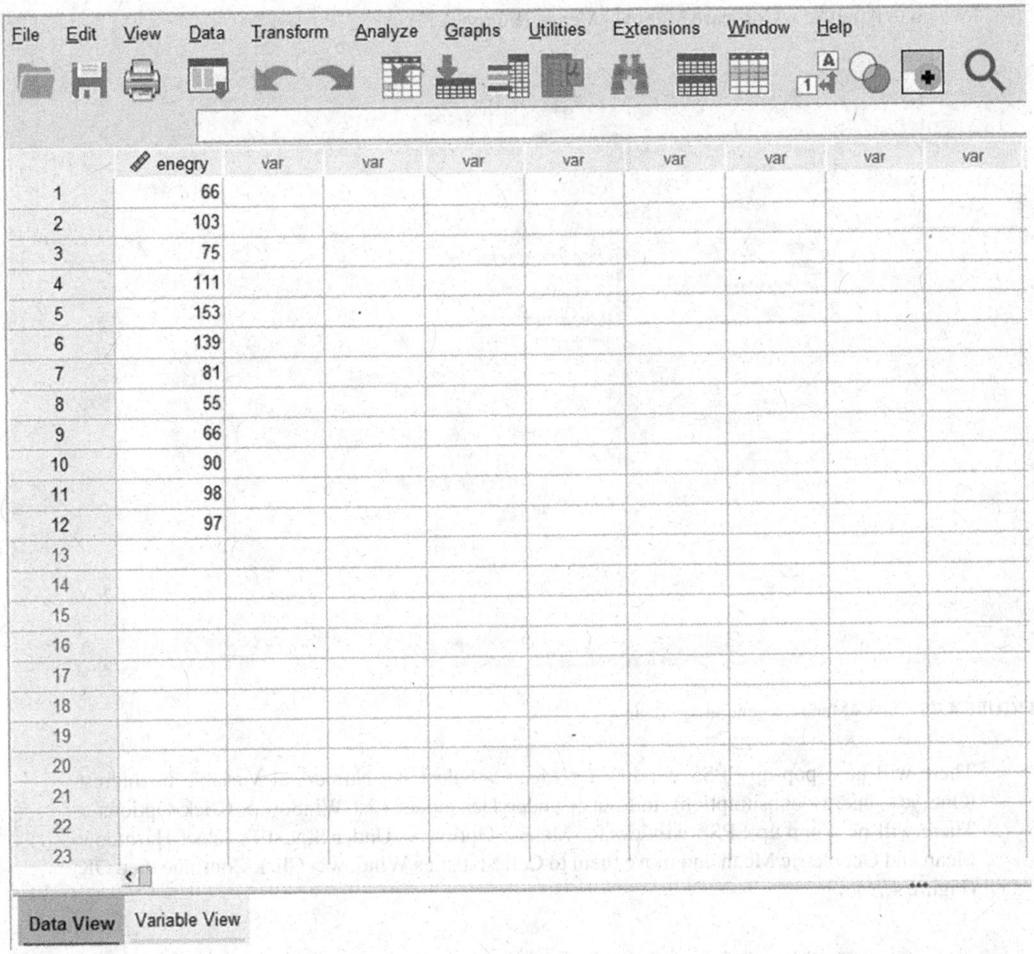

	⬗ enegry	var	var	var	var	var	var	var	var
1	66								
2	103								
3	75								
4	111								
5	153								
6	139								
7	81								
8	55								
9	66								
10	90								
11	98								
12	97								
13									
14									
15									
16									
17									
18									
19									
20									
21									
22									
23									

Data View Variable View

FIGURE 4.74 SPSS Data View Window for One Year Energy Consumption Data

We go to Analyze > Compare Means > Means (Figure 4.75)

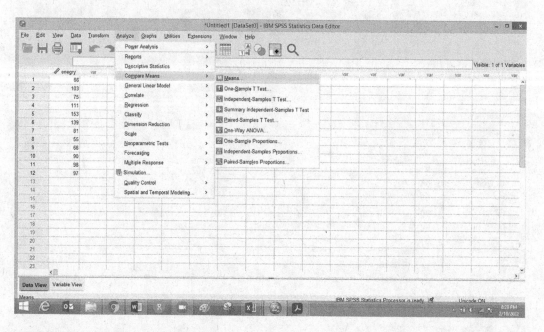

FIGURE 4.75   SPSS Menu for generating Means

There will be a pop-up SPSS window for Mean > Select the Numerical Variable of interest (One year energy consumption) and put it under Dependent List Window > Click Options > There will be a pop up SPSS window for Means: Options > Under Statistics select Harmonic Mean and Geometric Mean and move them to Cell Statistics Window > Click Continue then OK (Figure 4.76).

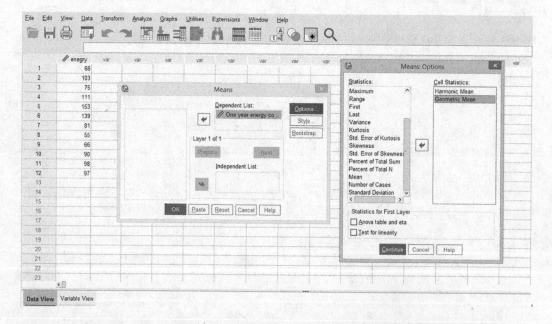

FIGURE 4.76   SPSS Pop Up Widow for generating Harmonic and Geometric Mean for One Year Energy Consumption Data

The SPSS output will give the Geometric mean and Harmonic mean.

Report
One-Year Energy Consumption

Harmonic Mean	Geometric Mean
86.81	90.52

FIGURE 4.77 Harmonic and Geometric Mean Values for the One Year Energy Consumption Data

Interpretation: Harmonic mean - the average one year energy consumption for 12 different households was about 86.81 units.

Geometric mean - the average one year energy consumption for 12 different households was about 90.52 units.

Harmonic and Geometric means are forms of averages.

ii. **Median**

Median of a distribution is the value of the variable, which divides it into two equal parts. To compute median value:

- Arrange data in ascending or descending order
- For an odd number of observations, the median is the middle value
- For an even number of observations, the median is the average of the two middle values
- Median can also be estimated by percentiles – it is the 50th percentiles

Median can also be computed using percentiles

Step 1: Arrange the data in ascending order

Step 2: Compute an index i

$$i = \left(\frac{p \times n}{100} \right)$$

where p is the percentile of interest and n is the number of observations

Step 3(a): If i is not an integer (i.e. with decimal), round up

Step 3(b): If i is an integer (whole number), the pth percentile is the average of the values in positions i and $i+1$

Example: Compute the median value for the one-year energy consumption data below

66	103	75	111	153	139	81	55	66	90	98	97

SOLUTION:

Step 1: Arrange the data in ascending order

55	66	66	75	81	90	97	98	103	111	139	153

Step 2: Compute an index i – median is the 50th percentile

$$i = \left(\frac{p \times n}{100} \right) = \left(\frac{50 \times 12}{100} \right) = 6$$

where p is the percentile of interest = 50 and n is the total number of observations = 12

Since $i = 6$, its an integer, so we go to step 3(b)

Step 3(a): If i is not an integer (i.e. with decimals), round up

Step 3(b): if i is an integer (whole number), the pth percentile is the average of the values in positions i and $i+1$

Position i is the 6th position, while position $i+1$ is the 7th position

The median value will be the average of values in position 6 and 7

$$Median = \frac{90+97}{2} = 93.5$$

iii. Mode

Mode is the value that occurs most frequently in the data set/set of observations.
Mode is the observation that appears several times.

Example: Compute the modal value for the one-year energy consumption data below

66   103   75   111   153   139   81   55   66   90   98   97

SOLUTION:

Arranging the data in ascending order allows one to visualize the most appearing value

55   66   66   75   81   90   97   98   103   111   139   153

In this case mode is 66

In SPSS, we can compute the Arithmetic Mean, Median and Mode using the path:
    Analyze > Descriptive Statistics > Frequencies > There will be a pop up SPSS window for Frequencies > Select the Numerical Variable of interest and put it in Variable(s) Window > Click Statistics > There will be a pop up SPSS window for Statistics > Under Central Tendency select Mean, Median and Mode then Continue and then OK.

Using the one-year energy consumption data:
    We go to Analyze > Descriptive Statistics > Frequencies > There will be a pop up SPSS window for Frequencies > select the Numerical Variable of interest (One year energy consumption) and put it in Variable(s) Window > Click Statistics > There will be a pop up SPSS window for Statistics > Under Central Tendency select Mean, Median and Mode then Continue and then OK (Figure 4.78).

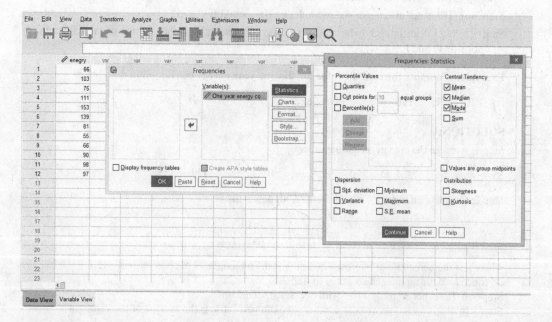

FIGURE 4.78   SPSS Pop Up Window for Frequencies displaying Measures of Central Tendency

The SPSS output will give the mean, median and modal values

Statistics		
One-Year Energy Consumption		
N	Valid	12
	Missing	0
Mean		94.50
Median		93.50
Mode		66

FIGURE 4.79 Mean, Median and Mode Values for the One Year Energy Consumption Data

Interpretation: the average energy consumption for 12 different households was 94.5. Half or 50% of for 12 different households consumed 93.5 and below. Most families from the 12 different households consumed energy amount of 66.

iv. **Partition values**

Partition values are the values which divide the series into a number of equal parts. They are Percentiles, Quartiles and Deciles. Ideally, the method for computing the partition values are the same as those of locating the median. Percentiles are usually used to estimate the partition values.

a. **Percentiles**

Percentiles are ninety nine point which divide the series into 100 equal parts. We can have 1st percentile to 99th percentile. For example, the forty seventh percentile, P_{47}, is the point which exceed 47% of the observations.

Example: Compute the 47th percentile for the one-year energy consumption data below

66   103   75   111   153   139   81   55   66   90   98   97.

SOLUTION:

Step 1: Arrange the data in ascending order

55   66   66   75   81   90   97   98   103   111   139   153

Step 2: Compute an index i – the 47th percentile

$$i = \left(\frac{p \times n}{100}\right) = \left(\frac{47 \times 12}{100}\right) = 5.64$$

Where p is the percentile of interest = 47 and n is the total number of observations = 12
    Since $i = 5.64$, it is not an integer, so we go to step 3(a)
Step 3(a): If i is not an integer (i.e. with decimals), round up
Step 3(b): if i is an integer (whole number), the pth percentile is the average of the values in positions i and $i+1$
    Rounding up 5.64 gives 6. Hence the value in position 6 is the 47th percentile.

$$P_{47} = 90$$

In SPSS, we can compute Percentiles using the path:

Analyze > Descriptive Statistics > Frequencies > There will be a pop up SPSS window for Frequencies > Select the Numerical Variable of interest and put it in Variable(s) Window > Click Statistics > There will be a pop up SPSS window for Frequencies: Statistics > Under Percentile Value click Percentile(s) then type the percentile value(s) of interest then click Add > then Continue and then OK.

Using the above example, we can compute the 47th percentile for the one-year energy consumption data using the path:

Analyze > Descriptive Statistics > Frequencies > There will be a pop up SPSS window for Frequencies > Select the Numerical Variable of interest (one year energy consumption) and put it in Variable(s) Window > Click Statistics > There will be a pop up SPSS window for Frequencies: Statistics > Under Percentile Value click Percentile(s) then type the percentile value(s) of interest (47) then click Add > then Continue and then OK (Figure 4.80).

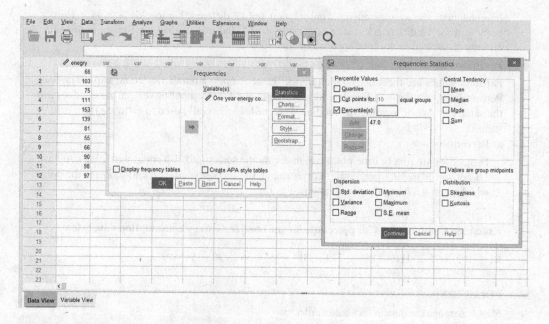

FIGURE 4.80   SPSS Pop Up Window for Frequencies displaying Percentiles

SPSS output will give the 47th percentile value

Statistics		
One-Year Energy Consumption		
N	Valid	12
	Missing	0
Percentiles	47	90.77

FIGURE 4.81   Forty Seventh Percentile Value for the One Year Energy Consumption Data

Interpretation: 47% of the 12 different households consumed 90.77 and below.

b. Quartiles

Quartiles are the three points which divide the series into four equal parts are called **quartiles**. We usually have three quartiles:

- The first quartile (Q1 = 25th percentile) is the value which exceed 25% of the observation and is exceeded by 75% of the observations
- The second quartile (Q2 = 50th percentile) coincides with median
- The third quartile (Q3 = 75th percentile) is the point which has 75% observations before it and 25% observations after it

To generate Quartiles we use the same path for generating percentiles but Under Percentile Value we click Percentile(s) then type the percentile values of interest (i.e. 25, 50 and 75) then click Add or select Quartiles > then Continue and then OK.

Example: Using the above data for one-year energy consumption,
we can generate quartiles using SPSS via the path

Analyze > Descriptive Statistics > Frequencies > There will be a pop up SPSS window for Frequencies > Select the Numerical Variable of interest (one year energy consumption) and put it in Variable(s) Window > Click Statistics > There will be a pop up SPSS window for Frequencies: Statistics > Under Percentile Value we click Percentile(s) then type the percentile values of interest (i.e. 25, 50 and 75) then click Add or select Quartiles or click > then Continue and then OK (Figure 4.82).

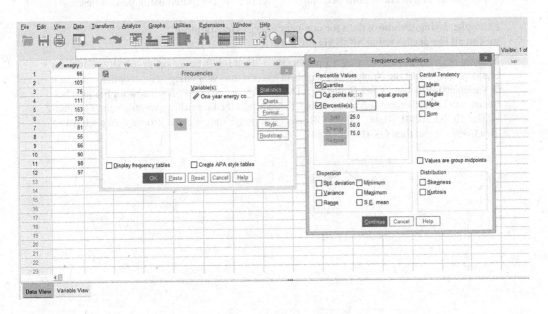

FIGURE 4.82   SPSS Pop Up Window for Frequencies displaying Quartiles

The SPSS output will give the quartile values as 25th percentile, 50th percentile and 75th percentile

Statistics		
One-Year Energy Consumption		
N	Valid	12
	Missing	0
Percentiles	25	68.25
	50	93.50
	75	109.00

FIGURE 4.83   Quartile Values for the One Year Energy Consumption Data

The 25th percentile is 68.25, 50th percentile is 93.5 and the 75th percentile is 109.0

Interpretation: 25th percentile (first quartile) is 68.25 implying that 25% or a quarter of the 12 different households had consumed about 68.25 units of the energy and below. 50th percentile (second quartile) is 93.5 implying that 50% or half of the 12 different households had consumed about 93.5 units of energy and below. 75th percentile (third quartile) or three-quarter of the 12 different households had consumed about 109 units of energy and below.

c. Deciles

Deciles are the nine points which divide the series into ten equal parts are called *decile*. We usually have 1st decile to 9th decile. Deciles are the percentiles of 10's, i.e. 1st decile = 10th percentile, 2nd decile = 20th percentile … Deciles are computed using percentiles.

Example: Using the above data for one-year energy consumption, we can all the deciles using SPSS via the path

Analyze > Descriptive Statistics > Frequencies > There will be a pop up SPSS window for Frequencies > Select the Numerical Variable of interest (one year energy consumption) and put it in Variable(s) Window > Click Statistics > There will be a pop up SPSS window for Frequencies: Statistics > Under Percentile Value we click Percentile(s) then type the percentile values of interest (i.e. 10, 20, 30, 40, 50, 60, …, 90) OR select Cut points for ten equal groups then click Add > then Continue and then OK (Figure 4.84).

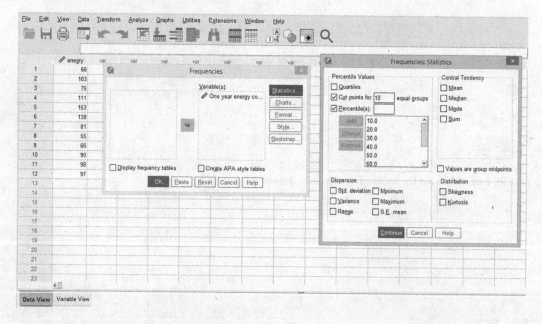

FIGURE 4.84   SPSS Pop Up Window for Frequencies displaying Deciles

The SPSS output will give the decile values as 10th percentile, 20th percentile, 30th percentile and so on

Statistics		
One-Year Energy Consumption		
N	Valid	12
	Missing	0
Percentiles	10	58.30
	20	66.00
	30	74.10
	40	82.80
	50	93.50
	60	97.80
	70	103.80
	80	122.20
	90	148.80

FIGURE 4.85   Decile Values for the One Year Energy Consumption Data

Interpretation: The interpretation of deciles is the same as that of the percentiles. For example, The second decile (or 20th percentile) was 66. This means that 20% of the 12 different households had consumed about 66 units of the energy and below.

4.3.2 Measures of Dispersion/Spread/Variability

Dispersion refers to the variability or spread in data. There are usually two types of measures of dispersion

- Absolute measures of dispersion – when dispersion is measured in original units, e.g. range
- Relative measures of dispersion – when dispersion is independent of original and are expressed in terms of ratio, percentage, etc. or when you use the already computed values, e.g. Variance

The most common measures of dispersions are Average deviation (Mean Absolute Deviation), Variance, Standard Deviation, Range, Interquartile Range, Quartile Deviation, Coefficient of Variation

i. **Average Deviation**

The Average Deviation is also called mean absolute deviation. The Average Deviation is computed as:

$$AD = \frac{\sum |X_i - \bar{X}|}{n}$$

Where X_i are the individual observation data while $\bar{X}$ is the mean

Example: Compute the Average Deviation or Mean Absolute Deviation for the one-year energy consumption data below

66   103   75   111   153   139   81   55   66   90   98   97

We first compute the mean value = $\bar{X} = 94.50$

We then subtract the mean value from each observation but ignoring the negative signs to the difference value. We then sum up the absolute difference values

$$AD = \frac{\sum |X_i - \bar{X}|}{n} = \frac{|66 - 94.5| + |103 - 94.5| + |75 - 94.5| + \cdots + |97 - 94.5|}{12} = \frac{268}{12} = 22.333$$

In SPSS, we can compute Average Deviation as follows:

We first compute the Mean value for the data using Analyze > Descriptive Statistics > Frequencies > There will be a pop up SPSS window for Frequencies > Select the Numerical Variable of interest and put it in Variable(s) Window > Click Statistics > There will be a pop up SPSS window for Frequencies: Statistics > Under Central Tendency select Mean then Continue and then OK.

Once you have the mean value go to Transform > Compute Variable > For the Target Variable type Absolute_Deviation > Under Function group click All > Under Functions and Special Variables double click Abs and it will appear in the Numeric Expression > Drag the numerical value of interest to the Numeric Expression window where there is Abs and subtract the mean value then click OK.

A new variable known as Absolute_Deviation will be created.

Go to Analyze > Descriptive Statistics > Frequencies > There will be a pop up SPSS window for Frequencies > Select the Absolute_Deviation and put it in Variable(s) Window > Click Statistics > There will be a pop up SPSS window for Frequencies: Statistics > Under Central Tendency select Mean then Continue and then OK.

Example

Generate the Average Deviation value for one-year energy consumption data

We go to Analyze > Descriptive Statistics > Frequencies > there will be a pop up SPSS window for Frequencies > Select one year energy consumption data and put it in Variable(s) Window > Click Statistics > There will be a pop up SPSS window for Frequencies: Statistics > Under Central Tendency select Mean then Continue and then OK (Figure 4.86).

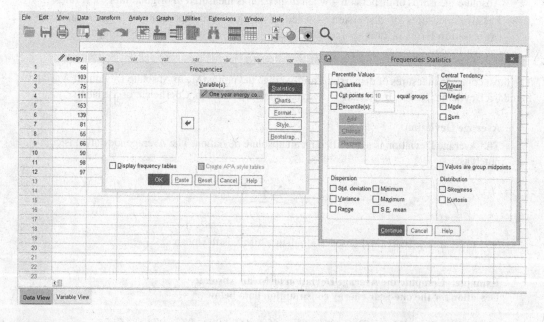

FIGURE 4.86 SPSS Pop Up Window for Frequencies displaying Mean

The SPSS output will give the mean value

Statistics		
One-Year Energy Consumption		
N	Valid	12
	Missing	0
Mean		94.50

FIGURE 4.87 Mean Value for the One Year Energy Consumption Data

The mean value is 94.5

Go to Transform > Compute Variable > For the Target Variable type Absolute_Deviation > Under Function group click All > Under Functions and Special Variables double click Abs and it will appear in the Numeric Expression > Drag Age in years to the Numeric Expression window where there is Abs and subtract the mean value of 47.66 then click OK (Figure 4.88).

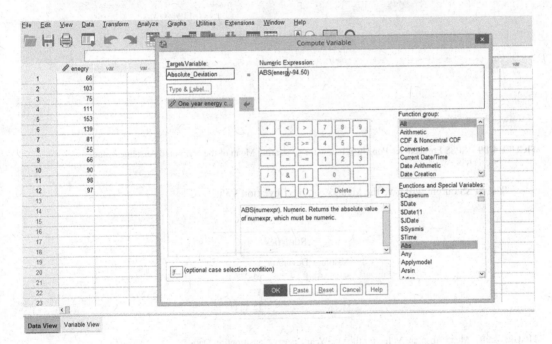

FIGURE 4.88 SPSS Pop Up Window for computing Absolute Values

Go to Analyze > Descriptive Statistics > Frequencies > There will be a pop up SPSS window for Frequencies > Select the Absolute_Deviation and put it in Variable(s) Window > Click Statistics > There will be a pop up SPSS window for Frequencies: Statistics > Under Central Tendency select Mean then Continue and then OK (Figure 4.89)

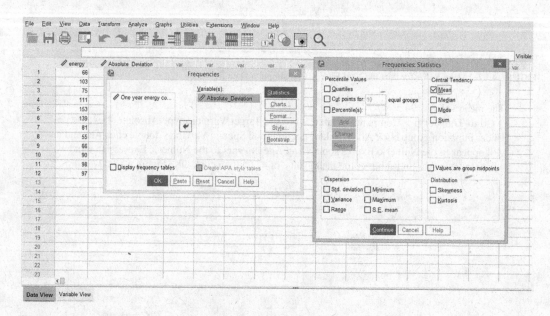

FIGURE 4.89 SPSS Frequencies Pop Up Window for generating Mean of the computed Absolute Values

The SPSS output will give the Mean Absolute Deviation Value

Statistics		
Absolute_Deviation		
N	Valid	12
	Missing	0
Mean		22.3333

FIGURE 4.90 Mean Absolute Value for the One Year Energy Consumption Data

The Average Deviation value will be 15.5460

Interpretation. Since the Mean Absolute Deviation is "quite high", it implies that there is variability in the one year energy consumption data i.e. that the one year energy consumption data is more spread out/ dispersed/ far from the mean.

ii. **Variance**

Variance, denoted by s^2, is computed as

$$s^2 = \frac{\sum(X_i - \bar{X})^2}{n-1}$$

Where X_i are the individual observation data, $\bar{X}$ is the mean and n is the sample

Example: Compute the Variance for the one year energy consumption data below

| 66 | 103 | 75 | 111 | 153 | 139 | 81 | 55 | 66 | 90 | 98 | 97 |

We first compute the mean value = $\bar{X} = 94.50$

We then subtract the mean value from each observation. We then square the differences, sum up the differences and divide the sum of differences by the sample

$$s^2 = \frac{\sum(X_i - \bar{X})^2}{n-1}$$

$$s^2 = \frac{(66-94.5)^2 + (103-94.5)^2 + \cdots + (97-94.5)^2}{12-1} = \frac{9533}{12} = 866.6364$$

iii. **Standard Deviation**

Standard deviation, denoted by s, is computed as

$$s = \sqrt{\frac{\sum(X_i - \bar{X})^2}{n-1}}$$

Using the above example, the standard deviation value will be

$$s = \sqrt{\frac{\sum(X_i - \bar{X})^2}{n-1}} = \sqrt{s^2} = \sqrt{866.6364} = 29.43869$$

iv. **Range**

Range is computed by estimating the difference between highest value and the lowest value

Using the above example on one-year energy consumption, the range will be given by

$$Range = Highest\ Value - Lowest\ Value = 153 - 55 = 98$$

In SPSS, we can compute Variance, Standard Deviation and Range using the path:

Analyze > Descriptive Statistics > Frequencies > There will be a pop up SPSS window for Frequencies > Select the Numerical Variable of interest and put it in Variable(s) Window > Click Statistics > There will be a pop up SPSS window for Statistics > Under Dispersion select Variance, Standard Deviation and Range then Continue and then OK.

Using the energy consumption data, we go to

Analyze > Descriptive Statistics > Frequencies > There will be a pop up SPSS window for Frequencies > Select the Numerical Variable of interest (one-year energy consumption) and put it in Variable(s) Window > Click Statistics > There will be a pop up SPSS window for Statistics > Under Dispersion select Variance, Standard Deviation and Range then Continue and then OK (Figure 4.91).

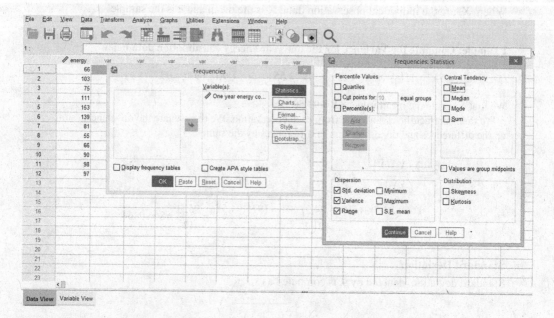

FIGURE 4.91   SPSS Pop Up Window for Frequencies displaying Measures of Dispersion

The SPSS output window will give the values of variance, standard deviation and range

Statistics	
One-Year Energy Consumption	
N Valid	12
Missing	0
Std. Deviation	29.439
Variance	866.636
Range	98

FIGURE 4.92   Standard Deviation, Variance and Range Values for the One Year Energy Consumption Data

Interpretation: Standard deviation value was 29.439. Since the standard deviation value is "quite high". it implies that there is variability in the one year energy consumption data i.e. that the one year energy consumption data is more spread out/ dispersed/ far from the mean.

We always interpret the standard deviation value instead of variance.

Range was 98, implying the difference between the highest and the lowest consumed energy among the 12 different households was 98 units.

v. **Inter-Quartile Range**

The Interquartile range is the difference between the third quartile (75th percentile) and the first quartile (25th percentile).

$$IQR = Q_3 - Q_1$$

Since we already know how to compute percentiles whether manually or using SPSS, we can go ahead and generate the interquartile range

In SPSS, we can compute the Interquartile Range using the path:

Analyze > Descriptive Statistics > Explore > There will be a pop up SPSS window for Explore > Select the Numerical Variable of interest and put it in the Dependent List Window > Click Statistics > There will be a pop up SPSS window for Explore: Statistics and select Descriptives > Click Continue > in the pop up SPSS window select Statistics and then OK.

**Example: Generate the Interquartile Range (IQR) for
the one-year energy consumption data below**

66   103   75   111   153   139   81   55   66   90   98   97

In SPSS, we can compute the Interquartile Range using the path:

Analyze > Descriptive Statistics > Explore > There will be a pop up SPSS window for Explore > Select the Numerical Variable of interest (one year energy consumption) and put it in the Dependent List Window > Click Statistics > There will be a pop up SPSS window for Explore: Statistics and select Descriptives > Click Continue > in the pop up SPSS window select Statistics and then OK (Figure 4.93).

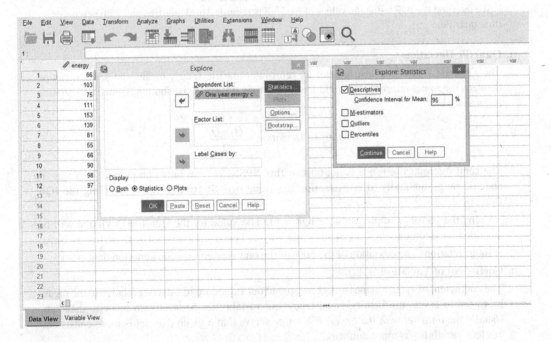

FIGURE 4.93   SPSS Explore Pop Up Window for Generating Descriptive Statistics Values

In the SPSS Data View Window, click Window then Output. From the SPSS output select the interquartile range value

Descriptives			
One-Year Energy Consumption			
Mean		**Statistic**	**Std. Error**
		94.50	8.498
95% Confidence Interval for Mean	Lower Bound	75.80	
	Upper Bound	113.20	
5% Trimmed Mean		93.44	
Median		93.50	
Variance		866.636	
Std. Deviation		29.439	
Minimum		55	
Maximum		153	
Range		98	
Interquartile Range		41	
Skewness		0.751	0.637
Kurtosis		0.088	1.232

FIGURE 4.94   Descriptive Statistics Values for the One Year Energy Consumption Data

Hence the IQR = 41

Interpretation: Since the IQR value is "quite high". it implies that the one year energy consumption data points are spread out/ dispersed. It also implies that range of the middle 50% (half) of the energy consumption data is between 109 and 68.25.

We interpret the IQR using the values of the 75th percentile (third quartile) and 25th percentile (first quartile).

vi. **Quartile Deviation**

The Quartile Deviation, also known as semi-interquartile range, means the semi variation between the upper quartiles (Q3) and lower quartiles (Q1) in a distribution. Quartile Deviation is computed as

$$Quartile\ Deviation = \frac{Q_3 - Q_1}{2}$$

The semi interquartile range is not built within SPSS. We can only compute the semi interquartile range manually after extracting the value of IQR from SPSS by dividing the IQR value by 2.

Using the above example, we have IQR = 41. The value for the Quartile Deviation will be the value for Interquartile Range and divide it by 2. Therefore the Quartile Deviation is 20.5.

Interpretation: The deviation of the 50% of the one year energy consumption data is 20.5.

vii. **Coefficient of Variation**

The coefficient of variation measures how spread out values are in a dataset relative to the mean. It is used to assess uniformity, consistency and variabilities between two or more data sets. Usually the *data set with the lowest C.V value* shows that a given data set is more consistent/ has lees variability/is more uniform.

The Coefficient of Variation (C.V.) is the ratio of the standard deviation to the mean multiplied by 100.

$$C.V. = \frac{s}{\bar{X}} \times 100$$

where s = standard deviation and $\bar{X}$ is the mean

Example: Generate the Coefficient of Variation (C.V.) value for the one year energy consumption data below

66   103   75   111   153   139   81   55   66   90   98   97

We have estimated the mean value as

$$\bar{X} = \frac{\sum X_i}{n} = \frac{66+103+75+111+153+139+\cdots+90+98+97}{12} = 94.50$$

We have also estimated the standard deviation value as

$$s^2 = \frac{\sum (X_i - \bar{X})^2}{n-1}$$

$$s^2 = \frac{(66-94.5)^2 + (103-94.5)^2 + \cdots + (97-94.5)^2}{12-1} = \frac{9533}{12} = 866.6364$$

$$s = \sqrt{\frac{\sum (X_i - \bar{X})^2}{n-1}} = \sqrt{s^2} = \sqrt{866.6364} = 29.43869$$

We can then compute the CV as

$$C.V. = \frac{s}{\bar{X}} \times 100 = \frac{29.43869}{94.5} \times 100 = 31.15205\%$$

In SPSS, we can compute the Coefficient of Variation (C.V.) using the path:
First create a column of 1's (one's) in the data set next to the numerical variable of interest > then go to Analyze > Descriptive Statistics > Ratio > there will be a pop up SPSS window for Ratio Statistics > Select the Numerical Variable of interest and put it in the Numerator and the variable of 1's into the Denominator > Click Statistics > There will be a pop up SPSS window for Ratio Statistics: Statistics and select Mean, Standard Deviation and Mean Centered COV > Click Continue then OK.

In the SPSS Data View Window, click Window then Output, the Coefficient of Variation for the dataset will be displayed

Using the above example,

Go to SPSS Data View window and create a column of 1's (one's) in the data set next to the numerical variable of interest (Figure 4.95)

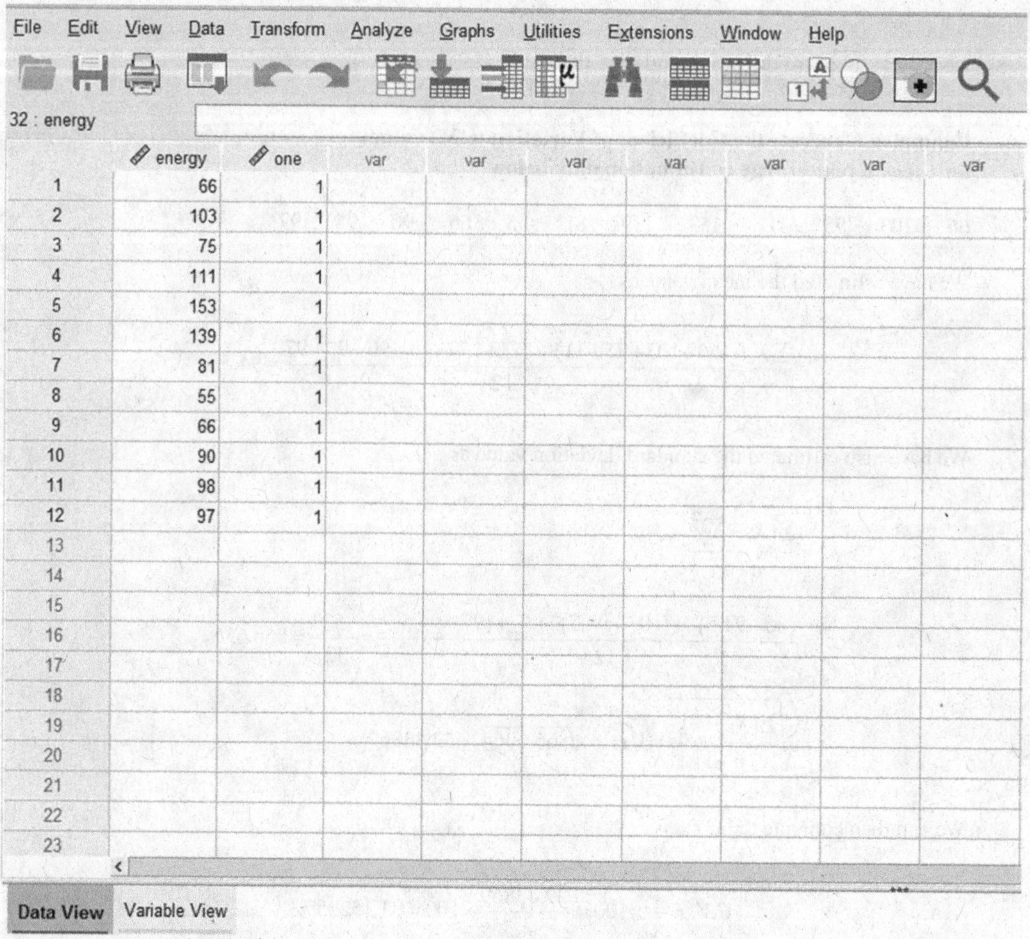

| File | Edit | View | Data | Transform | Analyze | Graphs | Utilities | Extensions | Window | Help |

32 : energy

	energy	one	var	var	var	var	var	var	var
1	66	1							
2	103	1							
3	75	1							
4	111	1							
5	153	1							
6	139	1							
7	81	1							
8	55	1							
9	66	1							
10	90	1							
11	98	1							
12	97	1							
13									
14									
15									
16									
17									
18									
19									
20									
21									
22									
23									

Data View   Variable View

FIGURE 4.95   SPSS Data View Window displaying One Year Energy Consumption Data and Variables of Ones

Go to Analyze > Descriptive Statistics > Ratio > there will be a pop up SPSS window for Ratio Statistics > Select the Numerical Variable of interest (one year energy consumption) and put it in the Numerator and the variable of 1's (one) into the Denominator > Click Statistics > There will be a pop up SPSS window for Ratio Statistics: Statistics and select Mean, Standard Deviation and Mean Centered COV > Click Continue then OK (Figure 4.96).

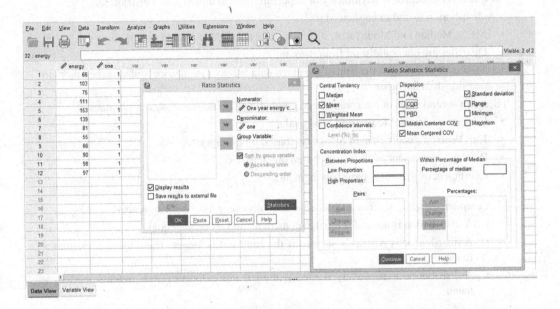

FIGURE 4.96 SPSS Ratio Pop Up Window for generating Coefficient of Variation Value

In the SPSS Data View Window, click Window then Output, the Coefficient of Variation for the dataset will be displayed

Ratio Statistics for One-Year Energy Consumption/One

Mean	Std. Deviation	Coefficient of Variation Mean Cantered
94.500	29.439	31.2%

FIGURE 4.97 Coefficient of Variation Value for One Year Energy Consumption Data

The value for the CV is 31.2%.

Interpretation: To interpret CV there has to be two sets of data that are being compared. The data set with the lowest CV value has lower variability/ is less dispersed and is consistent or uniform in terms of the observations in the dataset.

4.3.3 Measures of Shape

The shape of distributions refers to symmetry or lack of it (skewness) and flatness or peakedness (kurtosis). The measures of shape of a distribution can be described using Skewness and Kurtosis.

a. **Skewness**

Skewness is the degree of asymmetry or departure from symmetry, of a distribution.

A distribution is said to be skewed if:

- Mean, Median and Mode fall at different points, i.e. *Mean ≠ Median ≠ Mode*
- Quartiles are not equidistant from median
- The curve drawn with the help of the given data are not symmetrical but stretched more to one side than the other

There are several ways of interpreting and computing Coefficient of skewness values

- Comparing Mean, Median and Mode values
- Karl Pearson's coefficient of Skewness (based on averages)
- Bowley's coefficient of skewness (based on quartiles)
- Coefficient of Skewness based on moments

i. *Comparing Mean, Median and Mode values after they have been computed*

- A distribution is normal if the mean, median, and mode are equal
- A distribution is positively skewed if the right tail is longer. Then, mean > median > mode.
- A distribution is negatively skewed if the left tail is longer. Then, mode > median > mean

Example: interpret the skewness value for the one year energy consumption data below by comparing men, median and mode values

| 66 | 103 | 75 | 111 | 153 | 139 | 81 | 55 | 66 | 90 | 98 | 97 |

Since we have already computed the mean, median and mode values, the only thing we need to do is to compare the three different values and make interpretation. Otherwise, compute the mean, median and mode values to make the interpretation.

Mean = 94.5, Median = 93.5 and Mode = 66. We can see that mean value > median value > mode value, hence the one year energy consumption data are positively skewed.

ii. *Computing and Interpreting Karl Pearson coefficient of skewness*

- Skewness can be based on the averages in two ways:

$$S_k = \frac{Mean - Mode}{Standard\ Deviation}$$

or

$$S_k = \frac{3(Mean - Median)}{Standard\ Deviation}$$

Example: interpret the skewness value for the one year energy consumption data below using Karl Pearson coefficient of skewness

66   103   75   111   153   139   81   55   66   90   98   97

Mean = 94.5, Median = 93.5, Mode = 66 and Std Dev = 29.43869.

There are two ways of computing skewness value

Method 1: using mean, mode and standard deviation values

$$S_k = \frac{Mean - Mode}{Standard\ Deviation} = \frac{94.5 - 66}{29.43869} = 0.96811$$

Since $S_k > 0$, it implies that one year energy consumption data are positively skewed.

Method 2: using mean, median and standard deviation values

$$S_k = \frac{3(Mean - Median)}{Standard\ Deviation} = \frac{3(94.5 - 93.5)}{29.43869} = 0.1019$$

Since $S_k > 0$, it implies that one year energy consumption data are positively skewed.

iii. *Computing and Interpreting Skewness based on Quartiles which is computed using the quartiles*

$$S_k = \frac{Q_3 + Q_1 - 2Q_2}{Q_3 - Q_1}$$

Since we have already computed the quartiles, i.e. 25th percentile, 50th percentile and 75th percentile values, the only thing we need to do is to substitute the three different values into the skewness formula and make interpretation. Otherwise, compute the quartiles and substitute them into the skewness formula to make the interpretation.

We have $Q_1 = 68.25$, $Q_2 = 93.5$ and $Q_3 = 109$

Substituting the values into the formula

$$S_k = \frac{Q_3 + Q_1 - 2Q_2}{Q_3 - Q_1} = \frac{109 + 68.25 - (2 \times 93.5)}{109 - 68.25} = -0.2392$$

Since $S_k < 0$, it implies that one year energy consumption data are negatively skewed.

Interpretation of Coefficient of Skewness Value
- If $S_k = 0$, then we have normal distribution
- If $S_k < 0$, then we have negative distribution
- If $S_k > 0$, then we have positive distribution

iv. *Computing and interpreting the coefficient of Skewness based on moments*
Skewness value is computed using third (μ_3) and second (μ_2) central moments. Measure of Skewness based on moments are Beta coefficient and Gamma coefficients as follows:

$$\beta_1 = \frac{\mu_3^2}{\mu_2^3}$$

$$\gamma_1 = \sqrt{\beta_1} = \frac{\mu_3}{\mu_2^{3/2}}$$

where:

$$\mu_2 = \frac{\sum(x_i - \overline{x})^2}{n}$$

$$\mu_3 = \frac{\sum(x_i - \overline{x})^3}{n}$$

Interpretation

- If $\sqrt{\beta_1} = 0$ then the distribution is symmetrical, where mean, median and mode coincide
- If $\sqrt{\beta_1} < 0$ then the curve is negatively skewed. The left tail is longer
- If $\sqrt{\beta_1} > 0$ then the curve is positively skewed. The right tail is longer

Example: interpret the skewness value for the one year energy consumption data below using moments

66   103   75   111   153   139   81   55   66   90   98   97

We already have the mean value = 94.5. We will need to generate $(x_i - \overline{x})^2$ and $(x_i - \overline{x})^3$ values so that we can use them to generate μ_2 and μ_3 values

x	$(x - \overline{x})$	$(x_i - \overline{x})^2$	$(x_i - \overline{x})^3$
66	−28.5	812.25	−23149.1
103	8.5	72.25	614.125
75	−19.5	380.25	−7414.88
111	16.5	272.25	4492.125
153	58.5	3422.25	200201.6
139	44.5	1980.25	88121.13
81	−13.5	182.25	−2460.38
55	−39.5	1560.25	−61629.9
66	−28.5	812.25	−23149.1
90	−4.5	20.25	−91.125
98	3.5	12.25	42.875
97	2.5	6.25	15.625
Total		9533	175593

We can see that $\sum(x_i - \overline{x})^2 = 9533$ and $\sum(x_i - \overline{x})^3 = 175593$. Substituting these values gives μ_2 and μ_3 values

$$\mu_2 = \frac{\sum(x_i - \overline{x})^2}{n} = \frac{9533}{12} = 794.4167$$

$$\mu_3 = \frac{\sum(x_i - \overline{x})^3}{n} = \frac{175593}{12} = 14632.75$$

$$\beta_1 = \frac{\mu_3^2}{\mu_2^3} = \frac{14632.75^2}{794.4167^3} = 0.427077$$

$$\gamma_1 = \sqrt{\beta_1} = \sqrt{0.427077} = 0.6535$$

Since $\sqrt{\beta_1} > 0$ then it implies that one year energy consumption data are positively skewed.

b. **Kurtosis**

Kurtosis is a numerical measure for describing the "peakedness" or flatness of the probability distribution relative to a normal distribution based on second and fourth moments. A distribution maybe *"more peaked" than normal, normal* or *"flatter" than normal distribution*. Kurtosis values are computed using 2nd (μ_2) and 4th (μ_4) central moments

$$\beta_2 = \frac{\mu_4}{\mu_2^2}$$

$$\gamma_2 = \beta_2 - 3$$

where

$$\mu_4 = \frac{\sum (x_i - \bar{x})^4}{n}$$

Interpretation

- If $\beta_2 = 3$, or $\gamma_2 = 0$ then the distribution is mesokurtic (Normal curve)
- If $\beta_2 > 3$, or $\gamma_2 > 0$ then the distribution is leptokurtic (more peaked than normal curve)
- If $\beta_2 < 3$, or $\gamma_2 < 0$ then the distribution is platykurtic (flatter than normal curve)

Example: interpret the kurtosis value for the one year energy consumption data below using moments

66   103   75   111   153   139   81   55   66   90   98   97

We already have the mean value = 94.5 and $(x_i - \bar{x})^2$ we will need to compute $(x_i - \bar{x})^4$ values so that we can use them to generate μ_2 and μ_4 values. If these values are not given, then we will have to compute them.

Using the above table

x	$(x - \bar{x})$	$(x_i - \bar{x})^2$	$(x_i - \bar{x})^4$
66	−28.5	812.25	659750.1
103	8.5	72.25	5220.063
75	−19.5	380.25	144590.1
111	16.5	272.25	74120.06
153	58.5	3422.25	11711795
139	44.5	1980.25	3921390
81	−13.5	182.25	33215.06
55	−39.5	1560.25	2434380
66	−28.5	812.25	659750.1
90	−4.5	20.25	410.0625
98	3.5	12.25	150.0625
97	2.5	6.25	39.0625
Total		9533	19644810

We can see that $\sum(x_i - \bar{x})^2 = 9533$ and $\sum(x_i - \bar{x})^4 = 19644810$. Substituting these values gives μ_2 and μ_4 values

$$\mu_2 = \frac{\sum(x_i - \bar{x})^2}{n} = \frac{9533}{12} = 794.4167$$

$$\mu_4 = \frac{\sum(x_i - \bar{x})^3}{n} = \frac{19644810}{12} = 1637067$$

$$\beta_2 = \frac{\mu_4}{\mu_2^2} = \frac{1637067}{794.4167^2} = 2.594$$

$$\gamma_2 = \beta_2 - 3 = 2.594 - 3 = -0.406$$

Since $\beta_2 < 3$, or $\gamma_2 < 0$ then the one year energy consumption data are platykurtic (flatter than normal curve)

In SPSS, we can compute Skewness and Kurtosis values using the path:

Analyze > Descriptive Statistics > Frequencies > There will be a pop up SPSS window for Frequencies > Select the Numerical Variable of interest and put it in Variable(s) Window > Click Statistics > There will be a pop up SPSS window for Statistics > Under Distribution select Skewness and Kurtosis then Continue and then OK.

Using the above example for one year energy consumption data, we go to Analyze > Descriptive Statistics > Frequencies > There will be a pop up SPSS window for Frequencies > Select the one year energy consumption variable and put it in Variable(s) Window > Click Statistics > There will be a pop up SPSS window for Statistics > Under Distribution select Skewness and Kurtosis then Continue and then OK (Figure 4.98).

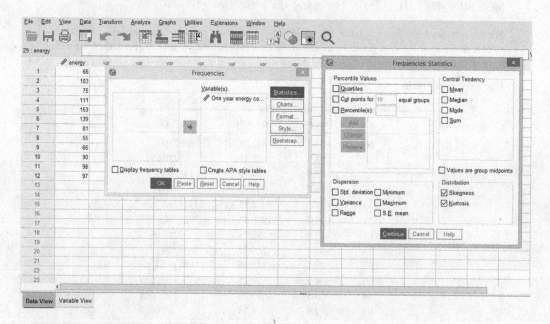

FIGURE 4.98   SPSS Frequencies Pop Up Window for generating Distribution Value

Statistics		
One year energy consumption		
N	Valid	12
	Missing	0
Skewness		.751
Std. Error of Skewness		.637
Kurtosis		.088
Std. Error of Kurtosis		1.232

FIGURE 4.99 Skewness and Kurtosis Values for One Year Energy Consumption Data

Interpretation: Skewness value is 0.751. This implies that the one year energy consumption data is positively skewed.

Kurtosis value is 0.088. This implies that the one year energy consumption data is flatter than a normal distribution.

4.3.4 Measures of Association

Association describes the relationships between two variables, i.e. a bivariate distribution. Correlation is used to describe relationships for a bivariate distribution where there are two variables.

There are three methods of studying correlation

- Generating and interpreting the scatter diagram (graphical method)
- Computing and interpreting Karl Pearson coefficient of correlation also known as product-moment correlation coefficient
- Computing and interpreting Spearman's Rank coefficient of correlation

For the bivariate distributions we should always have pairs of variables so that one can be able to do any bivariate analysis

i. *Using Scatter Diagram*

Scatter diagram is the simplest diagrammatic representation of bivariate data. Usually there are three types of graphs that will arise whenever you generate a scatter plot

- One showing positive relationship, i.e. an increase in one variable results into increase in the other variable.
- One showing negative relationship, i.e. an increase in one variable results into a decrease in another variable.
- One showing no relationship, i.e. an increase in one variable results into no increase nor decrease in the other variable.

In SPSS, we can generate a scatter plot using the path:

Graphs > Legacy Dialogs > Scatter/Dot > There will be a pop-up window for Scatter/Dot > Select Simple then Define > There will be a pop-up window for Simple Scatter Plot > select the numerical variable of interest and put it in Y Axis then select another numerical variable and put it in X Axis then click OK.

Once you have the scatter plot, you interpret using any of the above three ways.

Example: Generate a scatter plot for the following data on income versus the credit card debt for sampled individuals

Individual	Income	Credit Card Debt
1	1182	1116.5
2	1353	1005
3	1004	933.3
4	373	541.5
5	409	476.2
6	327	359.3

We create the variables in the Variable View Window then key the data into SPSS through the Data View Window (Figure 4.100)

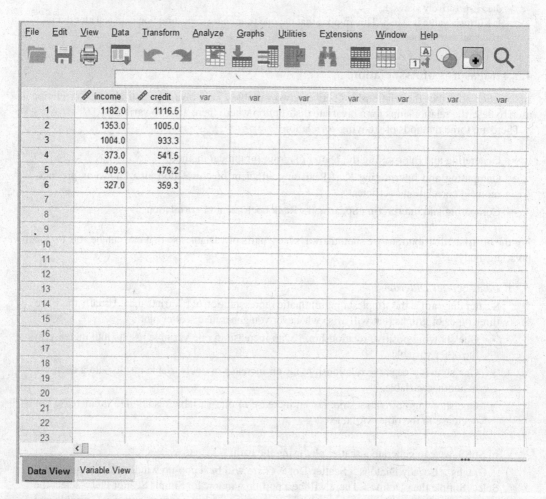

FIGURE 4.100 Data View displaying Income and Credit Card Debts

Go to Graphs > Legacy Dialogs > Scatter/Dot > There will be a pop-up window for Scatter/Dot > Select Simple then Define > There will be a pop-up window for Simple Scatter Plot > select income and put it in Y Axis then select credit and put it in X Axis then click OK (Figure 4.101).

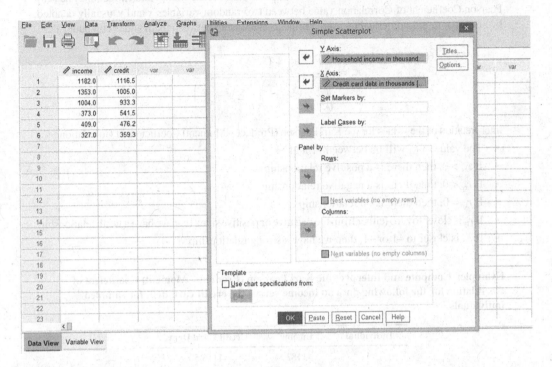

FIGURE 4.101 SPSS Pop Up Window for generating Simple Scatter Plot

The SPSS output will give the scatter plot (Figure 4.102)

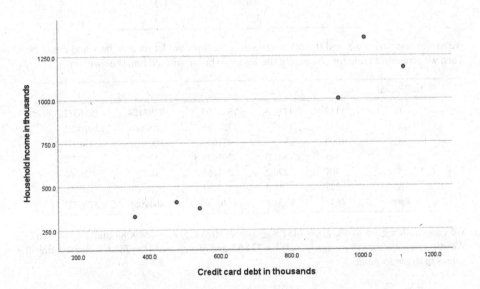

FIGURE 4.102 Scatter Plot for Household Income in Thousands and Credit card Debt in Thousands

From the above scatter plot we can see that there is a positive relationship between household income and credit card debts

ii. *Computing and Interpreting the Karl Pearson Coefficient of Correlation*

Karl Pearson Coefficient of Correlation value is *a numerical measure* describing the relationship between two numerical variable. The relationship is in terms of the *strength* and *direction*. The Karl Pearson Coefficient of Correlation value between two random variables x and y, usually denoted by $r(x, y)$ or r_{xy} is a numerical measure of linear relationship between them and is computed as:

$$r_{xy} = \frac{\frac{1}{n}\sum xy - \bar{x}\bar{y}}{\sqrt{\left(\frac{1}{n}\sum x^2 - \bar{x}^2\right)\left(\frac{1}{n}\sum y^2 - \bar{y}^2\right)}}$$

Interpretation of the values for the Karl Pearson (Product – Moment) Coefficient of Correlation (r_{xy})

- The value of r_{xy} will be between −1 and +1
- If $r_{xy} > 0$, then there is a positive relationship
- If $r_{xy} < 0$, then there is a negative relationship
- If $r_{xy} = 0$, then there is no relationship
- If r_{xy} is closer to zero (either from the negative or positive side), then we have a weak relationship
- If r_{xy} is closer to −1 or +1, then we have a strong relationship

Example: Compute and interpret the Karl Pearson (Product – Moment) Coefficient of Correlation for the following data on income versus the credit card debt for sampled individuals

Individual	Income	Credit Card Debt
1	1182	1116.5
2	1353	1005
3	1004	933.3
4	373	541.5
5	409	476.2
6	327	359.3

We assign one variable x and the other variable y. In this case let income be x and credit be y. Then we generate a table for computing the values to be substituted into the formula

Individual	x	y	xy	x^2	y^2
1	1182	1116.5	1319703	1397124	1246572
2	1353	1005	1359765	1830609	1010025
3	1004	933.3	937033.2	1008016	871048.9
4	373	541.5	201979.5	139129	293222.3
5	409	476.2	194765.8	167281	226766.4
6	327	359.3	117491.1	106929	129096.5
Total	4648	4431.8	4130738	4649088	3776731

We can see that $\sum x = 4648$, $\sum y = 4431.8$, $\sum xy = 4130738$, $\sum x^2 = 4649088$ and $\sum y^2 = 3776731$. We can compute means as $\bar{x} = \frac{\sum x}{n} = \frac{4648}{6} = 774.67$ and $\bar{y} = \frac{\sum y}{n} = \frac{4431.8}{6} = 738.63$. Substituting the values in to the formula

$$r_{xy} = \frac{\frac{1}{n}\sum xy - \bar{x}\bar{y}}{\sqrt{\left(\frac{1}{n}\sum x^2 - \bar{x}^2\right)\left(\frac{1}{n}\sum y^2 - \bar{y}^2\right)}} = \frac{\frac{1}{6}\times 4130738 - (774.67 \times 735.63)}{\sqrt{\left(\frac{1}{6}\times 4649088 - 774.67^2\right)\left(\frac{1}{6}\times 3776731 - 738.63^2\right)}} = 0.96033$$

In SPSS, we can generate Pearson Coefficient of Correlation values using the path:

In SPSS we go to Analyze > Correlate > Bivariate > There will be a pop up SPSS window for Bivariate Correlations > Select the two numerical variables of interest and drag them to Variables Window > Click Pearson Correlation Coefficient then click OK.

Using the above example we will go to Analyze > Correlate > Bivariate > There will be a pop up SPSS window for Bivariate Correlations > Select Household income in thousands and Credit card debt in thousand then drag them to Variables Window > Click Pearson Correlation Coefficient and Spearman Correlation Coefficients then click OK (Figure 4.103).

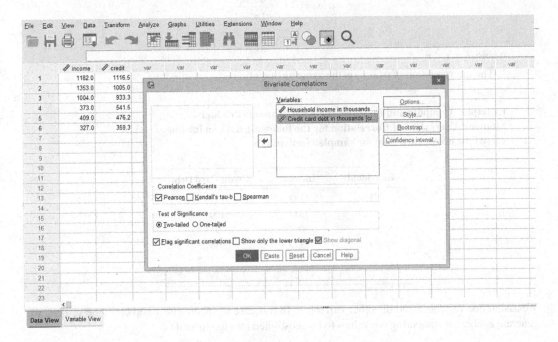

FIGURE 4.103   SPSS Pop Up Window for Pearson Correlation

The SPSS output will give the Pearson Correlation Coefficient value

Correlations		Household Income in Thousands	Credit Card Debt in Thousands
Household income in thousands	Person Correlation	1	0.960**
	Sig. (2-tailed)		0.002
	N	6	6
Credit card debt in thousands	Person Correlation	0.960**	1
	Sig. (2-tailed)	0.002	
	N	6	6

** Correlation is significant at the 0.01 level (2-tailed).

FIGURE 4.104   Pearson Correlation Value for household Income in Thousands and Credit Card Debt in Thousands

The Karl Pearson (Product – Moment) Coefficient of Correlation is 0.96 implying that there is a strong and positive relationship between household income in thousands and credit card debt in thousands.

iii. *Computing ad Interpreting Spearman Rank coefficient of Correlation*

Spearman rank coefficient of correlation is used when a group of n individuals are arranged in order or ranked. One will then compute the differences in the rankings of each group, $d_i = x_i - y_i$ and then apply the Spearman's formula. The Spearman Rank Coefficient of Correlation value between two random variables x and y, usually denoted by $\rho(x, y)$ or ρ_{xy} is a numerical measure of linear relationship between them and is computed as:

$$\rho(x, y) = 1 - \frac{6 \sum d_i^2}{n(n^2 - 1)}$$

where $d_i = x_i - y_i$

The interpretation of $\rho(x, y)$ is the same as that of r_{xy}

Example: Compute and interpret the Karl Pearson (Product – Moment) Coefficient of Correlation for the following data on income versus the credit card debt for sampled individuals

Individual	Income	Credit Card Debt
1	1182	1116.5
2	1353	1005
3	1004	933.3
4	373	541.5
5	409	476.2
6	327	359.3

We assign one variable x and the other variable y. In this case let income be x and credit be y. Then we generate a table for computing the values to be substituted into the formula

Individual	x	y	x_i	y_i	$d_i = x_i - y_i$	d_i^2
1	1182	1116.5	2	1	1	1
2	1353	1005	1	2	−1	1
3	1004	933.3	3	3	0	0
4	373	541.5	5	4	1	1
5	409	476.2	4	5	−1	1
6	327	359.3	6	6	0	0
Total						4

We can see that $\sum d_i^2 = 4$

Substituting the values into the formula

$$\rho(x, y) = 1 - \left(\frac{6 \sum d_i^2}{n(n^2 - 1)} \right) = 1 - \left(\frac{6 \times 4}{6(6^2 - 1)} \right) = 0.88571$$

In SPSS, we can generate Spearman Rank Coefficient of Correlation values using the path:

Analyze > Correlate > Bivariate > There will be a pop up SPSS window for Bivariate Correlations > Select the two numerical variables of interest and drag them to Variables Window > Click Pearson Correlation Coefficient and Spearman Correlation Coefficients then click OK.

Using the above example for one year energy consumption data we go to

Analyze > Correlate > Bivariate > There will be a pop up SPSS window for Bivariate Correlations > Select Household income in thousands and Credit card debt in thousand then drag them to Variables Window > Click Pearson Correlation Coefficient and Spearman Correlation Coefficients then click OK (Figure 4.105).

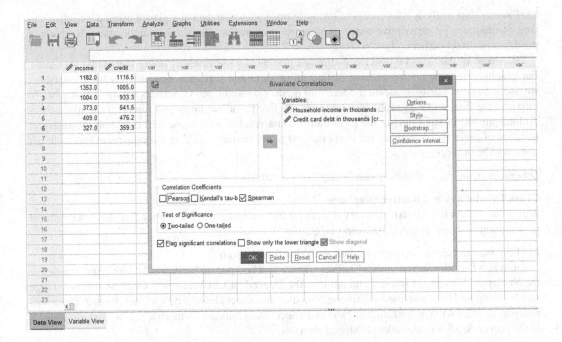

FIGURE 4.105 SPSS Pop Up Window for Spearman Correlation

The SPSS output will give the Pearson Correlation Coefficient value

Correlations			Household Income in Thousands	Credit Card Debt in Thousands
Spearman's rho	Household income in thousands	Correlation coefficient	1.000	0.886*
		Sig. (2-tailed)	.	0.019
		N	6	6
	Credit card debt in thousands	Correlation coefficient	0.886*	1.000
		Sig. (2-tailed)	0.019	.
		N	6	6

*Correlation is significant at the 0.05 level (2-tailed).

FIGURE 4.106 Spearman Correlation Value for household Income in Thousands and Credit Card Debt in Thousands

The Spearman Coefficient of Correlation is 0.886 implying that there is a strong and positive relationship between household income in thousands and credit card debt in thousands.

4.4 Exploratory Data Analysis

Exploratory data analysis is a process of trying to understand the data at a glance using descriptive statistics techniques, i.e. descriptive statistics measures and graphs. The aim is to understand the shape of the data and detect outliers.

The shape of data can be described through graphical techniques or using numerical measures. For graphical techniques we can use histogram or stem and leaf plot. For the numerical measures we use skewness, kurtosis, mean, median and mode values. To detect outliers we use a box plot. To interpret the Boxplot as a graph/chart one uses the five point summary and interpret the five points. To interpret the boxplot during the Exploratory Data Analysis, then one will only need to interpret in terms of outliers, i.e. if there are outliers, how many outliers, what is the minimum value/cut off point for the outliers.

Other techniques for exploratory data analysis for bivariate data

Cross Tabulations and Scatter Diagrams

A cross tabulation is a tabular summary of data for two variables while a Scatter diagram is a graphical presentation of the relationship between two quantitative variables.

In SPSS, we can conduct exploratory data analysis using the path:

Analyze > Descriptive Statistics > Explore > There will be a pop up SPSS window for Explore > Select the Numerical Variable of interest and put it in the Dependent List Window > Click Statistics > There will be a pop up SPSS window for Explore: Statistics and select Descriptives > Click Plots > There will be a pop up SPSS window for Explore: Plots and select Histogram and Stem-and-Leaf > Click Continue > in the pop up SPSS window select Both and then OK.

Example

Using *customer_subset.sav* data (inbuilt SPSS data), conduct Exploratory Data Analysis for Credit card debt in thousands.

To locate *customer_subset.sav* data we use the path: Click **Open data document** folder > drive (**C**): > **Program Files (x86)** > **IBM** > **SPS** > **Statistics** > **27** > **Samples** > **English** > customer_subset.sav data (In-built *SPSS data files*).

Analyze > Descriptive Statistics > Explore > there will be a pop up SPSS window for Explore > Select Age in years and put it in the Dependent List Window > Click Statistics > There will be a pop up SPSS window for Explore: Statistics and select Descriptives then click Continue (Figure 4.107)

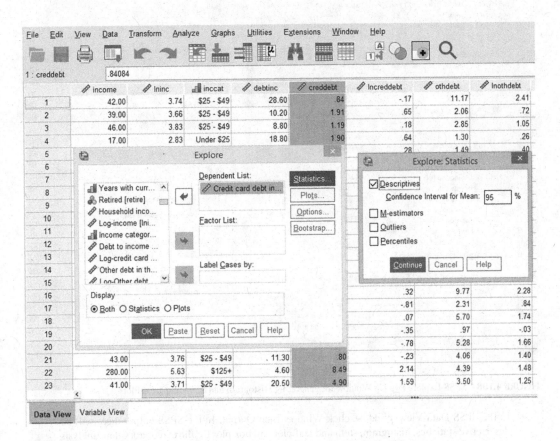

FIGURE 4.107 SPSS Explore Pop Up Window for generating Descriptive Statistics Values

In the SPSS pop up window for Explore Click Plots > There will be a pop up SPSS window for Explore: Plots and select Histogram and Stem-and-Leaf > Click Continue > in the pop up SPSS window select Both and then OK (Figure 4.108).

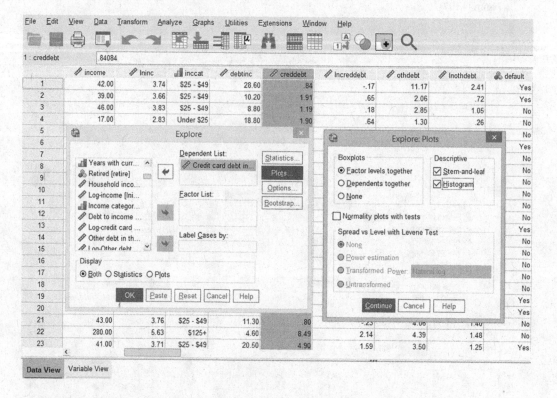

FIGURE 4.108   SPSS Explore Pop Up Window for generating Histogram and Stem-and-Leaf Plots

In the SPSS Data View Window, click Window then Output. In the SPSS output there will be descriptive statistics, histogram, stem and leaf plot and box plot for the exploratory data analysis for Credit card debt in thousands displayed.

Descriptives			**Statistic**	**Std. Error**
Credit card debt in thousands	Mean		1.8619	.24431
	95% Confidence Interval for Mean	Lower Bound	1.3756	
		Upper Bound	2.3482	
	5% Trimmed Mean		1.5944	
	Median		1.0688	
	Variance		4.775	
	Std. Deviation		2.18520	
	Minimum		0.09	
	Maximum		10.30	
	Range		10.21	
	Interquartile Range		1.87	
	Skewness		1.896	.269
	Kurtosis		3.274	.532

FIGURE 4.109   Descriptive Statistics values for credit Card Debt in Thousands

From the Descriptive Statistics output, interpret Skewness and Kurtosis values. The skewness value = 1.896 implying that credit card debt is positively skewed. The kurtosis value = 3.274 implying that the credit card debt data are more peaked than a normal distribution.

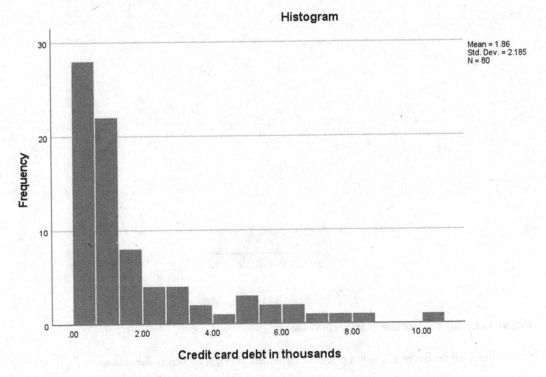

FIGURE 4.110   Histogram Plot for Credit Card Debt in Thousands

Interpret the shape of Histogram. It is evident that credit card debt in thousand is positively skewed with most of the households are having a credit card debt of between 0 and 4000.

Credit Card Debt in Thousands Stem-and-Leaf Plot			
Frequency	**Stem**	**&**	**Leaf**
24.00	0	.	011111222223333333344444
15.00	0	.	556677777778889
13.00	1	.	0000001222334
6.00	1	.	566689
4.00	2	.	1134
3.00	2	.	678
1.00	3	.	2
2.00	3	.	78
0.00	4	.	
4.00	4	.	6899
8.00 Extremes		.	(≥ 5.4)
Stem width:	1.00		
Each leaf:	1 case (s)		

FIGURE 4.111   Stem-and-Leaf Plot for Credit Card Debt in Thousands

Interpret the shape of the stem and leaf. It is evident that credit card debt in thousand is positively skewed with eight extreme values above 5400.

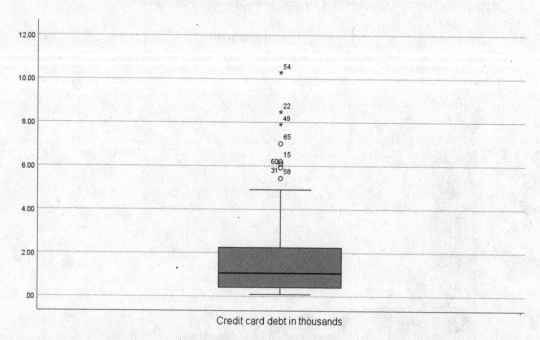

Credit card debt in thousands

FIGURE 4.112   Box Plot for Credit Card Debt in Thousands

Interpret the box plot. The box plot shows that there are eight outliers in the data set.

Practice Exercise

1. Using *customer_subset.sav* data (inbuilt SPSS data), consider Household income in thousands
   a.   Generate a grouped frequency table with intervals
   b.   Generate all the numerical measures of central tendency
   c.   Generate all the numerical measures of dispersion
   d.   Generate all the numerical measures of shape of a distribution
2. Using *customer_subset.sav* data (inbuilt SPSS data), generate and interpret all the graphs and charts using appropriate variables of interest
3. Using *customer_subset.sav* data (inbuilt SPSS data), generate and interpret the coefficients of correlation measures for variables household income in thousands and credit card debt
4. Using *customer_subset.sav* data (inbuilt SPSS data), conduct exploratory data analysis for Household income in thousands

Part II

Probability Concepts

5

Introduction to Probability

5.1 Basic Probability Concepts

Probability is a numerical measure of likelihood (certainty or uncertainty) that an event will occur. The value of probability lies between 0 (0%) and 1 (100%).

Experiment is a process that generates well-defined outcomes. Experiment is an act that can be repeated under given conditions. For example, tossing a coin will generate outcomes of either having head or tail, sitting for an exam, attending a job interview, etc.

A **sample space** for an experiment is the set of all experimental outcomes. By specifying all possible experimental outcomes, we identify the sample space for an experiment. *For example:* in the experiment of tossing a coin, the sample space is

$$S = \{Head, Tail\}$$

An **Events** is a collection of sample points. An **event** is a possible outcome of an experiment or a result of atrial or an observation.

Types of Events
The following are some of the types of events:

- **Elementary Events:** events that cannot be decomposed or broken down into other events They are events which contains only single outcome in the sample space. For example, tossing a six-sided dice will give elementary events which are 1,2,3,4,5,6.
- **Mutually Exclusive Events:** these are two or more events that cannot happen at the same time". For example, when one tosses a coin you can only get a head or a tail and not both at the same time.
- **Not Mutually Exclusive Events:** These are two or more events that can happen at the same time. For example, an individual can be reading and listening to music at the same time.
- **Independent Events:** two or more events are independent events if the occurrence or nonoccurrence of one of the events does not affect the occurrence or nonoccurrence of the other event(s).
- **Collectively Exhaustive Events:** a list of collectively exhaustive events contains all possible elementary events for an experiment. Thus, all sample spaces are collectively exhaustive lists.
- **Complementary Events:** the complement of event A is denoted A, pronounced "not A". All the elementary events of an experiment not in A comprise its complement.

Probability Notation
If E is an event, then $P(E)$ represents the probability that event E occurs. It is read "the probability of E".

5.2 Assigning Probabilities

Basic requirements for assigning probabilities are:

1. The probability assigned to each experimental outcome must be between 0 and 1, i.e. $0 \leq P(X) \leq 1$
2. The sum of the probabilities for all the experimental outcomes must be equal to 1, i.e. $\sum P(X) = 1$

DOI: 10.1201/9781003292654-7

There are three approaches for assigning probabilities: classical, relive frequency and subjective method.

The classical method of assigning probabilities is appropriate when all the experimental outcomes are equally likely:

$$P(A) = \frac{No. \ of \ observation \ of \ A}{Total \ number \ of \ outcomes} = \frac{N(A)}{N(S)}$$

Example: In tossing a coin experiment the two experimental outcomes are equally likely: head and tail. Therefore, the probability of having a head = 1/2 = 0.5 and the probability of having a tail = 1/2 = 0.5.

The relative frequency method is used when you have a frequency table whereby one computes the relative frequency values as follows:

$$PP(A) = \frac{No. \ of \ observation \ of \ A}{Sample \ size} = \frac{f}{n}$$

Example: Consider a study on waiting time in the X-ray department.

Number Waiting	Number of Days Outcome Occurred	Probability
0	2	2/20
1	5	5/20
2	6	6/20
3	4	4/20
4	3	3/20
Total	20	

The subjective method of assigning probabilities is used when one cannot assume that the experimental outcomes are equally likely and when little relevant data are available. Subjective probability comes from the person's intuition or reasoning (i.e., the probability of something happening is based on an individual's own experience or personal judgment). For example, betting for a particular football team to win before starting the game.

5.3 Tree Diagrams

Tree diagrams are used to represent the several outcomes that are occurring e.g. tossing two coins, fillip-ing dice two times or three times etc.

When you toss the first coin, there will be two possible outcomes (H,T).

Tossing the second coin, there will be two other possible outcomes (HT) and (HT).

The sample space will have four possible outcomes: $S = \{HH, HT, TH, TT\}$.

Example: Two coins are tossed, what is the probability of getting at least one head?

The sample space will be four total possible outcomes: $S = \{HH, HT, TH, TT\}$.

The probability of having at least one head –3 possible outcomes with at least one head $\{HH, HT, TH\}$ – reduced sample space.

To calculate probability,

$$P(A) = \frac{Outcomes}{Number \ of \ Outcomes} = \frac{3}{4} = 0.75$$

Interpretation: There is a 75% chance of getting at least one head.

5.4 Some Basic Relationships of Probability

To relate the two events we use Union, Intersection, Compliment and Null.

5.4.1 Union of Two Events

A union of two events E and F, denoted by $E \cup F$, will occur if either E or F occurs, i.e. combining the elements of two events without repetition. *For example:* If $E = \{g\}$ and $F = \{b\}$, then $E \cup F = \{g,b\}$ – the event containing all the sample points belonging to E, F or both. In probability, union is applicable for statements like at least, either, or in probability. Union will also be used in the additive law.

5.4.2 Intersection of Two Events

The intersection of two events E and F, denoted by $E \cap F$, will occur if both E and F occur, i.e. what is common in the two events. *Example:* If $E = \{2,5\}$ and $F = \{2,5,10\}$ then $E \cap F = \{2,5\}$ – the event containing the sample points belonging to both E and F. In probability, intersection will be applicable for statements like both, and in probability. An intersection will also be used in the multiplicative law.

5.4.3 Complement of an Event

A complement of an event E, denoted by E^c, is what is not in the event of interest but belongs to the sample space S, i.e. $E^c = S - E$. **The Complementation Rule**: For any event E, $P(E^c) = 1 - P(E)$.

5.4.4 Null/Empty

A null/empty set, denoted by $\varnothing$, is the set that has no element.

A Venn diagram is a graphical representation of events and it's very useful for illustrating logical relations among events.

5.5 Laws of Probability

5.5.1 Additive Law

The **addition law** provides a way to compute the probability that event A or event B or both occur.

5.5.1.1 Additive Law for Mutually Exclusive Events

When both events A and B are mutually exclusive then,

$$P(A \cup B) = P(A) + P(B)$$

5.5.1.2 Additive Law for Not-Mutually Exclusive Events

When both events are not mutually exclusive,

$$P(A \cup B) = P(A) + P(B) - P(A \cap B)$$

Mutually exclusive events are two events with no sample points in common, i.e. cannot happen at the same time.

5.5.2 Conditional Probability

If we obtain new information and learn that a related event, denoted by B, already occurred, we will want to take advantage of this information by calculating a new probability for event A. This new probability of event A is called a **conditional probability** of A given that B has occurred and is written $P(A\backslash B)$.

Let A and B be two events. The conditional probability of event A given that event B has occurred denoted by $P(A\backslash B)$ is defined as

$$P(A\backslash B) = \frac{P(A \cap B)}{P(B)}$$

To apply conditional probability, we will use joint probability table, i.e. one table combining two events.

5.5.3 Independent Events

An event B is said to be independent (or statistically independent) of event A, if the conditional probability of B given A, i.e. $P(B|A)$ is equal to the unconditional probability of B.

$$P(B|A) = P(B)$$

Two events A and B are independent if

$$P(A \cap B) = P(A) \times P(B)$$

For Independent and conditional events we use joint table. A joint table is one table that combines two events in one table. For example, employment status and gender has been combined into one table.

Gender	Employment Status		
	Employed	Unemployed	Total
Male	450	473	923
Female	293	1035	1328
Total	743	1508	2251

To create probabilities we divide each observation by the overall total, i.e. we convert the given values to be between 0 and 1.

We divide each number with the overall total = 2251 to get a joint probability table.

Gender	Employment Status		
	Employed	Unemployed	Total
Male	450/2251 = 0.20	473/2251 = 0.21	923/2251 = 0.41
Female	293/2251 = 0.13	1035/2251 = 0.46	1328/2251 = 0.59
Total	743/2251 = 0.33	1508/2251 = 0.67	2251/2251 = 1

The *probabilities of the totals* represents the *probability the respective event* for that given total. These are the marginal probabilities.

$$P(Male) = 0.41, \ P(Female) = 0.59, P(Employed) = 0.33, P(Unemployed) = 0.67$$

The probabilities of intersections will represent the probability of two intersecting events.

$$P(Female \cap Employed) = 0.13, P(Male \cap Employed) = 0.20,$$
$$P(Male \cap Unemployed) = 0.21, P(Female \cap Unemployed) = 0.46$$

We can generate a joint table in Statistical Package for Social Science (SPSS) using the path: Analyze > Descriptive Statistics > Crosstabs > there will be a pop up window for Crosstabs > Select the categorical variable for Column and another Categorical variable for Rows > Click Cells > Under Counts select Observed and under Percentages select Total then Continue and OK.

Example

Using customer_subset.sav data, generate a joint probability table for Gender and Size of hometown.

In SPSS we go to, Analyze > Descriptive Statistics > Crosstabs (Figure 5.1)

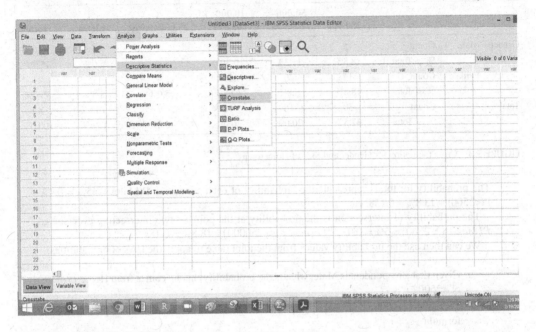

FIGURE 5.1 SPSS Menu for Crosstabs

There will be a pop up window for Crosstabs > Select the Size of hometown for Column and Gender for Rows > Click Cells > Under Counts select Observed and under Percentages select Total then Continue and OK (Figure 5.2).

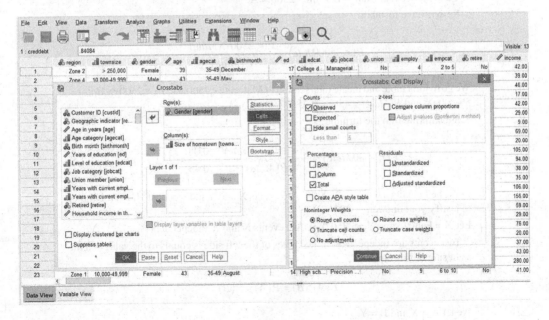

FIGURE 5.2 SPSS Crosstabs Pop Up Window displaying Cell Options

The SPSS Output will display the joint probability table values in form of percentages (Figure 5.3).

			>250,000	50,000–249,999	10,000–49,999	25,000–9,999	<2500	Total
					Size of Hometown			
Gender	Male	Count	15	9	5	3	4	36
		% of Total	18.8%	11.3%	6.3%	3.8%	5.0%	45.0%
	Female	Count	15	7	11	4	7	44
		% of Total	18.8%	8.8%	13.8%	5.0%	8.8%	55.0%
Total		Count	30	16	16	7	11	80
		% of Total	37.5%	20.0%	20.0%	8.8%	13.8%	100.0%

Gender × Size of Hometown Crosstabulation

FIGURE 5.3 Cross tabulation for Gender and Size of Hometown

The probability for totals represents the probability of each event, e.g. P(Male) = 45% = 0.45, P(<2500) = 13.8% = 0.138, etc.

The probability of intersection is the probability at the point of two intersecting events, e.g. $P(Male \cap < 2500) = 5\% = 0.05$, $P(Female \cap > 250000) = 18.8\% = 0.188$.

We can then use the probability values from the joint table to apply the laws of probabilities.

a. Given that a respondent is Male, what is the probability that he comes from Hometown of <2500?

This is a conditional probability question. We will use the conditional probability formula

$$P(A|B) = \frac{P(A \cap B)}{P(B)}$$

Statement is *given that the respondent is Male*, what is the probability that he is *from <2500?*

This implies that Male = B while <2500 = A.

Re-write the formula

$$P(< 2500|Male) = \frac{P(< 2500 \cap Male)}{P(Male)}$$

We go to joint table and locate intersection of Male and <2500, i.e. $P(< 2500 \cap Male) = 5\% = 0.05$ and $P(Male) = 45\% = 0.45$.

Substituting into the formula,

$$P(< 2500|Male) = \frac{P(< 2500 \cap Male)}{P(Male)} = \frac{0.05}{0.45} = 0.11$$

b. Let X be Female and Y be >250000. Are events X and Y independent? Use probability to justify your answer.

Let X = Female and Y => 250,000. Are events X and Y independent?

Two events are independent if the values of the left side is equals to the values on the right side of this equation.

$$P(A \cap B) = P(A) \times P(B)$$

We let A = X and B = Y

Substitute into the formula

$$P(X \cap Y) = P(X) \times P(Y)$$

We need to get the following values:

$$P(X \cap Y) = P(Female \cap > 250,000) = 18.8\% = 0.188$$

$$P(X) = P(Female) = 55\% = 0.55$$

$$P(Y) = P(> 250,000) = 37.5\% = 0.375$$

Multiplying P(X) and P(Y) we get

$$P(X) \times P(Y) = 0.55 \times 0.375 = 0.20625$$

Confirm if the values of the left side $[P(X \cap Y)]$ is equals to the values on the right side $[P(X) \times P(Y)]$.

Since the value of $P(X \cap Y)$ is not equal to the value of $P(X) \times P(Y)$, we conclude that X is not independent of Y.

Practice Exercise

1. Using *customer_subset.sav* data (inbuilt SPSS data), generate a joint probability table for Gender and Job category.

2. Using the joint probability table in (1) above, what is the probability that a respondent is a Male and is in Service Job Category?

3. Using the joint probability table in (1) above, what is the probability that a given respondent is not a Male and is not in Agricultural and Natural Resources Job Category?

4. Using the joint probability table in (1) above, given that a given respondent in Sales and Offices Job Category, what is the probability that a respondent is a Female?

5. Let event A be a Male respondent and event B be a respondent in Managerial and Professional Job Category, are events A and B independent? Use probabilities to justify your answer.

6

Random Variables and Probability Distributions

6.1 Random Variables

A random variable is a variable whose values are associated with some probability of being observed. A random variable represents outcomes of random experiments represented by uppercase variable such as X. *Examples of random variables:* a number of males in a class, height and weight.

6.2 Types of Random Variables

A random variable can be classified as being either discrete or continuous. *A discrete random variable* is one that can assume only finite number of values or an infinite sequence of values such as 0, 1, 2, ... For example, a number of examinations passed in an experiment of a student taking placement exams. A discrete random variable is a random variable whose possible values can be listed. A *continuous random variable* is one that can assume any numerical value in an interval or collections of intervals. For example, consider an experiment monitoring projects completed in on month, time between customer arrivals in minutes.

6.3 Describing Random Variables

Probability distribution is the set of all possible values of a random variable and its associated probabilities. If the random variable is discrete, then we have a *Probability mass function* denoted by $p(x) = p(X = x)$ for all possible values, x. If a random variable is continuous then we have a *Probability Density Functions (PDF)* denoted by $f(x)$.

6.3.1 Describing Discrete Random Variable

Properties of Discrete Random Variable

1. The probability assigned to each experimental outcome must be between 0 and 1, i.e. $0 \leq P(X) \leq 1$.
2. The sum of the probabilities for all the experimental outcomes must be equal to 1, i.e. $\sum P(X) = 1$.

Mean and Variance of a Discrete Random Variable/Distribution

Mean or Expected Value
The mean or expected value of discrete distribution is the long-run average of the occurrences and is defined using probability mass function. The expected value for a discrete random variable is computed as:

$$E[X] = \mu = \sum_{i=1}^{\infty} [x.P(x)]$$

DOI: 10.1201/9781003292654-8

where
$\mu = E[X] = $ long-run average
$x = $ an outcome
$P(x) = $ probability of that outcome

Variance of a Discrete Random Variable
The formula for computing the variance of a discrete random variable is as follows:

$$\sigma^2[X] = \sum_{i=1}^{\infty}\left[(x-\mu)^2.P(x)\right]$$

where
$x = $ an outcome
$P(x) = $ probability of a given outcome
$\mu = $ mean

Standard Deviation of a Discrete Random Variable
The standard deviation is then computed by taking the square root of the variance as:

$$\sigma = \sqrt{\sigma^2} = \sqrt{\sum_{i=1}^{\infty}\left[(x-\mu)^2.P(x)\right]}$$

Example: Find the expected value and standard deviation of *x*.

x	$P(x)$	$x.P(x)$
0	0.03125	0.0000
1	0.15625	0.15625
2	0.31250	0.62500
3	0.31250	0.93750
4	0.15625	0.62500
5	0.03125	0.15625
		$\sum x.P(x) = 2.5$

To find the standard deviation of *x*, we first compute the value of variance of *x*, then get the square root of variance of *x*.

x	$P(x)$	$(x-2.5)^2.P(x)$
0	0.03125	0.19531
1	0.15625	0.35156
2	0.31250	0.07812
3	0.31250	0.07812
4	0.15625	0.35156
5	0.03125	0.19531
		$\sum(x-2.5)^2.P(x) = 1.25$

$$\sigma = \sqrt{\sigma^2} = \sqrt{1.25} = 1.118034$$

Estimating Probabilities for a Discrete Random Variable
To find the required probability, we add the probabilities for the given limits

Example: Using the same example above, find $P(X \leq 2)$

$$P(X \leq 2) = P(X = 0) + P(X = 1) + P(X = 2)$$

$$= 0.03125 + 0.15625 + 0.31250 = 0.5$$

6.3.2 Describing Continuous Random Variable

Properties of Continuous Random Variable

1. $\int_{-\infty}^{\infty} f(x)dx = 1$, i.e. the total area under the curve/probabilities should be equal to 1.
2. $f(x) \geq 0$ for all x

Mean and Variance of a Continuous Random Variable
We use integration

NB: Review of Integration
Integration is the reverse of differentiation. This is also referred to as finding the area under the curve of a function $f(x)$. The integral symbol is $\int$.

The general (power) rule of integration

$$\int_{w}^{z} ax^b dx = \left[\frac{ax^{b+1}}{b+1} \right]_{w}^{z}$$

We then find the difference when the values of the limits (z and w) are substituted for x

$$\left(\frac{az^{b+1}}{b+1} \right) - \left(\frac{aw^{b+1}}{b+1} \right)$$

This is a case of *definite integral* which has limits of integration.

Example

Evaluate

$$\int_{0}^{2} 2xdx$$

This is a definite integral

$$\int_{0}^{2} 2xdx = \left[\frac{2x^{1+1}}{1+1} \right]_{0}^{2}$$

We then substitute 2 and 0, respectively, where there is x and find the difference

$$= \left[\frac{2x^2}{2} \right]_{0}^{2} - \left[\frac{2x^2}{2} \right]_{0}^{0}$$

$$= \left[\frac{2.2^2}{2} \right] - \left[\frac{2.0^2}{2} \right] = 4$$

The knowledge of integration will be used to estimate the mean and standard deviation of continuous random variables.

Mean or Expected Value

The mean or expected value of continuous random variable/distribution is defined using the probability density function and is computed as:

$$E[X] = \mu = \int_{-\infty}^{\infty} x.f(x)dx$$

where

x = an outcome

$f(x)$ = probability of a given outcome (continuous)

μ = mean

Variance of a Continuous Random Variable

The formula for computing the variance is as follows:

$$\sigma^2 = V[X] = E[X^2] - (E[X])^2$$

where

x = an outcome

$f(x)$ = probability of a given outcome

$\mu = \int_{-\infty}^{\infty} x.f(x)dx =$ mean

$E[X^2] = \int_{-\infty}^{\infty} x^2.f(x)dx$

Standard Deviation of a Continuous Random Variable

The standard deviation is then computed by taking the square root of the variance as:

$$\sigma = \sqrt{\sigma^2} = \sqrt{\left[E[X^2] - (E[X])^2 \right]}$$

Example: A random variable has a probability density function given by:

$$f(x) = 6x, \quad 0 \leq x \leq 2$$

Find mean and variance of X.

Mean of X is $E[X]$

$$E[X] = \int_{0}^{2} x.f(x)dx = \int_{0}^{2} x.6xdx = \int_{0}^{2} 6x^2 dx$$

$$= \left[\frac{6x^{2+1}}{2+1} \right]_{0}^{2} = \left[\frac{6x^3}{3} \right]_{0}^{2}$$

$$= \left[\frac{6x^3}{3} \right]^{2} - \left[\frac{6x^3}{3} \right]^{0}$$

$$= \left[\frac{6.2^3}{3} \right] - \left[\frac{6.0^3}{3} \right] = 16$$

Variance of X is given by $E[X^2] - (E[X])^2$

$$E[X] = 16$$

We have to estimate the value of $E\left[X^2\right]$ using $\int_0^2 x^2 . f(x) dx$

$$E\left[X^2\right] = \int_0^2 x^2 . f(x) dx = \int_0^2 x^2 . 6x \, dx = \int_0^2 6x^3 dx$$

$$= \left[\frac{6x^{3+1}}{3+1}\right]_0^2 = \left[\frac{6x^4}{4}\right]_0^2$$

$$= \left[\frac{6x^4}{4}\right]^2 - \left[\frac{6x^4}{4}\right]^0$$

$$= \left[\frac{6.2^4}{4}\right] - \left[\frac{6.0^4}{4}\right] = 24$$

Hence,

$$Var[X] = E\left[X^2\right] - \left(E[X]\right)^2 = 24 - 16^2 = -232$$

This means that $f(x) = 6x$ is not a valid probability density function since it has a negative variance.

Estimating Probabilities for a Continuous Random Variable
To find the probability, we integrate within the given limits.

Example: Using the above $f(x)$ function, find $P(X \le 1)$

$$P(X \le 1) = \int_0^1 f(x) dx = \int_0^1 6x \, dx$$

$$= \left[\frac{6x^{1+1}}{1+1}\right]_0^1 = \left[\frac{6x^2}{2}\right]_0^1$$

$$= \left[\frac{6x^2}{2}\right]^1 - \left[\frac{6x^2}{2}\right]^0$$

$$= \left[\frac{6.1^2}{2}\right] - \left[\frac{6.0^2}{2}\right] = 3$$

Hence, $f(x) = 6x$ is not a valid probability density function because the probability is greater than 1.

6.4 Common Distributions

The outcomes of random variables and their associated probabilities can be organized into *distributions*. The two types of distributions are discrete distributions constructed from discrete random variables and continuous distributions based on continuous random variables. Discrete distributions include: *Binomial distribution*, *Poisson distribution* and hypergeometric distribution. Continuous distribution includes *Uniform distribution*, *Normal distribution*, *Normal Approximation of Binomial distribution*, Exponential distribution, t distribution, Chi-square distribution and F distribution. Other distributions are Weibull, Gamma, Beta, Multivariate Normal and Multivariate Normal t Distribution.

6.4.1 Binomial Distribution

Binomial distribution is one of the discrete probability distributions. Binomial distribution is used when there are only two outcomes, e.g. pass or fail or when dealing with percentages. Binomial distribution is described by n (sample/number of trials) and p (percentage/probability of event of interest). We denote the *binomial distribution* as b (n, p)

The Binomial distribution formula is given by:

$$P(X = x) = C_x^n p^x (1-p)^{n-x}, \ for \ x = 1, 2, \ldots, n$$

where x are the successes in n trials with p probability of success in n trials and $(1-p)$ is the probability of getting a failure of getting a failure on any one trial.

We can use formula, Binomial Table, Excel or Statistical Package for Social Science (SPSS) to compute Binomial Probabilities

The mean and variance of a binomial random variable is given by:

$$E[X] = \mu = np$$

$$\sigma^2 = V[X] = np(1-p)$$

In SPSS, we use Cumulative Distribution Function (CDF). Binom to compute $\leq$ probabilities and PDF. Binom to compute exact probabilities.

The SPSS path for computing Binomial probabilities

When computing exact probability, e.g. $P(X = x)$ with n and p values: **Transform > Compute Variable**. Enter a value of zero in the top left cell. In the Compute Variable window, type pdfbinomial as **Target Variable**. Since we are calculating $P(X = x)$, under Function group, select **PDF & Noncentral PDF** and under **Functions and Special Variables**, double-click on **Pdf.Binom**. In the Numeric Expression box enter the values for the three question marks in the following order: quant = value of x, n = sample and prob = p then click OK. Go to Data View Window and you have the results of the computed probability.

When computing $\leq$ probabilities (cumulative probabilities), e.g. $P(X \leq x)$ with n and p values: **Transform > Compute Variable**. In the Compute Variable window, type cdfbinomial as **Target Variable**. Since we are calculating $P(X \leq x)$, under **Function group** select **CDF & Noncentral CDF**, and under **Functions and Special Variables** double-click on **Cdf.Binom**. In the Numeric Expression box enter the values for the three question marks in the following order: quant = value of x, n = sample and prob=p then click OK. Go to Data View Window and you have the results of the computed probability.

Example

A certain interactive survey reported that 55% of Nairobi residents were unemployed. Suppose five residents are selected randomly and will be interviewed about their employment status.

    a. Is the selection of five residents a binomial experiment? Explain.

      Yes because we have percentage ($p = 55\% = 0.55$) and sample ($n = 5$)

        We also have percentage of unemployment = 55% and we can compute percentage of employment = 45%

b. What is the probability that 3 of the 5 residents will say that they are unemployed?

This is a Binomial Distribution: $n = 5$ and $p = 0.55$ and X is a discrete random variable representing a number of employees.

We want to compute the probability that X is equals to 3.

Using the formula

$$P(X = x) = C_x^n p^x (1-p)^{n-x}, \ for \ x = 1, 2, \ldots, n$$

Substituting the values for $x = 3$, $n = 5$ and $p = 0.55$

$$P(X = 3) = C_3^5 0.55^3 (1 - 0.55)^{5-3} = 0.3369$$

In SPSS, to find the probability that X is 3: we use PDF.Binom since we want to compute exact probabilities

$$P(X = 3)$$

Go to SPSS Data View Window > Click 0 in the first cell then press enter > Go to Transform > Compute Variable > there will be a pop up window for Compute Variable > type pdfbinomial in the Target Variable > select PDF and Noncentral PDF under Function group > Double click Pdf.Binom under Functions and Special Variables > In the Numeric Expression box enter the values for the three question marks in the following order: quant=value of x (3), n=sample (5) and prob=p (0.55) then click OK (Figure 6.1).

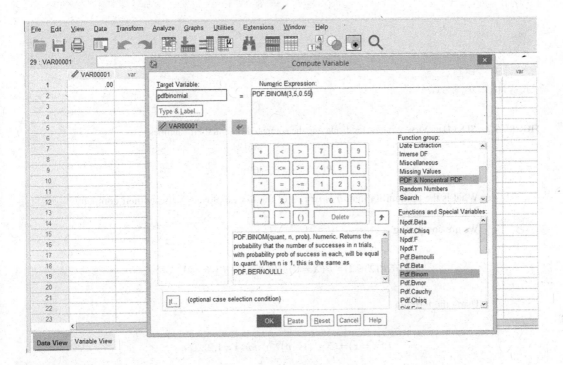

FIGURE 6.1   SPSS Pop Window for computing Binomial Probabilities

Go to Data View Window and you have the results. Hence $P(X=3)=0.33691$ (Figure 6.2)

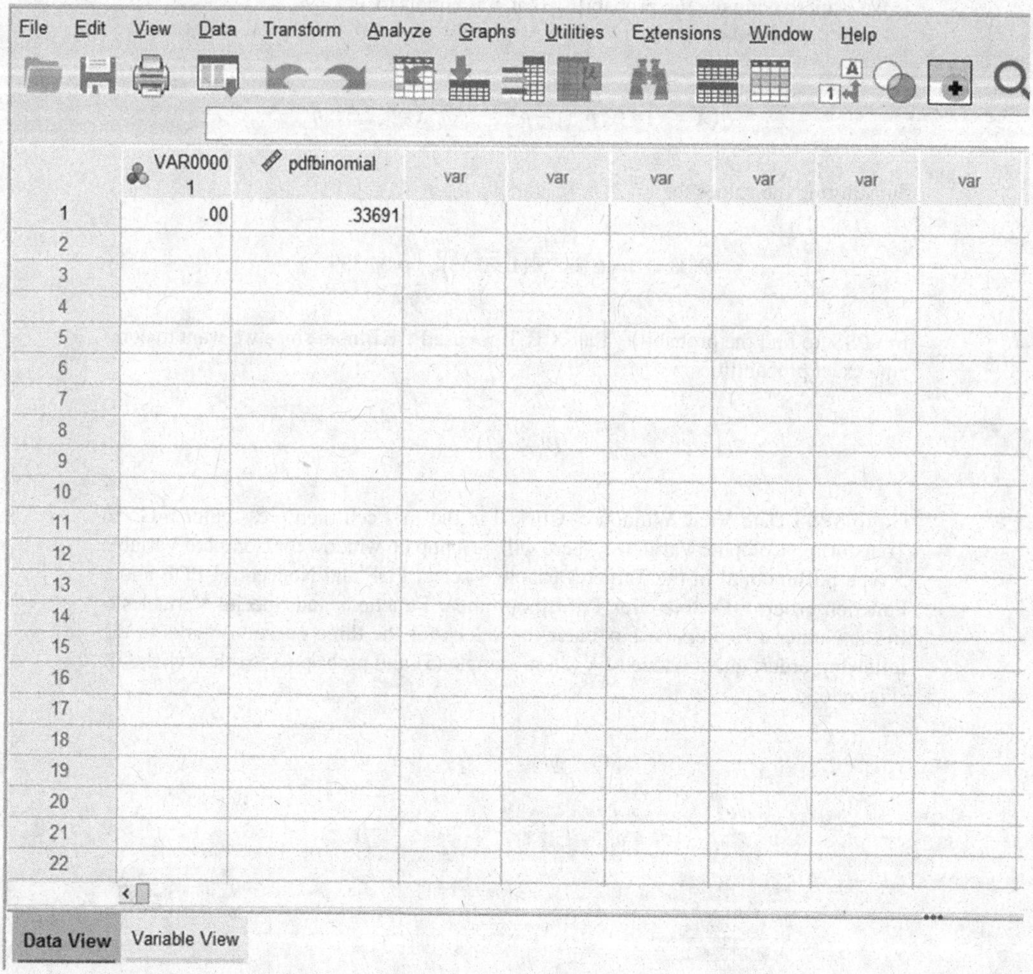

FIGURE 6.2 SPSS Data View Window for displaying the computed Binomial Probabilities

$$P(X=3)=0.33691$$

c. What is the probability that *at most two* of the five employees will say their company is loyal to them?

We are computing

$$P(X \le 2) = P(X=0) + P(X=1) + P(X=2)$$

Using the formula

$$P(X=x) = C_x^n p^x (1-p)^{n-x}, \; for \; x=1,2,\dots,n$$

Substituting the values for $x = 0,1$ and 2, $n = 5$ and $p = 0.55$

$$P(X = 0) = C_0^5 0.55^0 (1 - 0.55)^{5-0} = 0.018453$$

$$P(X = 1) = C_1^5 0.55^1 (1 - 0.55)^{5-1} = 0.112767$$

$$P(X = 2) = C_2^5 0.55^2 (1 - 0.55)^{5-2} = 0.275653$$

$$P(X \leq 2) = 0.018453 + 0.112767 + 0.275653 = 0.406873$$

We are computing at most probabilities. This is the same as $\leq$ probabilities. We use CDF.Binom to compute $\leq$

We are computing $P(X \leq 2)$, so we use CDF.Binomial.

Go to SPSS Data View Window > Click 0 in the first cell then press enter > Go to Transform > Compute Variable > there will be a pop up window for Compute Variable > type cdfbinomial in the Target Variable > select CDF and Noncentral CDF under Function group > Double click Cdf.Binom under Functions and Special Variables > In the Numeric Expression box enter the values for the three question marks in the following order: quant = value of x (2), n = sample (5) and prob = p (0.55) then click OK (Figure 6.3).

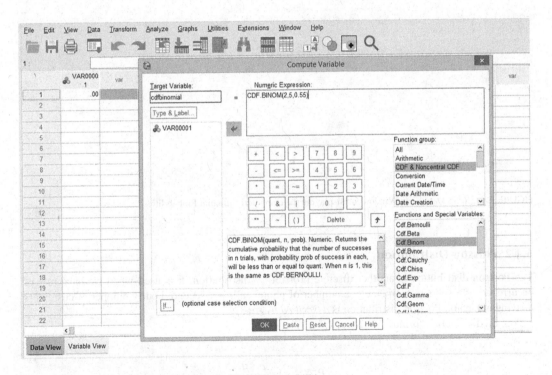

FIGURE 6.3 SPSS Pop Window for computing Binomial Probabilities

Go to Data View Window and you have the results. Hence $P(\leq 2) = 0.40687$ (Figure 6.4).

$$P(\leq 2) = 0.40687$$

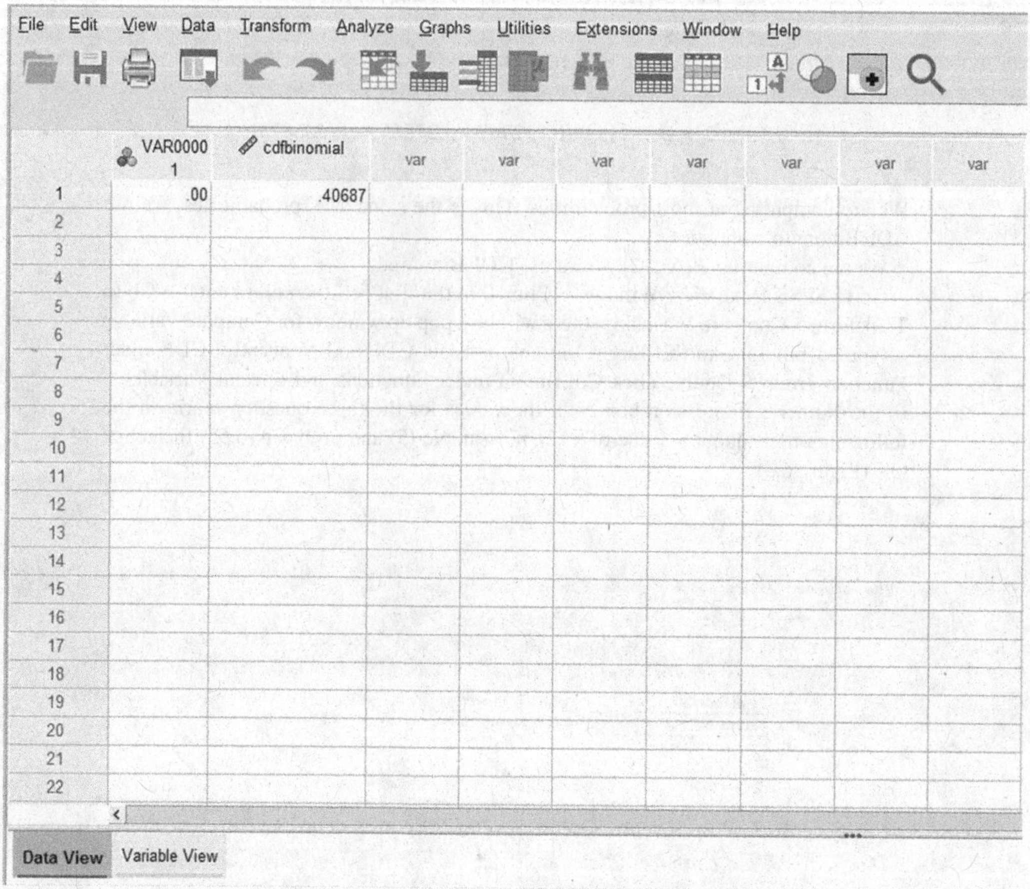

	VAR0000 1	cdfbinomial	var	var	var	var	var	var	var
1	.00	.40687							
2									
3									
4									
5									
6									
7									
8									
9									
10									
11									
12									
13									
14									
15									
16									
17									
18									
19									
20									
21									
22									

FIGURE 6.4   SPSS Data View Window for displaying the computed Binomial Probabilities

6.4.2 Poisson Distribution

The Poisson distribution is another discrete probability distribution. It is used when we have rate of occurrences of events over time, e.g. a number of phone calls per second, hourly number of customers arriving at a bank. Poisson distribution is *described by rate* μ.

Poisson distribution formula is given by:

$$P(X = x) = \frac{e^{-\mu}\mu^x}{x!}$$

where

x is the designated number of successes (number of times and event occurs).

$P(X)$ is the probability of X number of successes.

μ is the average number of successes per unit of time (rate of occurrences of events over time).
e is the base of the natural logarithmic system, or 2.71828.

The mean and variance of a Poisson random variable is given by:

$$E[X] = \mu.$$

$$\sigma^2 = V[X] = \mu$$

We can use formula, Poisson Table, Excel or SPSS to compute Poisson Probabilities
   In SPSS, we use CDF.Poisson to compute $\leq$ probabilities and PDF.Poisson to compute exact probabilities

The SPSS path for computing Poisson probabilities
   When computing exact probability, e.g. $P(X = x)$ *when given rate* μ: **Transform > Compute Variable**. Type pdfpoisson as **Target Variable**. Since we are calculating $P(X = 2)$, under **Function group** under **Function group** select **PDF & Noncentral PDF**, and under **Functions and Special Variables** double-click on **Pdf.Poisson**. In the Numeric Expression box enter the values for the two question marks in the following order: quant = value of x and mean = rate then click OK. Go to Data View Window and you have the results of the computed probability.
   When computing $\leq$ *probabilities (cumulative probabilities), e.g.* $P(X \leq 3)$ *when given rate* μ: **Transform > Compute Variable**. Type cdfpoisson as **Target Variable**. Since we are calculating $P(X \leq 3)$, under **Function group** select **CDF & Noncentral CDF**, and under **Functions and Special Variables**, double-click on **Cdf.Poisson**. In the Numeric Expression box enter the values for the two question marks in the following order: quant=value of x and mean=rate then click OK. Go to Data View Window and you have the results of the computed probability.

Example

Madaraka express train passengers arrive randomly and independently at the passenger-screening facility at a major train terminal. The mean arrival rate is nine passengers per minute.

   a. Compute the probability of no arrivals in a one-minute period.
      This is a Poisson distribution since we have rates of occurrences of events over time.
      The mean rate is 9 per one minute, i.e. $\mu = 9$
      We are computing $P(X = 0)$
      Using the formula

$$P(X = x) = \frac{e^{-\mu}\mu^x}{x!}$$

Substituting the value of $x = 0$ and $\mu = 9$ into the formula

$$P(X = 0) = \frac{e^{-9}9^0}{0!} = 0.0001234$$

Using SPSS to compute probabilities: we use CDF.Poisson to compute $\leq$ probabilities and PDF.Poisson to compute exact probabilities
   We are computing $P(X = 0)$, we will use PDF.Poisson
   Go to SPSS Data View Window > Click 0 in the first cell then press enter > Go to Transform > Compute Variable > there will be a pop up window for Compute

Variable > type pdfpoisson in the Target Variable > select PDF and Noncentral PDF under Function group > Double click Pdf.Poisson under Functions and Special Variables > In the Numeric Expression box enter the values for the two question marks in the following order: quant = value of x (0) and mean = rate (9) then click OK (Figure 6.5).

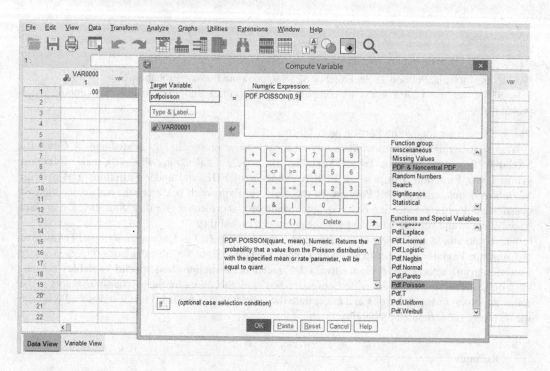

FIGURE 6.5   SPSS Pop Window for computing Poisson Probabilities

Go to Data View Window and you have the results. Hence $P(X = 0) = 0.000123$ (Figure 6.6).

$$P(X = 0) = 0.000123$$

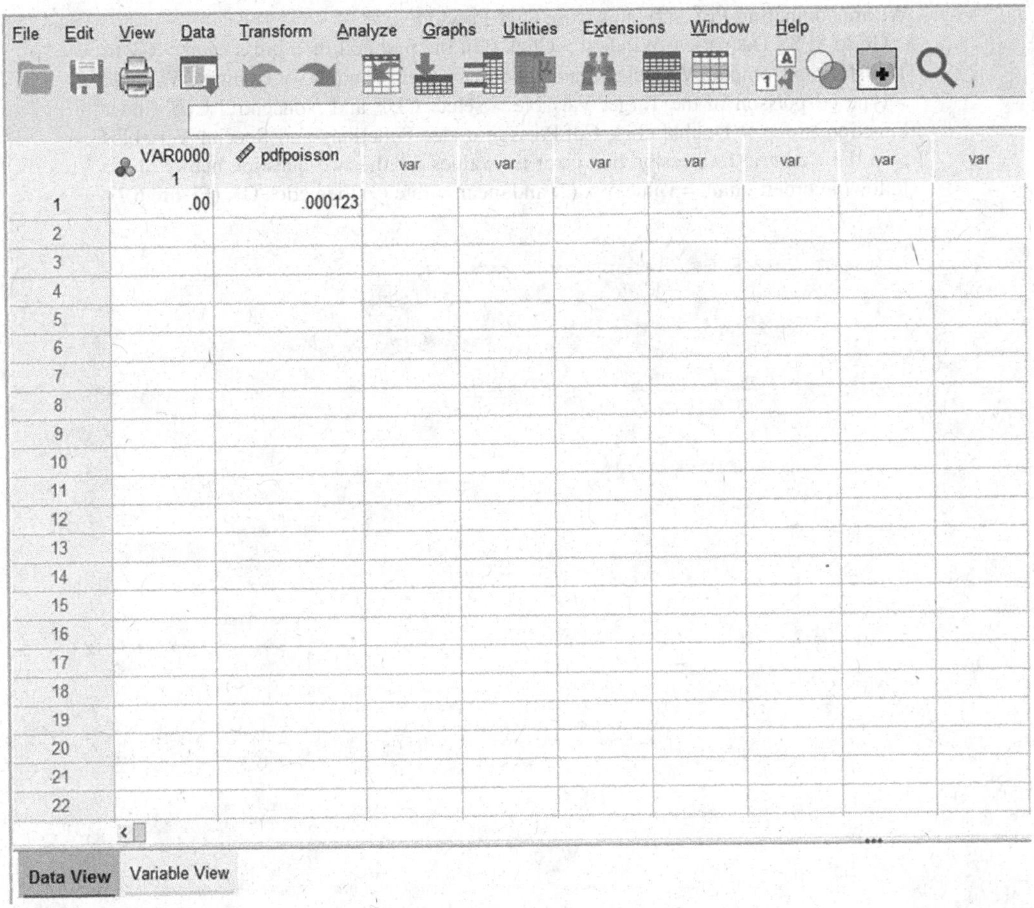

	& VAR0000 1	⬙ pdfpoisson	var	var	var	var	var	var	var	var
1	.00	.000123								
2										
3										
4										
5										
6										
7										
8										
9										
10										
11										
12										
13										
14										
15										
16										
17										
18										
19										
20										
21										
22										

Data View Variable View

FIGURE 6.6 SPSS Data View Window for displaying the computed Poisson Probabilities

b. Compute the probability of at most two arrivals in one-minute period
We are computing $P(X \le 2)$

$$P(X \le 2) = P(X = 0) + P(X = 1) + P(X = 2)$$

Using the formula

$$P(X = x) = \frac{e^{-\mu}\mu^x}{x!}$$

Substituting the value of x = 0, 1 and 2 and $\mu = 9$ into the formula

$$P(X = 0) = \frac{e^{-9}9^0}{0!} = 0.0001234$$

$$P(X = 1) = \frac{e^{-9}9^1}{1!} = 0.001110$$

$$P(X = 2) = \frac{e^{-9}9^2}{2!} = 0.004998$$

$$P(X \le 2) = 0.0001234 + 0.001110 + 0.004998 = 0.0062314$$

We are computing $P(X \leq 2)$, so we use CDF.Poisson

Go to SPSS Data View Window > Click 0 in the first cell then press enter > Go to Transform > Compute Variable > there will be a pop up window for Compute Variable > type cdfpoisson in the Target Variable > select CDF and Noncentral CDF under Function group > Double click Cdf.Poisson under Functions and Special Variables > In the Numeric Expression box enter the values for the two question marks in the following order: quant = value of x (2) and mean = rate (9) then click OK (Figure 6.7).

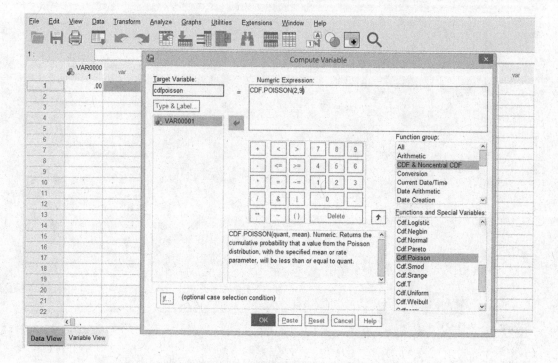

FIGURE 6.7   SPSS Pop Window for computing Poisson Probabilities

Go to Data View Window and you have the results. Hence $P(X \leq 2) = 0.00623$ (Figure 6.8).

$$P(X \leq 2) = 0.00623$$

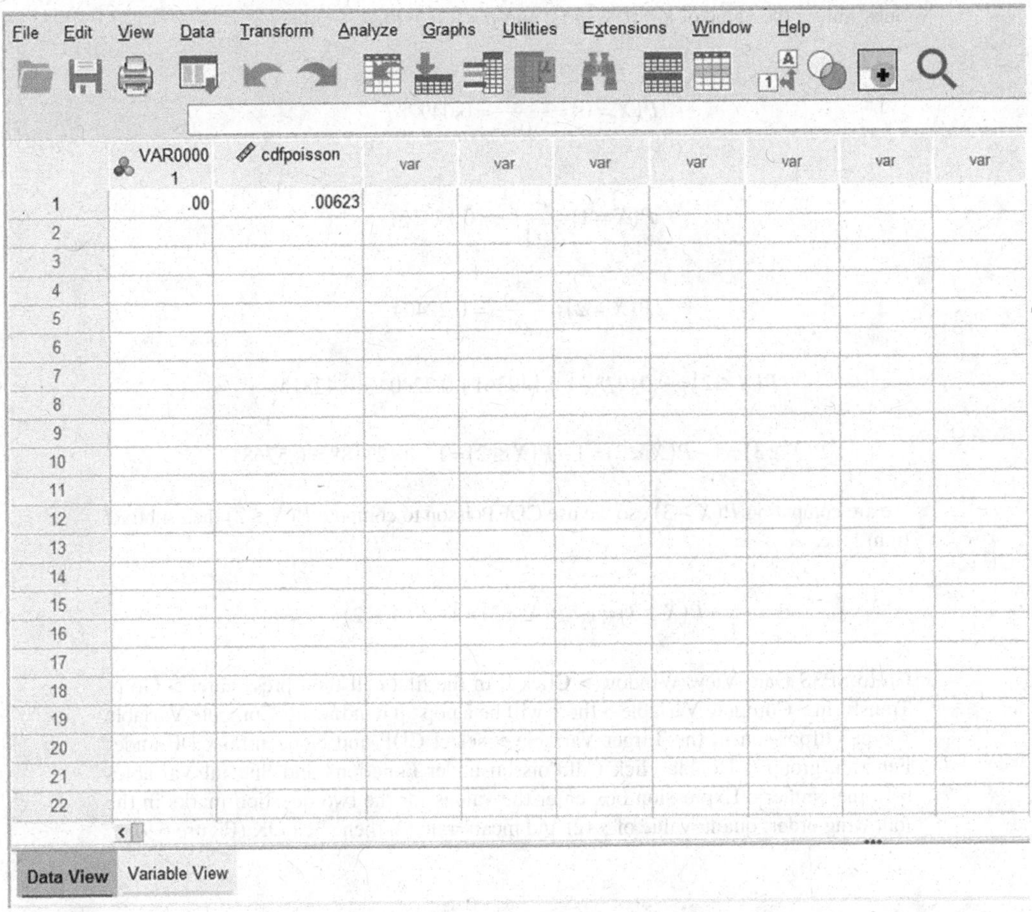

FIGURE 6.8   SPSS Data View Window for displaying the computed Poisson Probabilities

c. Compute the probability that three or more passengers arrive in a twenty seconds period.

We first convert the rate to twenty seconds since we were given the rate per minute

9 = 1 minute/60 seconds

What about 20 seconds = $\frac{9 \times 20}{60} = 3$. The rate becomes 3 per 20 seconds

We are computing $P(X \geq 3)$, so we use compute $P(X \leq 2)$ then subtract from 1, i.e.

$$P(X \geq 3) = 1 - P(X < 3) = 1 - P(X \leq 2)$$

We are computing $P(X \leq 2)$

$$P(X \leq 2) = P(X = 0) + P(X = 1) + P(X = 2)$$

Using the formula

$$P(X = x) = \frac{e^{-\mu}\mu^x}{x!}$$

Substituting the value of x = 0, 1 and 2 and $\mu = 3$ into the formula

$$P(X=0) = \frac{e^{-3}3^0}{0!} = 0.049787$$

$$P(X=1) = \frac{e^{-3}3^1}{1!} = 0.149361$$

$$P(X=2) = \frac{e^{-3}3^2}{2!} = 0.22404$$

$$P(X \leq 2) = 0.049787 + 0.149361 + 0.22404 = 0.423188$$

$$P(X \geq 3) = 1 - P(X < 3) = 1 - P(X \leq 2) = 1 - 0.423188 = 0.576812$$

We are computing $P(X \geq 3)$, so we use CDF.Poisson to compute $P(X \leq 2)$ then subtract from 1, i.e.

$$P(X \geq 3) = 1 - P(X < 3) = 1 - P(X \leq 2)$$

Go to **SPSS** Data View Window > Click 0 in the first cell then press enter > Go to Transform > Compute Variable > there will be a pop up window for Compute Variable > type cdfpoisson in the Target Variable > select CDF and Noncentral CDF under Function group > Double click Cdf.Poisson under Functions and Special Variables > In the Numeric Expression box enter the values for the two question marks in the following order: quant=value of x (2) and mean=rate (3) then click OK (Figure 6.9).

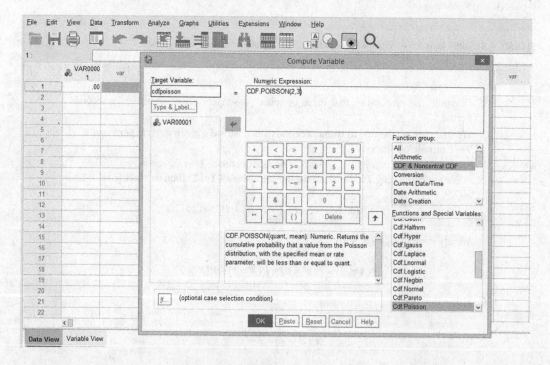

FIGURE 6.9   SPSS Pop Window for computing Poisson Probabilities

Go to Data View Window and you have the results. Hence $P(X \leq 2) = 0.4232$ (Figure 6.10).

$$P(X \geq 3) = 1 - P(X \leq 2) = 1 - 0.4232 = 0.5768$$

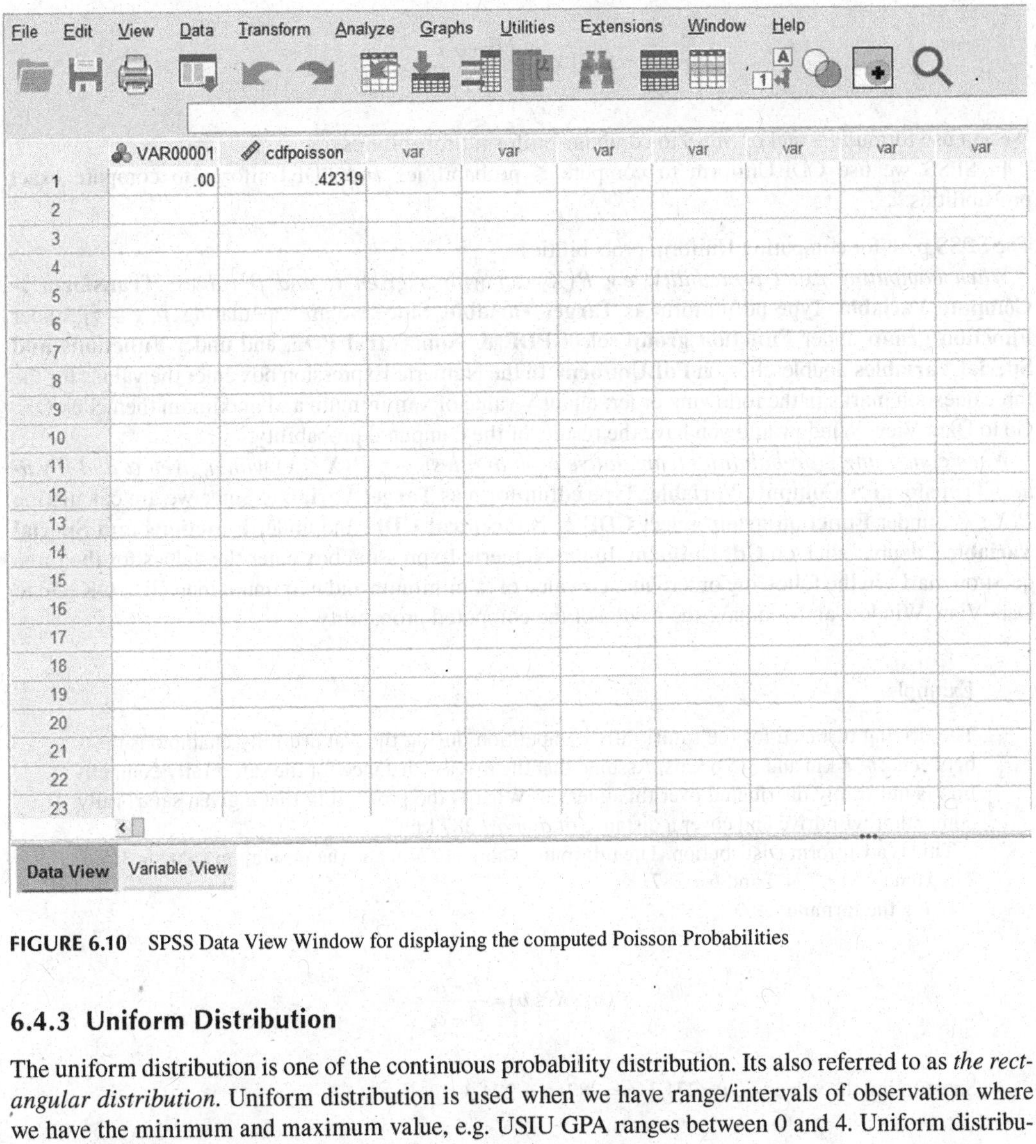

FIGURE 6.10 SPSS Data View Window for displaying the computed Poisson Probabilities

6.4.3 Uniform Distribution

The uniform distribution is one of the continuous probability distribution. Its also referred to as *the rectangular distribution*. Uniform distribution is used when we have range/intervals of observation where we have the minimum and maximum value, e.g. USIU GPA ranges between 0 and 4. Uniform distribution is *described by α (mimimum) and β (maximum)*.

A continuous random variable X is said to be uniformly distributed if its density function is given by:

$$f(x) = \frac{1}{\beta - \alpha}, \ for -\infty < \alpha \leq x \leq \beta < \infty$$

The values α and β are the parameters of the uniform distribution.

To find the probability for uniform distribution we estimate the area under the cure, i.e. we first estimate a and b values within the interval of α (mimimum) and (maximum), the substitute into the fomula

$$P(a \leq X \leq b) = \frac{b - a}{\beta - \alpha}$$

The mean and variance of a uniform random variable is given by:

$$E[X] = \frac{\alpha + \beta}{2}$$

$$Var[X] = \frac{(\beta - \alpha)^2}{12}$$

We can use formula, Excel or SPSS to compute Uniform Probabilities

In SPSS we use CDF.Uniform to compute $\leq$ probabilities and PDF.Uniform to compute exact probabilities

The SPSS path for computing Uniform probabilities

When computing exact probability, e.g $P(X = x)$ *with a given* α *and* β *values:* **Transform > Compute Variable**. Type pdfuniform as **Target Variable**. Since we are calculating $P(X = 2)$, under **Function group** under **Function group** select **PDF & Noncentral PDF**, and under **Functions and Special Variables** double-click on **Pdf.Uniform**. In the Numeric Expression box enter the values for the three question marks in the following order: quant = value of x, minimum and maximum then click OK. Go to Data View Window and you have the results of the computed probability.

When computing $\leq$ *probabilities (cumulative probabilities), e.g.* $P(X \leq x)$ *with a given* α *and* β *values:* **Transform > Compute Variable**. Type cdfuniform as **Target Variable**. Since we are calculating $P(X \leq 3)$, under **Function group** select **CDF & Noncentral CDF**, and under **Functions and Special Variables**, double-click on **Cdf.Uniform**. In the Numeric Expression box enter the values for the three question marks in the following order: quant = value of x, minimum and maximum then click OK. Go to Data View Window and you have the results of the computed probability.

Example

The driving distance for the safari rally competition during the Nakuru rally championship is between 274.3 km and 315.6 kms. Assume that the driving distance for the safari rally competition is uniformly distributed over this interval. What is the probability that a given safari rally competitor will drive and cover a distance of *at most* 287 kms?

This is a Uniform Distribution. The minimum value is 274.3 and the maximum value is 315.6. The value of $a = 274.3$ and $b = 287$

Using the formula

$$P(a \leq X \leq b) = \frac{b - a}{\beta - \alpha}$$

Substituting the values of $a = 274.3$, $b = 287$, $\alpha = 274.3$ and $\beta = 315.6$

$$P(287 \leq X \leq 315) = \frac{b - a}{\beta - \alpha} = \frac{287 - 274.3}{315.6 - 274.3} = 0.307506$$

We will use SPSS to compute probabilities: we use CDF.Uniform to compute $\leq$ probabilities and PDF.Uniform to compute exact probabilities

We are computing $P(X \leq 287)$, so we use CDF.Uniform

Go to SPSS Data View Window > Click 0 in the first cell then press enter > Go to Transform > Compute Variable > there will be a pop up window for Compute Variable > type cdfuniform in the Target Variable > select CDF and Noncentral CDF under Function group > Double click Cdf.Uniform under Functions and Special Variables > In the Numeric Expression box enter the values for the three question marks in the following order: quant = value of x (287), minimum (274.3) and maximum (315.6) then click OK (Figure 6.11).

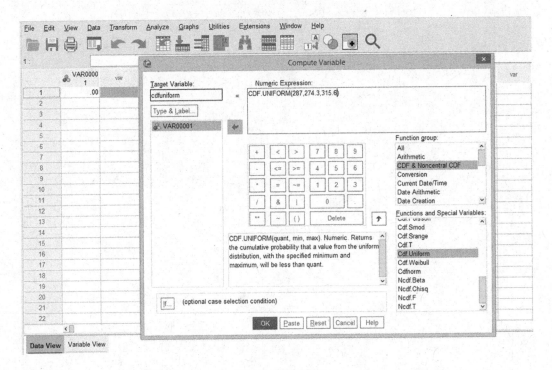

FIGURE 6.11 SPSS Pop Window for computing Uniform Probabilities

Go to Data View Window and you have the results. Hence $P(\leq 287) = 0.3075$ (Figure 6.12)

$$P(X \leq 287) = 0.3075$$

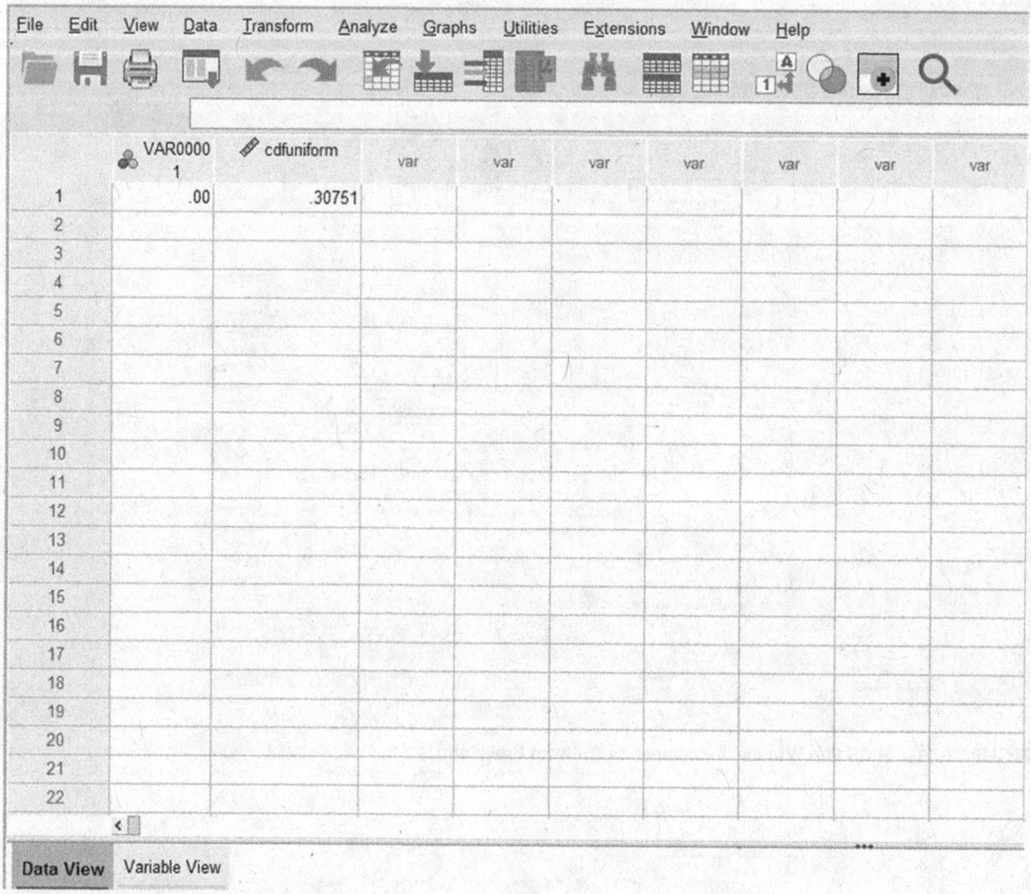

FIGURE 6.12   SPSS Data View Window for displaying the computed Uniform Probabilities

6.4.4 Normal Probability Distribution

The normal distribution is a continuous probability distribution and the most commonly used distribution in statistical analysis. Also called the Gaussian distribution. A normal distribution is a bell-shaped distribution that is symmetrical about the center (mean). Normal distribution is *described by mean (μ) and standard deviation (σ)*.

A random variable X is said to have the normal distribution with parameters μ and σ if its probability density function is given by:

$$f(x) = \frac{1}{\sigma\sqrt{2\pi}} e^{\left\{-\frac{1}{2}\left(\frac{x-\mu}{\sigma}\right)^2\right\}}, \; for -\infty < x < \infty$$

where
  μ = mean of x
  σ = standard deviation of x
  π = 3.14159
  e = 2.71828

Traditionally, to compute probabilities for a normal distribution we convert the X random variable to Z (standard normal random variable) and then use the Z table to read probabilities as explained in the steps below:

Steps to be followed when estimating probabilities under normal distribution using Z table

- Estimate the mean and standard deviation
- Convert the *X* value into its corresponding *Z* value (standard normal random variable), as follows:

$$Z = \frac{X - \mu}{\sigma}, \sigma \neq 0$$

- We then draw the area of interest under the normal curve using the Z values.
- Then we look up for the z value in Z-table. This gives the proportion of the area (probability) included under the curve between the mean (0) and that z value.

In SPSS we use CDF.Normal to compute ≤ probabilities and PDF.Normal to compute exact probabilities or Idf.Normal to estimate X values when probabilities are given.

The SPSS path for computing Normal probabilities

When computing exact probability, e.g. $P(X = x)$ given that μ and σ: **Transform > Compute Variable**. Type pdfnormal as **Target Variable**. Since we are calculating $P(X = x)$, under **Function group** under **Function group** select **PDF & Noncentral PDF**, and under **Functions and Special Variables** double-click on **Pdf.Normal** > In the Numeric Expression box enter the values for the three question marks in the following order: quant=value of x, mean and standard deviation then click OK. Then click **OK**. Go to Data View Window and you have the results of the computed probability.

When computing ≤ probabilities (cumulative probabilities), e.g. $P(X \leq x)$ given μ and σ: **Transform > Compute Variable**. Type cdfnormal as **Target Variable**. Since we are calculating $P(X \leq x)$, under **Function group** select **CDF & Noncentral CDF**, and under **Functions and Special Variables**, double-click on **Cdf.Normal**. In the Numeric Expression box enter the values for the three question marks in the following order: quant=value of x, mean and standard deviation then click OK. Then click **OK**. Go to Data View Window and you have the results of the computed probability.

The Inverse Transformation is used if we are interested to get the cumulative X values for a given probability. For example, find X *given the p, μ and σ values:* From the menu choose **Transform > Compute Variable**. In the Numeric Expression box enter the values for the three question marks in the following order: prob=cumulative probability, mean and standard deviation then click OK. Go to Data View Window and you have the results of the computed X value.

Example

The early morning trading volumes (millions of shares) at Nairobi Securities Exchange for 18 days in January are as shown below.

216, 143, 221, 169, 198, 202, 210, 122, 187, 212, 232, 181, 199, 220, 258, 233, 219, 221

Assuming that the trading volume is normally distributed with mean and standard deviation.

- a. What is the probability that, on a randomly selected day, the early morning trading volume *will be at most 189* million shares?

    This is a normal distribution. We need to compute *mean* and *standard deviation*
    We create the variables in the Variable View Window and key the data in the Data View Window (Figures 6.13 and 6.14).

Variable View

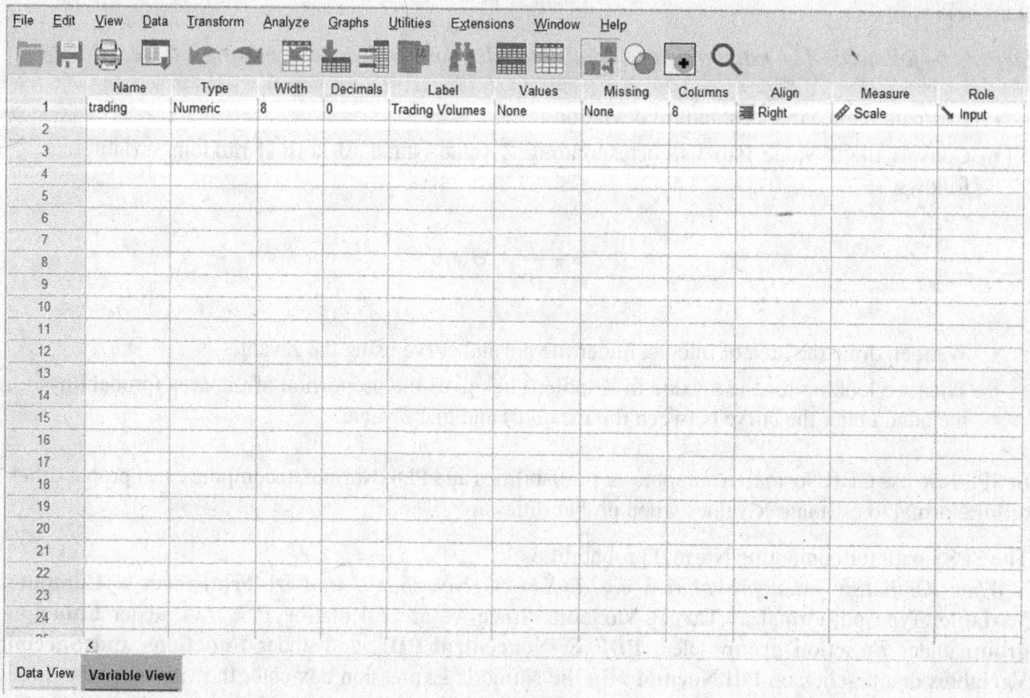

	Name	Type	Width	Decimals	Label	Values	Missing	Columns	Align	Measure	Role
1	trading	Numeric	8	0	Trading Volumes	None	None	8	Right	Scale	Input
2											
3											
4											
5											
6											
7											
8											
9											
10											
11											
12											
13											
14											
15											
16											
17											
18											
19											
20											
21											
22											
23											
24											

Data View   **Variable View**

FIGURE 6.13   Variable View Window for Trading Volumes Data

Variable View

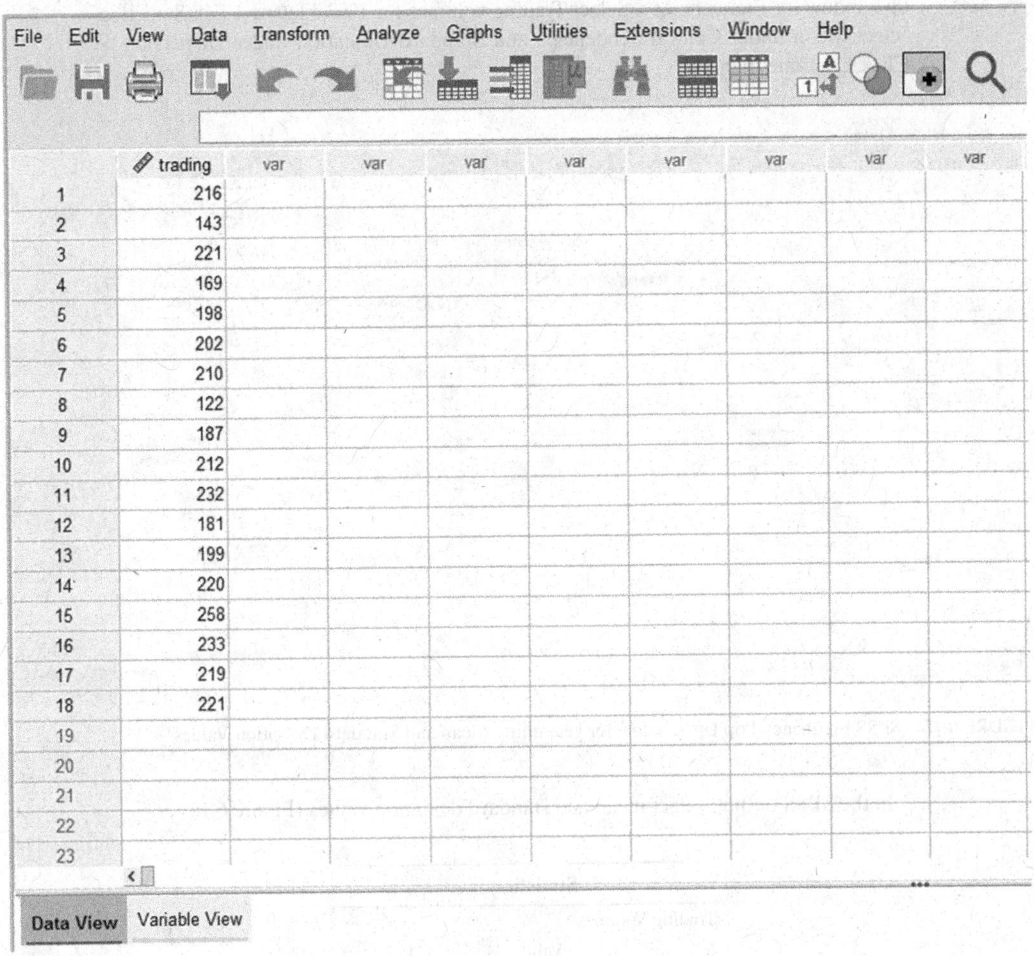

	trading	var	var	var	var	var	var	var	var
1	216								
2	143								
3	221								
4	169								
5	198								
6	202								
7	210								
8	122								
9	187								
10	212								
11	232								
12	181								
13	199								
14	220								
15	258								
16	233								
17	219								
18	221								
19									
20									
21									
22									
23									

Data View   Variable View

FIGURE 6.14 Data View Window for Trading Volumes Data

Step 1: We compute mean and standard deviation values (Figure 6.15)

In SPSS we to Analyze > Descriptive Statistics > Frequencies > there will be a pop up window for Frequencies > Select Trading volumes for Variable(s) > Click Statistics select Mean under Central Tendencies and Standard Deviation under Dispersion > Click Continue then OK

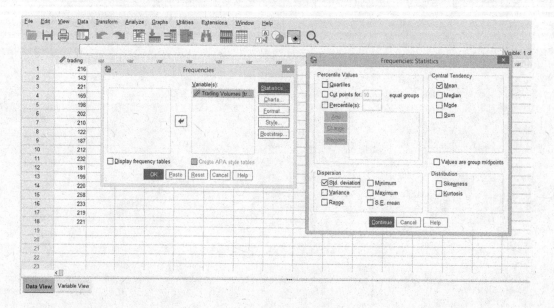

FIGURE 6.15   SPSS Frequency Pop Up Window for generating Mean and Standard Deviation Values

In the SPSS output, select mean and standard deviation values (Figure 6.16)

Statistics		
Trading Volumes		
N	Valid	18
	Missing	0
Mean		202.39
Std. Deviation		32.855

FIGURE 6.16   Mean and Standard Deviation Values for Trading Volumes

$$\mu = 205.39$$

$$\sigma = 32.855$$

We are computing $P(X \leq 189)$

Step 2: Convert X to Z

$$P(X \leq 189) = P\left(\frac{X - \mu}{\sigma} \leq \frac{189 - 205.39}{32.855}\right) = P(Z \leq -0.50)$$

Step 3: Draw the area of interest under the normal curve using Z value. Since z is negative, we draw it on the left side and shade the area of interest using the inequality

Step 4: We read the area from the center to $z = -0.50$ which is 0.1915

Step 5: Find the total shaded area = 0.5−0.1915 = 0.3085

We are computing $P(X \leq 189)$, so we use CDF.Normal

Go to SPSS Data View Window > Click 0 in the first cell then press enter > Go to Transform > Compute Variable > there will be a pop up window for Compute Variable > type cdfnormal in the Target Variable > select CDF and Noncentral CDF under Function group > Double click Cdf.Normal under Functions and Special Variables > In the Numeric Expression box enter the values for the three question marks in the following order: quant=value of x (189), mean (202.39) and standard deviation (32.8555) then click OK (Figure 6.17).

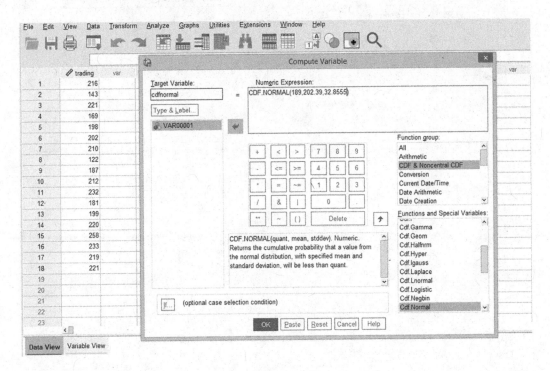

FIGURE 6.17 SPSS Pop Window for computing Normal Probabilities

Go to Data View Window and you have the results. Hence $P(X \leq 189) = 0.3418$
The SPSS output will give the probability value (Figure 6.18).

$$P(X \leq 189) = 0.3418$$

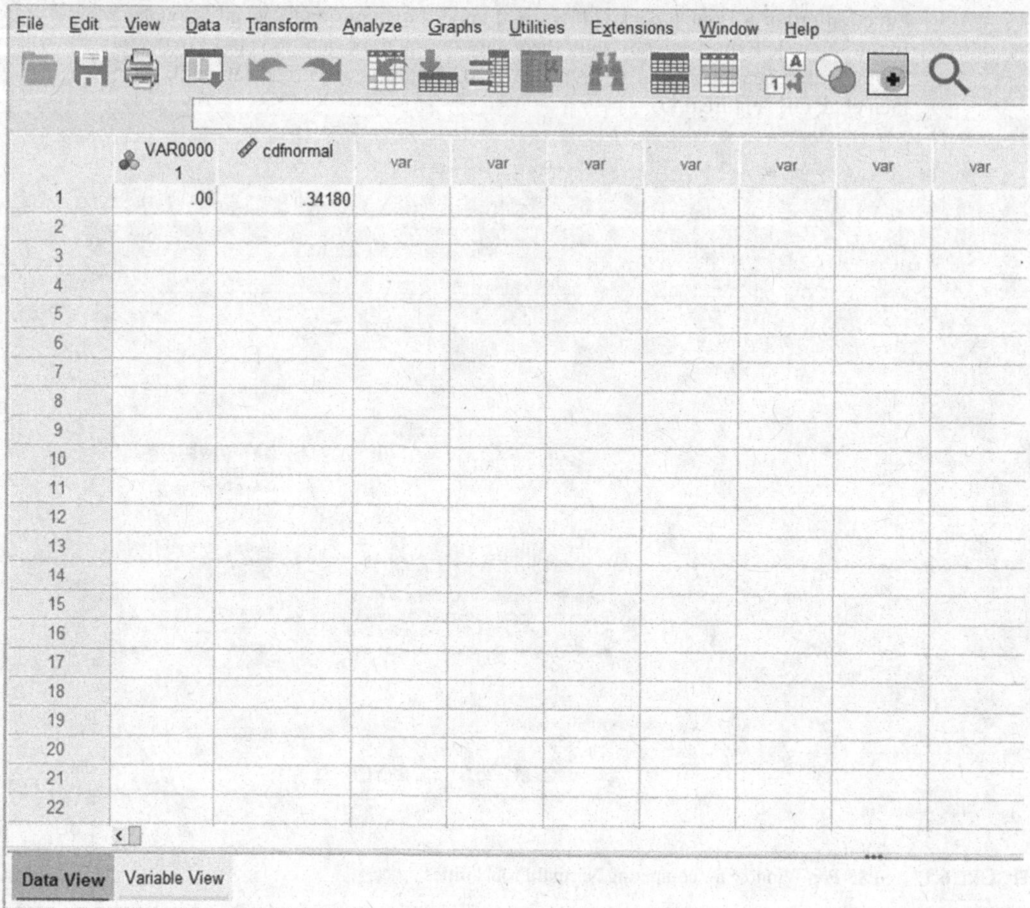

FIGURE 6.18 SPSS Data View Window for displaying the computed Normal Probabilities

b. How many trading volumes should the company have to be in the top 40%?

We have been given the area to the extreme right. We follow the following steps to get X value of interest.

Step 1: Draw the area of interest (top 40%) on the normal curve. We estimate the area from center to the 40% on the right by subtracting 40% from 50%, i.e. $0.5 - 0.4 = 0.1$. This represents the area from the center to Z.

Step 2: Find Z that is equal to the area from the center to z (0.10), which is $z = 0.25$. Z will be positive since its on the right side.

Step 3: Find the value of X using the Z formula

$$Z = \frac{X - \mu}{\sigma}$$

$$0.25 = \frac{X - 205.39}{32.8555}$$

$$X = (0.25 \times 32.8555) + 205.39 = 213.603$$

In SPSS, we first estimate the cumulative area/probability we subtract 40% = 0.40 from 1, i.e. 1 − 0.40 = 0.60

If it was bottom percentage, then you use the given percentage for cumulative probability

Once we have the cumulative percentage/probability we use Idf.Normal to get the X values corresponding to the cumulative percentage/probability

Go to SPSS Data View Window > Click 0 in the first cell then press enter > Go to Transform > Compute Variable > there will be a pop up window for Compute Variable > type idfnormal in the Target Variable > select Inverse DF under Function group > Double click Idf.Normal under Functions and Special Variables > In the Numeric Expression box enter the values for the three question marks in the following order: prob=cumulative probability (0.60), mean (202.39) and standard deviation (32.8555) then click OK (Figure 6.19).

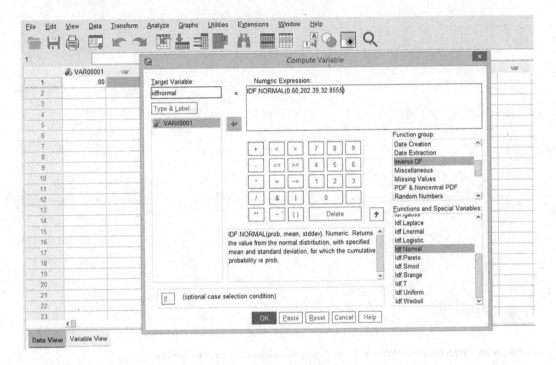

FIGURE 6.19   SPSS Pop Window for computing Normal Probabilities

Go to Data View Window and you have the results. Hence a probability value of 0.60 gives X value of 210.71 (Figure 6.20).

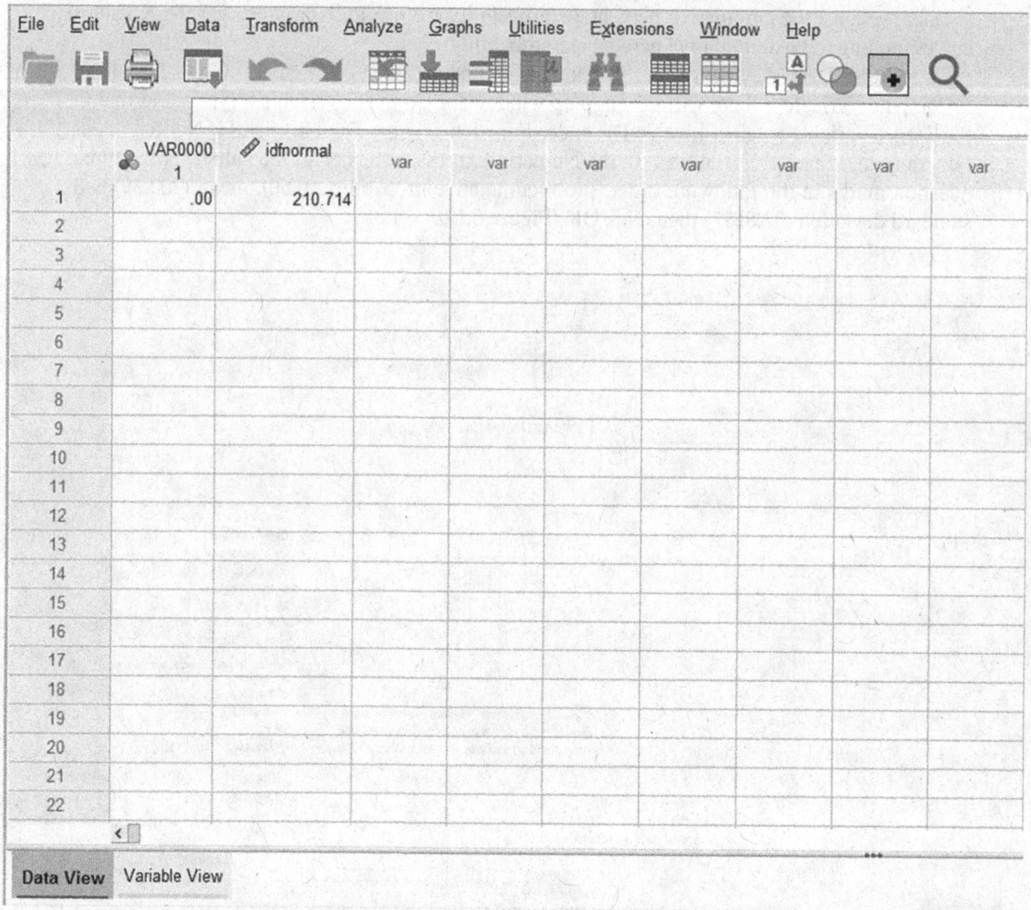

File	Edit	View	Data	Transform	Analyze	Graphs	Utilities	Extensions	Window	Help

	VAR0000 1	idfnormal	var	var	var	var	var	var	var
1	.00	210.714							
2									
3									
4									
5									
6									
7									
8									
9									
10									
11									
12									
13									
14									
15									
16									
17									
18									
19									
20									
21									
22									

Data View   Variable View

FIGURE 6.20   SPSS Data View Window for displaying the computed Normal Probabilities

Hence, a probability value of 0.60 gives X value of 210.71

6.4.4.1 Checking for Normality

Tests for Normality are used to assess if a given data set comes from a normal distribution. Parametric statistical test requires that data are normally distributed or approximately normally distributed. There are several test that can be used to assess for normality:

1. Graphical methods – use Q-Q plots, shape of a Histogram and Stem and Leaf plot, Boxplot
2. Use statistical tests for normality such as Shapiro-Wilk test. One may also interpret the Skewness and Kurtosis values

To test for normality in SPSS we use the path:

Analyze > Descriptive Statistics > Explore > There will be SPSS pop up window for Explore > Choose the numerical variable of interest and put it under Dependent List > Click Plots > there will be SPSS pop up window for Explore: Plots > Select Stem-and-leaf, Histogram and Normality plots with test > Click then in the Explore pop up window select Both then Continue then OK.

Example: Assess if trading volume is normally distributed

Analyze > Descriptive Statistics > Explore > There will be SPSS pop up window for Explore > Choose the Trading volume and put it under Dependent List > Click Statistics and select Descriptives then Continue (Figure 6.21).

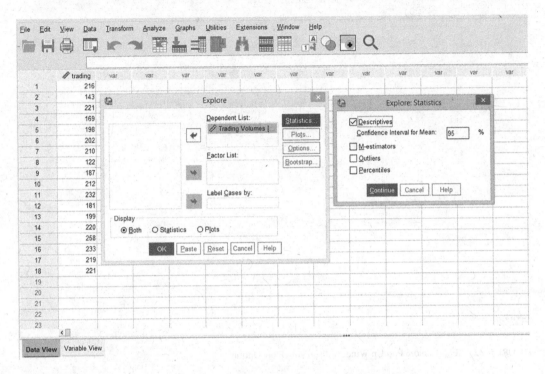

FIGURE 6.21 SPSS Explore Pop Up Window for generating Descriptive Statistics Values

Click Plots > there will be SPSS pop up window for Explore: Plots > Select Stem-and-leaf, Histogram and Normality plots with test > Click then in the Explore pop up window select Both then Continue then OK (Figure 6.22).

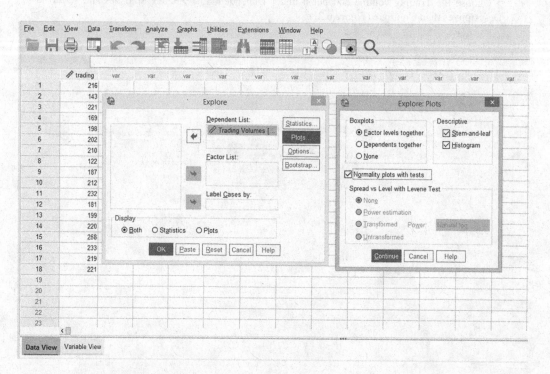

FIGURE 6.22 SPSS Explore Pop Up Window displaying Plots Options

The SPSS Output Window will give the descriptive statistics table, test for normality table, histogram, stem and leaf plot, QQ norm plot and Box plot (Figure 6.23).

Descriptives			Statistic	Std. Error
Trading Volumes	Mean		202.39	7.744
	95% Confidence	Lower Bound	186.05	
	Interval for Mean	Upper Bound	218.73	
	5% Trimmed Mean		203.77	
	Median		211.00	
	Variance		1079.428	
	Std. Deviation		32.855	
	Minimum		122	
	Maximum		258	
	Range		136	
	Interquartile Range		36	
	Skewness		−0.982	0.536
	Kurtosis		1.235	1.038

FIGURE 6.23 Descriptive Statistics values for Trading Volumes

The skewness value = −0.982 indicating that trading volume data are left/negatively skewed. Kurtosis value = 1.235 indicating that trading volume data are flatter than normal distribution. The SPSS output will also give a Table for Tests of Normality. Interpret the *p*-values for Kolmogorov-Smirnov and Shapiro-Wilk tests statistics (Figure 6.24).

	Tests of Normality					
	Kolmogorov-Smirnov[a]			Shapiro-Wilk		
	Statistic	df	Sig.	Statistic	df	Sig.
Trading Volumes	.169	18	.187	.926	18	.164

[a] Lilliefors significance correction.

FIGURE 6.24 Normality Tests for Trading Volumes

The p-value for both Kolmogorov-Smirnov (*p* = 0.187) and Shapiro-Wilk (*p* = 0.164) is more than 0.5. This shows that trading volume data are approximately normally distributed.

The SPSS output will also give a Histogram (Figure 6.25). Interpret the shape of the histogram.

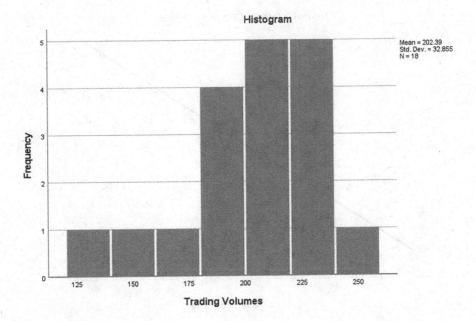

FIGURE 6.25 Histogram for Trading Volumes

The shape of a histogram shows that trading volume data are negatively skewed

The SPSS output will also give a Stem and Leaf Plot (Figure 6.26). Interpret the shape of the stem and leaf plot.

Trading Volumes Stem-and-Leaf Plot

Frequency	Stem	&	Leaf
1.00	Extremes		(≤122)
1.00	1	.	4
5.00	1	.	68899
10.00	2	.	0111122233
1.00	2	.	5
Stem width:	100		
Each leaf:	1 case (s)		

FIGURE 6.26   Stem-and-Leaf Plot for Trading Volumes

The shape of a stem and leaf plot shows that trading volume data are slightly negatively skewed and has one outlier below 122.

The SPSS output will also give Normal Q-Q Plot (Figure 6.27). Interpret the graph.

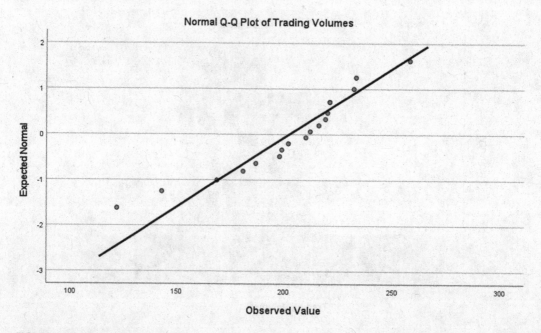

FIGURE 6.27   Normal Q-Q plot for Trading Volumes

The Normal QQ plot shows that trading volume data are approximately normally distributed.

The SPSS output will give the Boxplot (Figure 6.28). Interpret the boxplot by indicating the presence or absence of the outliers.

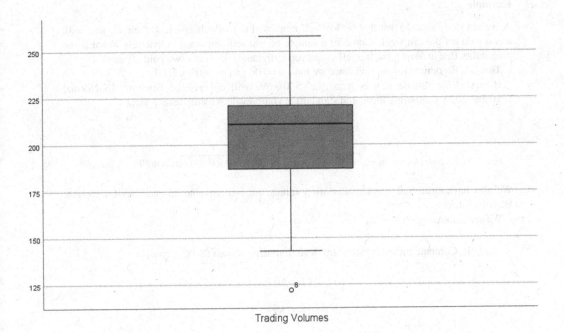

Trading Volumes

FIGURE 6.28   Box Plot for Trading Volumes

The boxplot shows that trading volume has one outlier.

6.4.5 Normal Approximation of Binomial Probabilities

When the number of trials becomes large, i.e. $n \geq 30$, evaluating the binomial probability function by hand or with a calculator becomes tedious and difficult. Binomial distribution are converted to Normal distribution by computing the mean and standard deviation.

When using the normal approximation to the binomial, we set $\mu = np$ and $\sigma = \sqrt{np(1-p)}$ is the definition of a normal curve using the following steps

Step 1: Compute mean and standard deviation using $\mu = np$ and $\sigma = \sqrt{np(1-p)}$

Step 2: Convert X to Z

$$Z = \frac{X - \mu}{\sigma}$$

Step 3: Draw the area of interest under the normal curve using Z value and shade the area of interest using the inequality

Step 4: We read the area from the center to z from the Z table

Step 5: Find the total shaded area

Then the probabilities are computed in SPSS the same way we compute probabilities for normal distribution using CDF.Normal or PDF. Normal or Idf.Normal to estimate X values when probabilities are given.

Example

A certain study showed that that 60% of self-employed individuals provided their workers with a two paid off days in week. Consider a sample of 100 self-employed individuals. What is the probability that *at least 65* of the self-employed individuals provide a two paid off days?

This is a Binomial Distribution since we have $n = 100$ and $p = 60\% = 0.60$

However, the sample size is large (i.e. >30). We will convert the Binomial to Normal Distribution by computing mean and standard deviation values using n and p values

$$Mean = n \times p = 100 \times 0.60 = 60$$

$$Std\ Deviation = \sqrt{n \times p \times (1-p)} = \sqrt{100 \times 0.60 \times (1-0.60)} = 4.899$$

We now have mean and standard deviation values, we can compute probabilities for Normal Distribution

We are computing $P(X \geq 65)$

Step 1: Compute mean and standard deviation using $\mu = np$ and $\sigma = \sqrt{np(1-p)}$

$$\mu = 60$$

$$\sigma = 4.899$$

Step 2: Convert X to Z

$$P(X \geq 65) = P\left(\frac{X-\mu}{\sigma} \geq \frac{65-60}{4.899} = P(Z \geq 1.02)\right)$$

Step 3: Draw the area of interest under the normal curve using Z value. Since z is positive, we draw it on the right side and shade the area of interest using the inequality

Step 4: We read the area from the center to $z = 1.02$ which is 0.3438

Step 5: Find the total shaded area = $0.5 - 0.3438 = 0.1562$

Using SPSS,

We are computing $P(X \geq 65)$, so we use CDF.Normal to compute $P(X \leq 64)$ then subtract from 1, i.e.

$$P(X \geq 65) = 1 - P(X < 65) = 1 - P(X \leq 64)$$

We are computing $P(X \leq 64)$, so we use CDF.Normal

Go to SPSS Data View Window > Click 0 in the first cell then press enter > Go to Transform > Compute Variable > there will be a pop up window for Compute Variable > type cdfnormal in the Target Variable > select CDF and Noncentral CDF under Function group > Double click Cdf.Normal under Functions and Special Variables > In the Numeric Expression box enter the values for the three question marks in the following order: quant=value of x (64), mean (60) and standard deviation (4.899) then click OK (Figure 6.29).

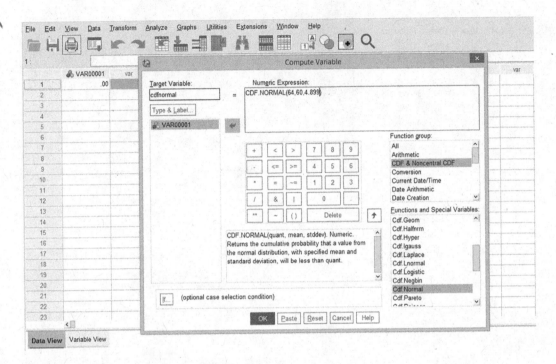

FIGURE 6.29 SPSS Pop Window for computing Normal Probabilities

Go to Data View Window and you have the results. Hence $P(X \leq 64) = 0.79289$
The SPSS Data View Window will give the probability value (Figure 6.30).

$$P(X \leq 64) = 0.79289$$

$$P(X \geq 85) = 1 - P(X < 85) = 1 - P(X \leq 84) = 1 - 0.79289 = 0.20711$$

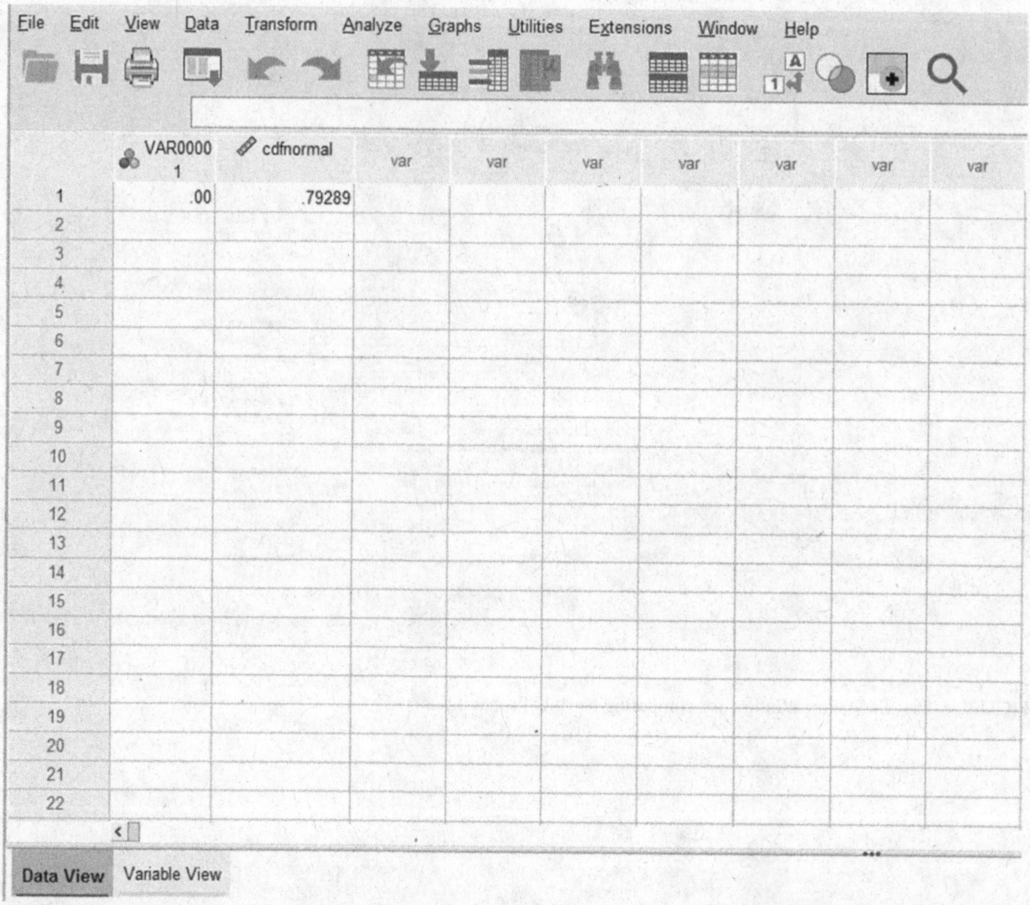

FIGURE 6.30   SPSS Data View Window for displaying the computed Normal Probabilities

6.4.6 Student's t Distribution

The Student's t distribution is the ratio of a standard normal random variable Z to the square root of an independently distributed chi square random variable U divided by its degree of freedom v.

$$X = \frac{Z}{\sqrt{U/_v}}$$

Student's t distribution is one of the continuous probability distributions that arise when estimating the mean of a normally distributed population in situations where the sample size is small (<30) and the population standard deviation is unknown. The probability density function for t distribution is given by

$$f(x:v) = \frac{1}{\sqrt{\pi v}} \frac{\Gamma\left(\frac{v+1}{2}\right)}{\Gamma\left(\frac{v}{2}\right)} \left(1 + \frac{x^2}{v}\right)^{-(v+1)/2}$$

t-statistics is defined by

$$t = \frac{\bar{x} - \mu}{s/\sqrt{n}}$$

6.4.7 F-Distribution

The F-distribution, also known as Fisher–Snedecor (F-distribution) is another continuous probability distribution. The F statistics is defined by

$$F = \frac{s_1^2}{s_2^2}$$

The probability density function for F-distributed random variable has degree of freedom n (numerator df = n1 − 1) and m (denominator df = n2 − 1) is given by

$$f(x) = \left[\frac{\Gamma\left(\frac{n+m}{2}\right)}{\Gamma\left(\frac{n}{2}\right)\Gamma\left(\frac{m}{2}\right)}\right] (n)^{\frac{n}{2}} (m)^{\frac{m}{2}} \frac{x^{\frac{n}{2}-1}}{(m+nx)^{\frac{n+m}{2}}}.$$

where > 0, $\Gamma(n)$ is a Gamma function, n1 is sample for group 1 and n2 is the sample for group 2.

6.4.8 Chi-Square Distribution

The chi-square (χ^2) distribution is a distribution that is used to solve statistical tests/inference involving contingency tables and can be used to test if a sample of data came from a population with a specific distribution. The probability density function (pdf) for a chi-square random variable with k degrees of freedom is given by:

$$f(x; k) = \frac{1}{\Gamma\left(\frac{k}{2}\right)} \left(\frac{1}{2}\right)^{k/2} x^{\frac{k}{2}-1} e^{-\frac{1}{2}x} ; x \geq 0$$

where $x > 0$ and Γ is the Gamma function

Practice Exercise

Use SPSS to solve the following questions

1. A certain study showed that the employment age was between 18 years and 60 years. Let the employment age be represented by uniform random variable X.
   a. What is the probability that a person will work for at most 53 years?
   b. What is the probability that a person will work for between 33 and 58 years?

2. The following data represents students heights (in cm) during a class exercise

   116, 143, 121, 159, 178, 102, 110, 187, 112, 132, 171, 169, 120, 158, 133, 119, 121, 131, 145, 106, 139, 138, 132, 145, 111

   Let student height be approximately normally distributed.
   a. What is the probability that the student weight will be at least 117 cm?
   b. What weight must a student have to be in the top 25%?

3. A certain survey found that 36% of employees were loyal to their employers. For a sample of 78 employees, what is the probability that at most 48 employees will be loyal to their employers?

4. A job seeker has done several job interviews with the hope of getting a job. The chance of getting a job for nay job interview done is 62%. Consider a sample of 12 job interviews done:
   a. What type of a distribution can be related to the above description? Explain
   b. What is the probability of getting a job having attended between 8 and 10 job interviews?

5. At a certain reservation desk, phone calls arrived at rate of 24 phone calls per hour.
   a. With reasons, which probability distribution can be related with the above description?
   b. What is the probability that at there will be at most three phone calls in 25 minutes?
   c. What is the probability that at least three phone calls will arrive in 1 hour?
   d. What is the probability that at least three phone calls will arrive 35 minutes?

7

Sampling and Sampling Distribution

7.1 Sampling Terminologies and Concepts

An *element* is the entity on which data are collected. For example, employees, students, etc.

A *population* is the collection of all the elements of interest. For example, employees from Kenya Commercial Bank (KCB), students from United States International University – Africa (USIU-A). Denoted by N if it's known. There are two types of population infinite and finite. *Infinite population* is a population that is not known/given. For example, the total number of people living in a given location as at today is not known. *Finite population* is a population that is known/given. For example, the total number of students in a given class is known.

A *sample* is a subset of the population. Denoted by n. For example, KCB employees from finance department, USIU-A graduate students, etc.

A *statistic* is a numerical characteristic of a sample such as sample mean $\bar{x}$

A *parameter* is a numerical characteristic of a population, such as a population mean μ, a population standard deviation σ, a population proportion p etc.

Sampling is the process/an act of selecting a representative sample from the population. There are non-probabilistic and probabilistic methods of sampling. There are several advantages of sampling. Selecting a sample is less time consuming than selecting every item in the population (census). Selecting a sample is also less costly than selecting every item in the population. Conducting an analysis of a sample is less cumbersome and more practical than an analysis of the entire population.

7.2 Sampling Methods

There are two types of sampling methods, probability and nonprobability sampling methods. In *probability sampling* every element has equal chance of being selected. Probability sampling is also known as *random sampling*. Probability sampling methods include: Simple random sampling, Systematic sampling, Stratified sampling and Cluster sampling. In *nonprobability sampling* every element has no equal chance of being selected (the method is biased with more errors). The nonprobability sampling method is also known as *nonrandom sampling*. The nonprobability sampling methods include: judgmental sampling, convenience sampling, snowball sampling and quota sampling.

Simple random sampling is a probability sampling method in which a sample is designed in such a way as to ensure that every member of the population or item from the frame has an equal chance of being selected/chosen. Random numbers for selecting the samples are obtained from table of random numbers or computer generated random numbers. There are two types of simple random sampling: sampling without replacement and sampling with replacement. In *sampling without replacement,* once an element has been included in the sample, it is removed from the population and cannot be selected a second time. In *sampling with replacement,* once an element has been included in the sample, it is returned to the population. A previously selected element can be selected again and therefore may appear in the sample more than once.

In *systematic random sampling*, elements from the population are selected by following a given pattern or in a systematic way. It's a probability sampling method in which we randomly select one of the first k elements and then select every kth element thereafter, where $k = \frac{N}{n}$. First one decides the sample to be selected (n) from the population (N).

DOI: 10.1201/9781003292654-9

Stratified random sampling is a probability sampling method in which the population is first divided into strata and a simple random sample is then taken from each stratum with sample sizes proportional to strata sizes. Samples from subgroups are combined into one. This is a common technique when sampling population of voters, stratifying across racial or socio-economic lines.

Cluster sampling is a probability sampling method in which the population is first divided into clusters and then a simple random sample of the clusters is taken. All items in the selected clusters can be used, or items can be chosen from a cluster using another probability sampling technique. A common application of cluster sampling involves election exit polls, where certain election districts are selected and sampled.

Convenience sampling is a nonprobability method of sampling whereby elements are selected for the sample on the basis of convenience. In convenience sampling, items are selected based only on the fact that they are easy, inexpensive or convenient to sample.

Judgment sampling is a nonprobability method of sampling whereby elements are selected for the sample based on the judgment of the person doing the study. In a judgment sample, you get the opinions of pre-selected experts in the subject matter.

Quota sampling is a nonrandom sampling technique which appears to be similar to stratified random sampling. Certain population subclasses, such as age group, gender, or geographic region, are used as strata. However, instead of randomly sampling from each stratum, the researcher uses a nonrandom sampling method to gather data from one stratum until the desired quota of samples is filled.

Snowball sampling is a nonrandom sampling technique in which survey subjects are selected based on referral from other survey respondents. The researcher identifies a person who fits the profile of subjects wanted for the study. The researcher then asks this person for the names and locations of others who would also fit the profile of subjects wanted for the study.

The reason we select a sample is to collect data to make an inference and/or answer a research question about a population.

7.2.1 Sampling Error: Need for Sampling Distribution

Using a sample to acquire information about a population is often preferable to conducting a census. Generally, sampling is less costly and can be done more quickly than a census; it is often the only practical way to gather information. Since a sample provides data for only a portion of an entire population, we cannot expect the sample to yield perfectly accurate information about the population. Thus, we should anticipate that a certain amount of error called *sampling error* simply because we are sampling. *Sampling error* is the error resulting from using a sample to estimate a population characteristic. To answer questions about sampling error, we need to know the distribution of all possible sample that could be obtained by sampling n individuals. That distribution is called *the sampling distribution*.

7.3 Sampling Distribution

A probability distribution that describes a *statistic* used to estimate a *parameter*. Describes how precise and accurate a statistic is for measuring a population parameter.

There are two types of sampling distribution: *sampling distribution of sample mean* and *sampling distribution of sample proportion*.

7.3.1 Sampling Distribution of the Sample Mean ($\bar{x}$)

If the sample mean ($\bar{x}$) is a random variable and a given sample size n, the probability distribution of $\bar{x}$ is called the sampling distribution of the sample mean $\bar{x}$. The sampling distribution of $\bar{x}$ is the probability distribution of all possible sample means for samples of given sample size being used to estimate the population mean. The knowledge of this sampling distribution and its properties will enable us to make probability statements about how close the sample mean $\bar{x}$ is to the population mean μ.

We generally do not know the sampling distribution of the sample mean exactly. However, we can often approximate that sampling distribution by *a normal distribution*; that is, under certain conditions, the variable $\bar{x}$ is approximately normally distributed. A variable is *normally distributed* if its distribution has the shape of a normal curve and that a normal distribution is determined by *the mean* and *standard deviation*.

7.3.1.1 The Mean and Standard Deviation of the Sample Mean

The first step in learning how to approximate the sampling distribution of the sample mean by a normal distribution is to obtain the mean and standard deviation of the sample mean, that is, of the variable $\bar{x}$.

7.3.1.1.1 The Mean of the Sample Mean

For any particular sample size n, the mean of all possible sample means equals the population mean. This equality holds regardless of the size of the sample.

$$E(\bar{x}) = \mu_{\bar{x}} = \mu$$

$\bar{x}$ is the sample mean and μ is the population mean.

When the expected value of a point estimator equals the population parameter, then the point estimator is *unbiased*.

7.3.1.1.2 Standard Deviation of Sample Mean

The standard deviation of the sampling distribution of $\bar{x}$ is given by:

1. For Finite Population

$$\sigma_{\bar{x}} = \sqrt{\frac{N-n}{N-1}} \left(\frac{\sigma}{\sqrt{n}} \right)$$

   This is used when sampling is done without replacement from a finite population

2. For Infinite Population

$$\sigma_{\bar{x}} = \frac{\sigma}{\sqrt{n}}$$

where
  $\sigma_{\bar{x}}$ is the standard deviation of $\bar{x}$
  σ is the standard deviation of the sample
  n is the sample size
  N is the population size

This is used when sampling is done with replacement from a finite population or when it is done from an infinite population.

The standard deviation of sample mean ($\sigma_{\bar{x}}$) is also referred to as the *standard error of sample mean*.

Hence, the sampling distribution of the sample mean for n selected samples is a normal distribution with mean (μ) and standard deviation ($\sigma_{\bar{x}}$).

In Statistical Package for Social Sciences (SPSS), we generate the mean value and standard error or mean using the path:

Analyze > Descriptive Statistics > Frequencies > There will be a pop up SPSS window for Frequencies > Select the Numerical Variable of interest and put it in Variable(s) Window > Click Statistics > There will be a pop up SPSS window for Statistics > Under Central Tendency select Mean and under Dispersion select S.E. Mean then Continue and then OK.

7.3.1.2 Sampling Distribution of the Sample Mean for Normally Distributed Variables

If the variable under consideration is normally distributed, so is the variable $\bar{x}$. If a variable x of a population is normally distributed with mean μ and standard deviation σ. Then, for samples of size n, the variable $\bar{x}$ is also normally distributed and has mean μ and standard deviation $\sigma/\sqrt{n}$.

7.3.1.3 Central Limit Theorem

In selecting random samples of size n from a population, the sampling distribution of the sample mean $\bar{x}$ can be approximated by a normal distribution as the sample size becomes large, i.e. the sampling distribution of the mean of a random sample drawn from any population is approximately normal for a sufficiently large sample size ($n > 30$).

The practical reason we are interested in the sampling distribution of $\bar{x}$ is that it can be used to provide probability information about the difference between the sample mean and the population mean, i.e. we will use this distribution to answer the probability question.

We follow the same steps as for estimating probabilities for a normal distribution.

Step 1: We compute mean and standard error of mean values.

In SPSS, we have to Analyze > Descriptive Statistics > Frequencies > there will be a pop up window for Frequencies > Select numerical variable of interest for Variable(s) > Click Statistics select Mean under Central Tendencies and S.E. Mean Deviation under Dispersion > Click Continue then OK.

In the SPSS output, select mean and standard error of mean values.

Step 2: Convert $\bar{X}$ to Z as follows:

$$Z = \frac{\bar{X} - \mu}{\sigma_{\bar{x}}}$$

Step 3: Draw the area of interest under the normal curve using Z value and shade the area of interest using the inequality.

Step 4: We read the area from the center to Z value of interest.

Step 5: Find the total shaded area of interest.

In SPSS, once we have the mean and standard error of mean values, we can use those values to estimate probabilities for a normal distribution by using CDF.Normal to compute $\leq$ probabilities and PDF.Normal to compute exact probabilities or Idf.Normal to estimate $\bar{X}$ values when probabilities are given.

When computing exact probability, e.g. $P\left(\bar{X} = \bar{x}\right)$ *given that μ and $\sigma_{\bar{x}}$:* In SPSS Data View Window, go to **Transform > Compute Variable**. Type pdfnormal as **Target Variable**. Since we are calculating $P(\bar{X} = \bar{x})$, under **Function group** under **Function group** select **PDF & Noncentral PDF**, and under **Functions and Special Variables** double-click on **Pdf.Normal** > In the Numeric Expression box enter the values for the three question marks in the following order: quant=value of x, mean and standard deviation then click OK. Then click **OK**. Go to Data View Window and you have the results of the computed probability.

When computing $\leq$ probabilities (cumulative probabilities), e.g. $P\left(\bar{X} \leq \bar{x}\right)$ *given μ and σ:* In SPSS Data View Window, go to **Transform > Compute Variable**. Type cdfnormal as **Target Variable**. Since we are calculating $P(\bar{X} \leq \bar{x})$, under **Function group** select **CDF & Noncentral CDF**, and under

Functions and Special Variables, double click on **Cdf.Normal**. In the Numeric Expression box enter the values for the three question marks in the following order: quant=value of *x*, mean and standard deviation then click OK. Then click **OK**. Go to Data View Window and you have the results of the computed probability.

The Inverse Transformation is used if we are interested to get the cumulative X values for a given probability. For example, find $\bar{X}$ *given the p*, μ *and* σ *values:* From the menu click **Transform > Compute Variable**. In the Numeric Expression box enter the values for the three question marks in the following order: prob=cumulative probability, mean and standard deviation then click OK. Go to Data View Window and you have the results of the computed X value.

Example

Consider the following set of data for mean trading volumes:

3012, 1504, 2902, 5115, 7174, 1445, 3335, 2456, 3963, 2404, 6025, 6398, 1923, 6332, 3331, 4231, 2237

a. Show the sampling distribution of $\bar{x}$, sample mean trading volume.
   Sampling distribution of sample mean is a probability distribution of selected samples that is a Normal Distribution.
   We create the variables in the Variable View window and key the data in the Data View window (Figures 7.1 and 7.2).
   Variable View window

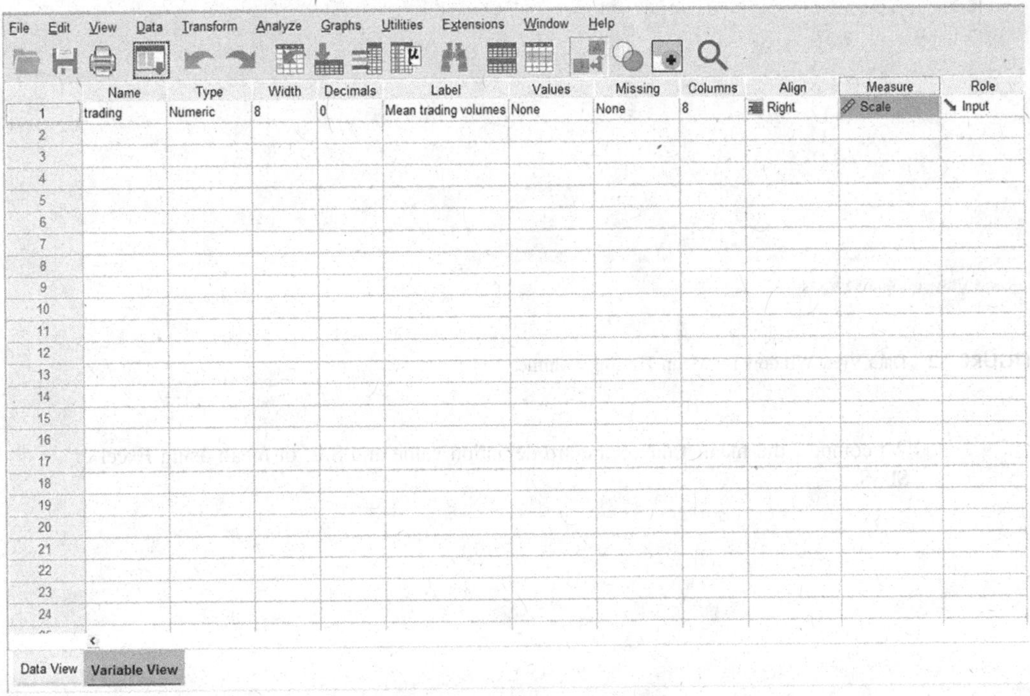

FIGURE 7.1 Variable View Window for Mean Trading Volumes

Data View window

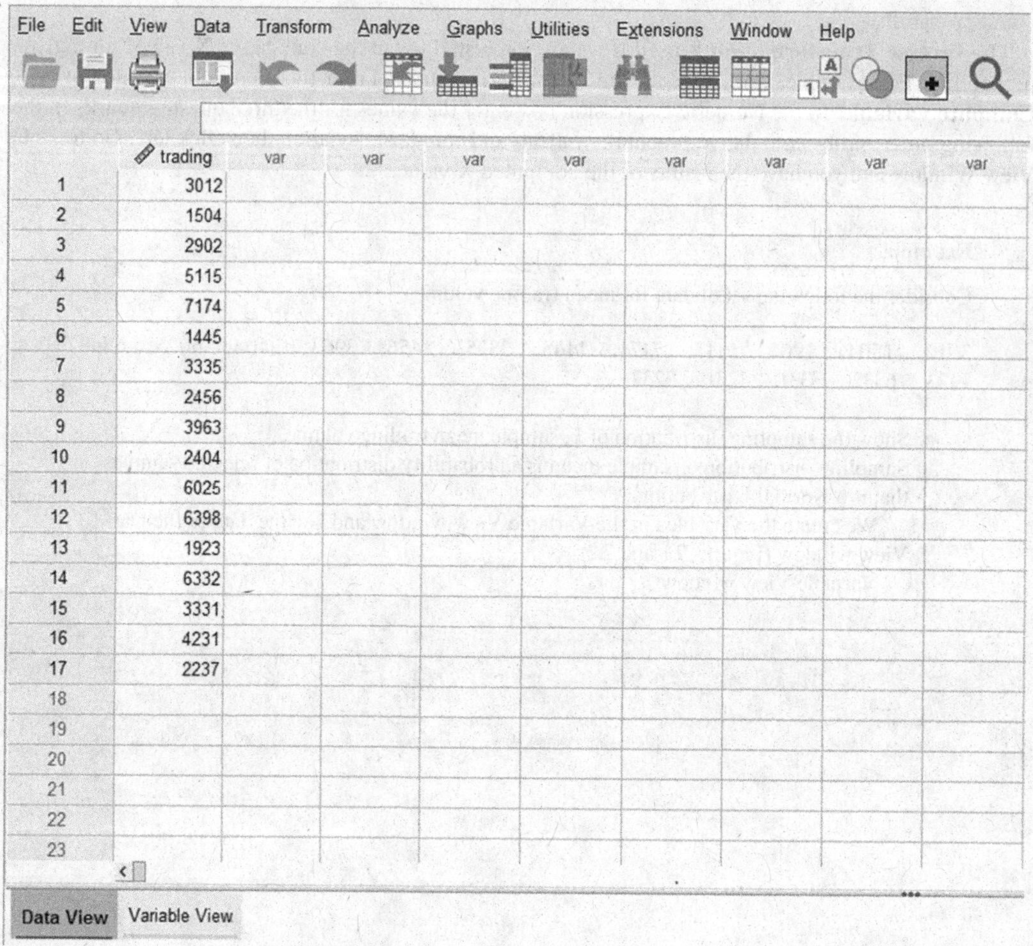

File	Edit	View	Data	Transform	Analyze	Graphs	Utilities	Extensions	Window	Help

	trading	var	var	var	var	var	var	var	var
1	3012								
2	1504								
3	2902								
4	5115								
5	7174								
6	1445								
7	3335								
8	2456								
9	3963								
10	2404								
11	6025								
12	6398								
13	1923								
14	6332								
15	3331								
16	4231								
17	2237								
18									
19									
20									
21									
22									
23									

Data View   Variable View

FIGURE 7.2   Data View Window for Mean Trading Volumes

We compute the mean value, standard deviation value and S.E. of mean using Excel or SPSS.

Analyze > Descriptive Statistics > Frequencies > Select mean trading volumes > Under Statistics select Mean and under Dispersion select Standard Deviation and S.E. Mean then click Continue then OK (Figure 7.3).

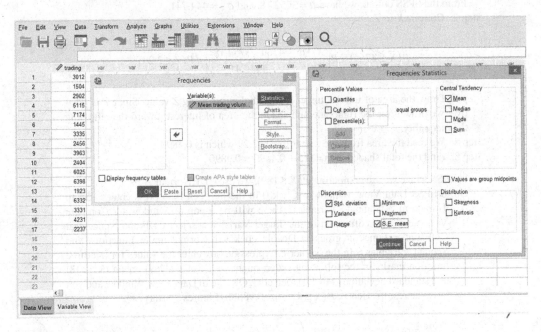

FIGURE 7.3 SPSS Frequency Pop Up Window for generating Mean, Standard Deviation and S.E. mean Values

SPSS output will give the mean, standard deviation and standard error of mean values (Figure 7.4).

Statistics		
Mean trading volumes		
N	Valid	17
	Missing	0
Mean		3752.18
Std. Error of Mean		444.741
Std. Deviation		1833.716

FIGURE 7.4 Mean, Standard Deviation and S.E. mean Values for Mean Trading Volumes Data

From the SPSS output we have $n = 17$, $\mu = 3752.18$, $\sigma = 1833.716$ and $\sigma_{\bar{X}} = 444.741$.

$$E\left(\bar{X}\right) = \mu = 3752.18$$

$$\sigma_x = \frac{\sigma}{\sqrt{n}} = \frac{1833.716}{\sqrt{17}} = 444.741$$

Sampling distribution of the sample mean for 17 selected samples is a normal distribution with mean 3752.18 and standard deviation 444.741

b. What is the probability that the sample mean trading volumes will be at most 4769?

$$P\left(\bar{X} \leq 4769\right)$$

We are computing $P(X \le 4769)$

Step 1: We compute mean and standard error of mean values
From the SPSS output, we have $\mu = 3752.18$, and $\sigma_{\bar{x}} = 444.741$.
Step 2: Convert $\bar{X}$ to Z

$$P(\bar{X} \le 4769) = P\left(\frac{\bar{X} - \mu}{\sigma_{\bar{x}}} \le \frac{4769 - 3752.18}{444.741}\right) = P(Z \le 2.29)$$

Step 3: Draw the area of interest under the normal curve using Z value. Since Z is positive, we draw it on the right side and shade the area of interest toward the left using the inequality.
Step 4: We read the area from the center to $Z = 2.29$ which is 0.4890
Step 5: Find the total shaded area $= 0.5 + 0.4890 = 0.9890$

In SPSS, since we are computing $P(X \le 189)$, we use CDF.Normal
Go to SPSS Data View Window > Click 0 in the first cell then press enter > Go to Transform > Compute Variable > there will be a pop up window for Compute Variable > type cdfnormal in the Target Variable > select CDF and Noncentral CDF under Function group > Double click Cdf.Normal under Functions and Special Variables > In the Numeric Expression box enter the values for the three question marks in the following order: quant=value of x (4769), mean (3752.18) and standard deviation (444.741) then click OK. Go to Data View Window and you have the results. Hence, $P(X \le 4769) = 0.98888$ (Figures 7.5 and 7.6).

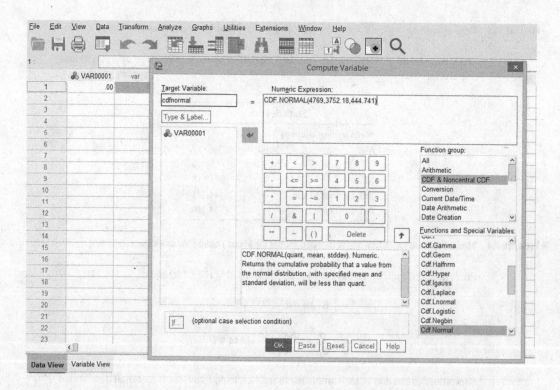

FIGURE 7.5   SPSS Pop Window for computing Normal Probabilities

The SPSS output will give the probability value

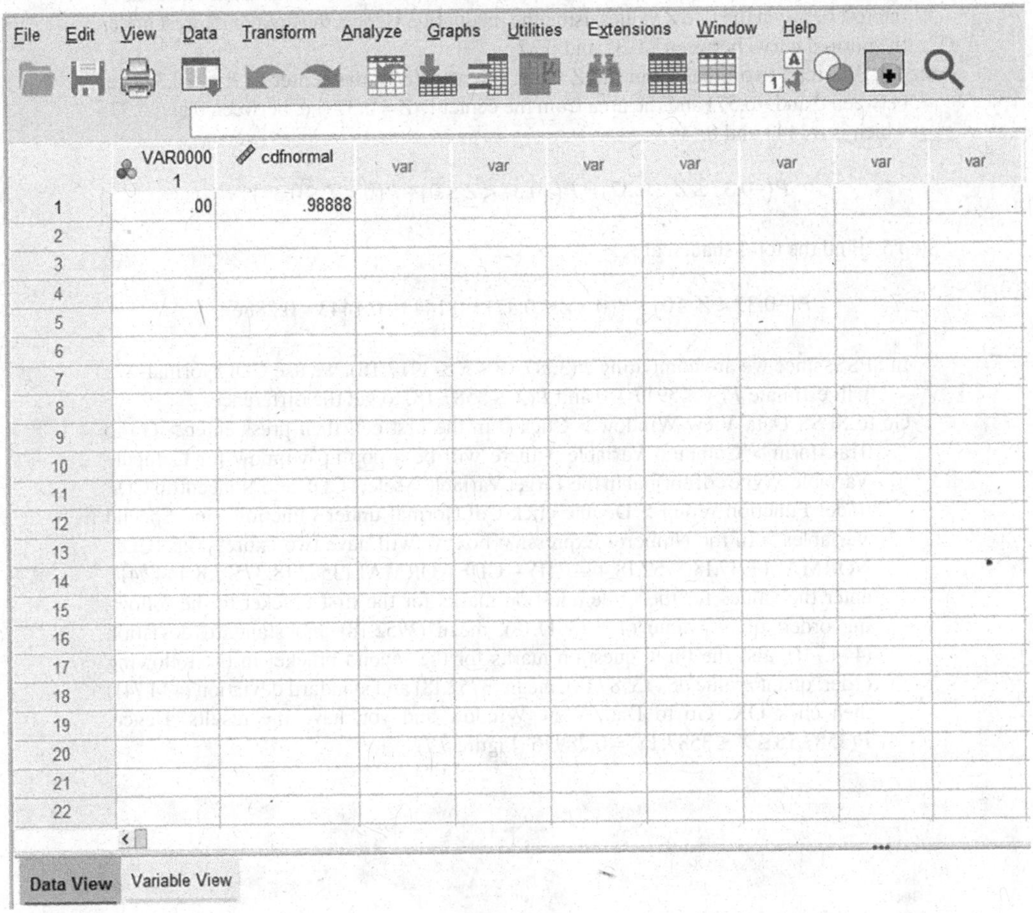

FIGURE 7.6 SPSS Data View Window for displaying the computed Normal Probabilities

$$P(X \leq 4769) = 0.98888$$

c. What is the probability that the sample mean trading volumes will be within 165 points of the population mean?

$$P(\bar{X} \pm 165) = P(3752.18 - 165 \leq \bar{X} \leq 3752.18 + 165)$$

$$P(3587.18 \leq \bar{X} \leq 3917.18)$$

We have

$$E(\bar{X}) = \mu = 3752.18$$

$$\sigma_x = \frac{\sigma}{\sqrt{n}} = 444.741$$

Step 1: We compute mean and standard error of mean values
From the SPSS output, we have $\mu = 3752.18$, and $\sigma_{\bar{x}} = 444.741$.
Step 2: Convert $\bar{X}$ value into its corresponding Z value

$$P(3587.18 \leq \bar{X} \leq 3917.18) = P\left(\frac{3587.18 - 3752.18}{444.741} \leq \frac{\bar{X} - \mu}{\sigma_x} \leq \frac{3917.18 - 3752.18}{444.741}\right) = P(-0.37 \leq Z \leq 0.37)$$

Step 3: Draw the area of interest under the normal curve using Z value. Since Z is positive and negative, we draw it on both the right side and the left side then shade the area of interest between the two Z values using the inequality. Hence, the area of interest under the normal curve between -0.37 and 0.37

Step 4: We read two areas from the Z table: the area from the center to $Z = -0.37$ (i.e. between 0 and -0.37) and the area from the center to $Z = 0.37$ (i.e. between 0 and 0.37) which is 0.1443 and 0.1443

$$P(-0.37 \leq Z \leq 0.37) = P(-0.37 < Z \leq 0) + P(0 < Z \leq 0.37)$$

Step 5: Find the total shaded area

$$P(-0.37 < Z \leq 0) + P(0 < Z \leq 0.37) = 0.1443 + 0.1443 = 0.2886$$

In SPSS, since we are computing $P(3587.18 \leq \bar{X} \leq 3917.18)$, we use CDF.Normal. We will estimate $P(\bar{X} \leq 3917.18)$ and $P(\bar{X} \leq 3587.18)$ to get the difference.

Go to SPSS Data View Window > Click 0 in the first cell then press enter > Go to Transform > Compute Variable > there will be a pop up window for Compute Variable > type cdfnormal in the Target Variable > select CDF and Noncentral CDF under Function group > Double click Cdf.Normal under Functions and Special Variables > In the Numeric Expression box we will have two expressions (CDF.NORMAL(3917.18,3752.18,444.741) − CDF.NORMAL(3587.18,3752.18,444.741): enter the values for the three question marks for the first bracket in the following order: quant=value of x (3917.18), mean (3752.18) and standard deviation (444.741), and the three question marks for the second bracket in the following order: quant=value of x (3587.18), mean (3752.18) and standard deviation (444.741) then click OK. Go to Data View Window and you have the results. Hence, $P(3587.18 \leq \bar{X} \leq 3587.18) = 0.28936$ (Figure 7.7).

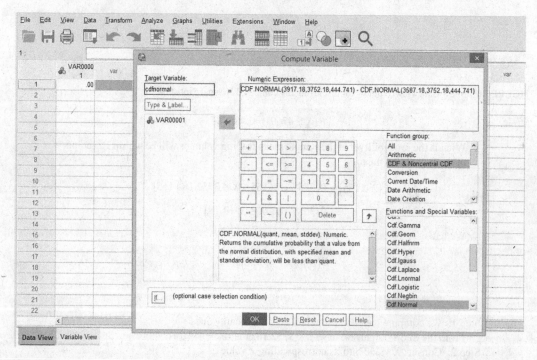

FIGURE 7.7   SPSS Pop Window for computing Normal Probabilities

The SPSS output will give the probability value (Figure 7.8).

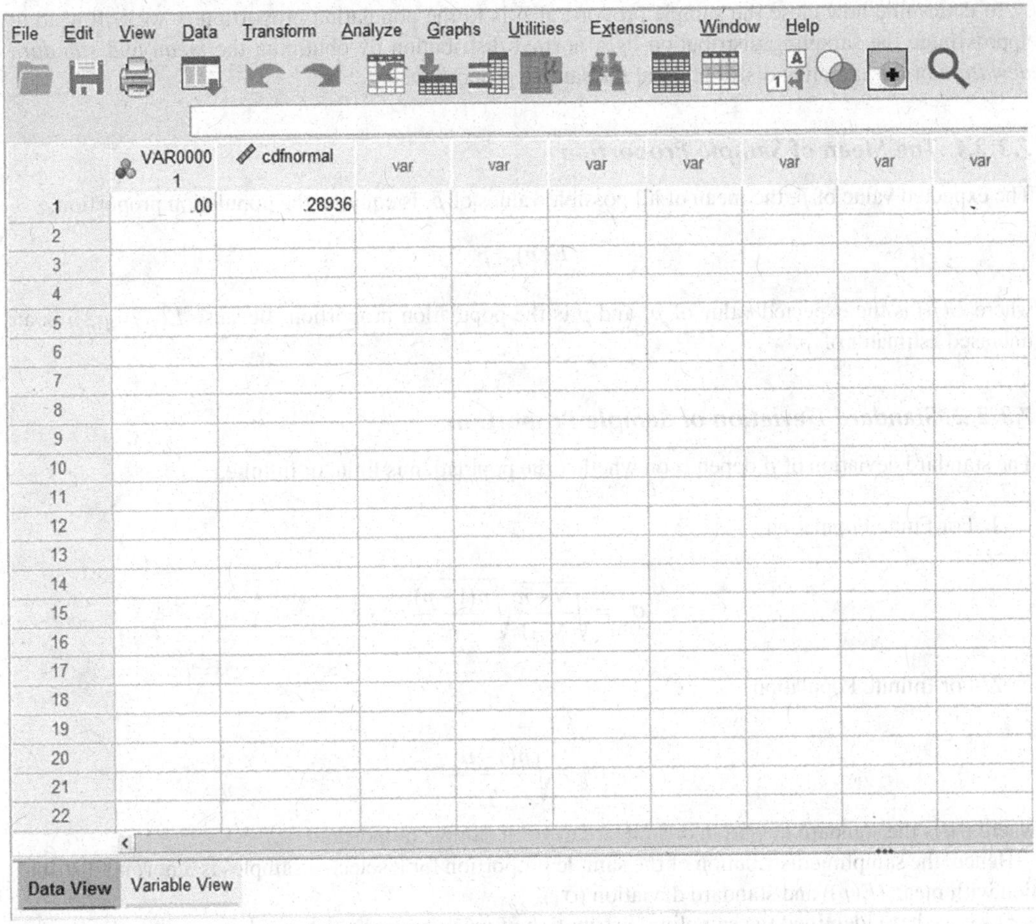

	VAR0000 1	cdfnormal	var	var	var	var	var	var	var
1	.00	.28936							
2									
3									
4									
5									
6									
7									
8									
9									
10									
11									
12									
13									
14									
15									
16									
17									
18									
19									
20									
21									
22									

Data View   Variable View

FIGURE 7.8   SPSS Data View Window for displaying the computed Normal Probabilities

Consequently, 28.936% of all samples of 17 observations have mean values within 165 units of the population mean value of all the observations.

Interpretation: There is about a 28.936% chance that the sampling error made in estimating the mean value of all observations by that of a sample of 17 observations will be within 165 units.

7.3.2 Sampling Distribution of the Sample Proportion ($\bar{p}$)

Probability distribution of sample proportion $\bar{p}$ of sample size n. If the sample proportions $\bar{p}$ is a random variable, the sampling distribution of $\bar{p}$ is the probability distribution of all possible values of the sample proportion $\bar{p}$. The sampling distribution of $\bar{p}$ is the probability distribution of all possible sample proportions for samples of given sample size being used to estimate the population proportion. The sample proportion $\bar{p}$ is the point estimator of the population proportion p. The formula for computing the sample proportion is

$$\bar{p} = \frac{x}{n}$$

where x is the number of elements in the sample that possess the characteristic of interest, and n is the sample size.

To determine how close the sample proportion $\bar{p}$ is to the population proportion p, we will need to approximate the sampling distribution by a normal distribution by obtaining the *mean* and *standard deviation* of the sampling distribution of the sample proportion.

7.3.2.1 The Mean of Sample Proportion

The expected value of $\bar{p}$, the mean of all possible values of $\bar{p}$, is equal to the population proportion p.

$$E(\bar{p}) = p$$

where $E(\bar{p})$ is the expected value of $\bar{p}$, and p is the population proportion. Because $E(\bar{p}) = p$, $\bar{p}$ is an unbiased estimator of p.

7.3.2.2 Standard Deviation of Sample Proportion

The standard deviation of $\bar{p}$ depends on whether the population is finite or infinite.

1. For Finite Population

$$\sigma_{\bar{p}} = \sqrt{\frac{N-n}{N-1}} \sqrt{\frac{p(1-p)}{n}}$$

2. For Infinite Population

$$\sigma_{\bar{p}} = \sqrt{\frac{p(1-p)}{n}}$$

where $\sigma_{\bar{p}}$ is the standard deviation or standard error of sample proportion.

Hence, the sampling distribution of the sample proportion for n selected samples is a normal distribution with mean $(E(p))$ and standard deviation (σ_p).

Once we have identified the sampling distribution of mean or proportion we then use this distribution to answer probability questions by first calculating z then reading the probability from z table just as for the normal distribution. We convert the $\bar{p}$ value into its corresponding Z value (standard normal random variable), as follows:

$$Z = \frac{\bar{p} - E(p)}{\sigma_{\bar{p}}}$$

We follow the same steps for computing probabilities of a normal distribution. In SPSS we can use CDF. Normal to compute cumulative probabilities ($\leq$) or PDF.Normal to compute exact probabilities.

Example

A certain report on spending amounts by families showed that 73% of households spend more than \$100 per week on family dinners. Assume the population proportion is $p = 0.73$ and a sample of 650 households will be selected from the population.

a. Show the sampling distribution of $\bar{p}$, the sample proportion of households spending more than \$100 per week on family dinners.

Sampling distribution of sample proportion is a probability distribution of selected samples that is a Normal Distribution.

We compute the mean value, standard deviation value and S.E. of mean using the formula

$$E(\bar{p}) = p = 0.73$$

$$\sigma_{\bar{p}} = \sqrt{\frac{p(1-p)}{n}} = \sqrt{\frac{0.73(1-0.73)}{650}} = 0.0174$$

Sampling distribution of the sample proportion for 650 selected samples is a normal distribution with mean $(E(p) = 0.73)$ and standard deviation $(\sigma_p = 0.0174)$.

b. What is the probability that the sample proportion will be within 0.05 of the population proportion?

$$P(\bar{p} \pm 0.05) = P(0.73 - 0.05 \le \bar{p} \le 0.73 + 0.05)$$

$$P(0.68 \le \bar{p} \le 0.78)$$

We have

$$E(\bar{p}) = p = 0.73$$

$$\sigma_{\bar{p}} = \sqrt{\frac{p(1-p)}{n}} = \sqrt{\frac{0.73(1-0.73)}{650}} = 0.0174$$

We convert the $\bar{p}$ value into its corresponding Z value

$$P(0.68 \le \bar{p} \le 0.78) = P\left(\frac{0.68 - 0.73}{0.0174} \le \frac{\bar{p} - E(\bar{p})}{\sigma_{\bar{p}}} \le \frac{0.78 - 0.73}{0.0174}\right) = P(-2.87 \le Z \le 2.87)$$

We then draw the area of interest under the normal curve between −2.87 and 2.87.

Using the Z tables, we find two probabilities: between 0 and −2.87 and between 0 and 2.87.

$$P(-2.87 \le Z \le 2.87) = P(-2.87 < Z \le 0) + P(0 < Z \le 2.87)$$

$$P(-2.87 < Z \le 0) + P(0 < Z \le 2.87) = 0.4979 + 0.4979 = 0.9958$$

In SPSS, since we are computing $P(0.68 \leq \bar{p} \leq 0.78)$, we use we use CDF.Normal. We will estimate $P(\bar{p} \leq 0.78)$ and $P(\bar{p} \leq 0.68)$ the get the difference.

Go to SPSS Data View Window > Click 0 in the first cell then press enter > Go to Transform > Compute Variable > there will be a pop up window for Compute Variable > type cdfnormal in the Target Variable > select CDF and Noncentral CDF under Function group > Double click Cdf.Normal under Functions and Special Variables > In the Numeric Expression box we will have two expressions (CDF.NORMAL(0.78,0.73,0.0174) – CDF.NORMAL(0.68,0.73,0.0174)): enter the values for the three question marks for the first bracket in the following order: quant=value of x (0.78), mean (0.73) and standard deviation (0.0174), and the three question marks for the second bracket in the following order: quant=value of x (0.68), mean (0.73) and standard deviation (0.0174) then click OK. Go to Data View Window and you have the results. Hence, $P(0.68 \leq \bar{p} \leq 0.78) = 0.99594$ (Figure 7.9).

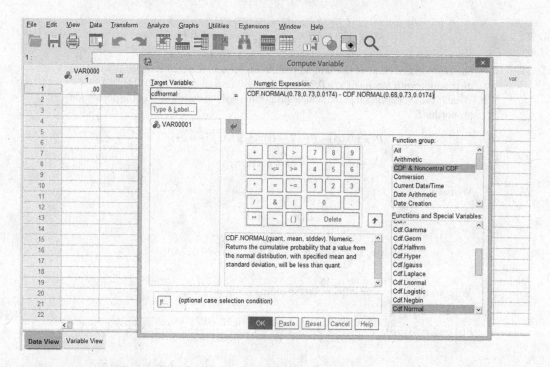

FIGURE 7.9 SPSS Pop Window for computing Normal Probabilities

The SPSS Data View window will give the probability value (Figure 7.10)

$$P(0.68 \leq \bar{p} \leq 0.78) = 0.99594$$

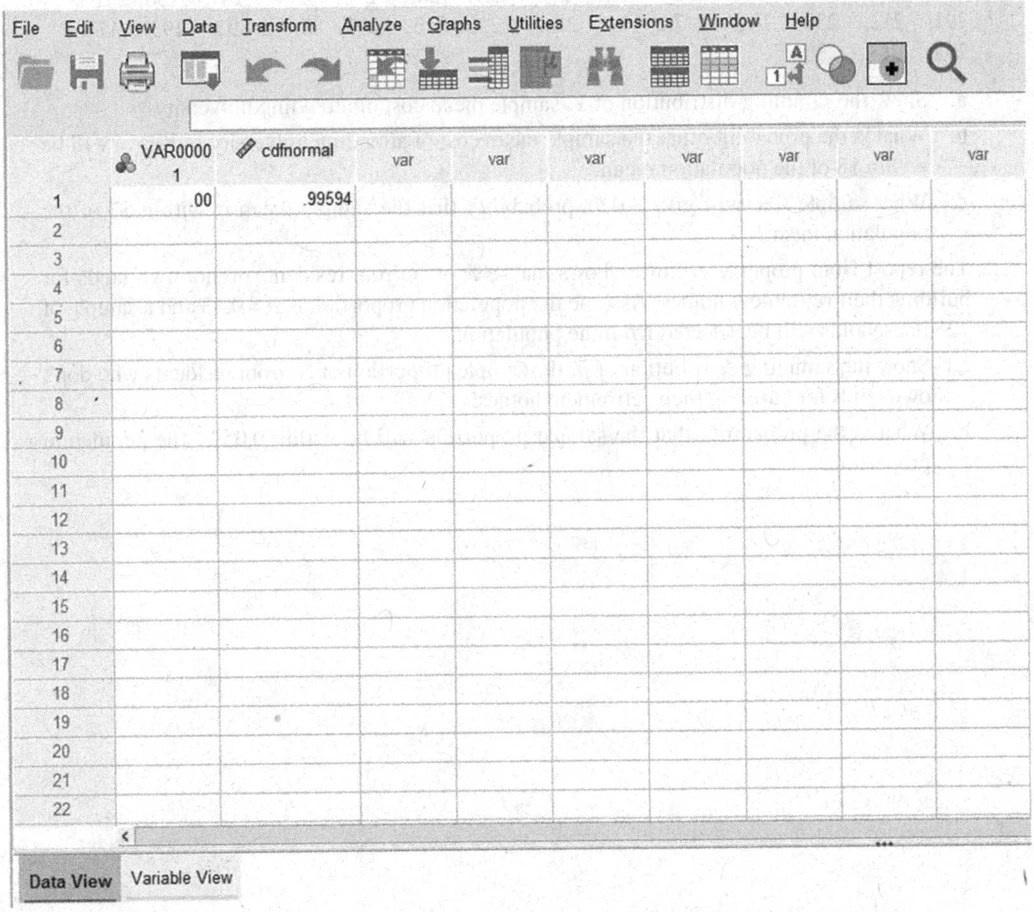

FIGURE 7.10   SPSS Data View Window for displaying the computed Normal Probabilities

Practice Exercise

1. The following is a sample of selected data showing the average costs of attending Universities in Kenya in US dollars:

   220, 184, 167, 213, 238, 154, 296, 211, 261, 241, 194, 187, 158, 253, 291, 212, 233, 201, 217, 179, 214, 222, 213, 232, 217, 897, 193, 173, 297, 215, 230

   a. Show the sampling distribution of $\bar{x}$, sample mean cost of attending university
   b. What is the probability that the sample mean cost of attending university in Kenya will be within $8 of the population mean?
   c. What sample size will give a 0.98 probability that the sample mean is within $7 of the population mean?

2. The report from property ventures shows that 43% of Nairobi residents do not own lands for building their retirement homes. Assume the population proportion is $p = 0.43$ and a sample of 725 households will be selected from the population.

   a. Show the sampling distribution of $\bar{p}$, the sample proportion of Nairobi residents who don't own lands for building their retirement homes.
   b. What is the probability that the sample proportion will be within 0.05 of the population proportion?

Part III

Introduction to Statistical
Inference Concepts and Methods

8

Statistical Inference: Estimation

8.1 Statistical Inference

Inferential statistics is part of statistics that enables one to generalize about a population based upon information obtained from a sample. This is done through *Estimation* and *Test of Hypothesis*, which are the two major branches of inferential statistics. Inferential statistics has two branches: *parametric statistics* and *nonparametric statistics*. *Parametric statistics* requires certain assumptions about the distribution of the data. It requires interval or ratio data. Parametric statistical analyses require that data must meet some assumptions including normality, data collected randomly and independently, and assumptions of equal variances when dealing with two or more groups. There are several ways of assessing the assumptions. For the *nonparametric statistics*, there is no assumption about the distribution of the data. It requires data that are in nominal and ordinal.

8.2 Estimation

Estimation is the process of inferring or estimating a population parameter from the corresponding statistic of a sample drawn from the population. An estimate of a population parameter may be expressed in two ways: *Point Estimate* and *Interval Estimate (confidence interval estimation)*. A good estimator should be unbiased, consistent and efficient. An unbiased estimator of a population parameter is an estimator whose expected value of the sample mean equals the population mean, i.e. $E(\bar{X}) = \mu$. An unbiased estimator is said to be a consistent estimator if the difference between the estimator and the parameter grows smaller as the sample size grows larger. As n grows larger, the variance of the sample mean grows smaller. In other words, the sample mean $\bar{x}$ is a consistent estimator of the population mean, μ, with variance given by $Var(\bar{x}) = \frac{\sigma^2}{n}$. If there are two unbiased estimators of a parameter, the one whose variance is smaller is said to be the more efficient estimator. In the estimation process, we will consider *mean* and *proportions* both for one group (population) and two groups (populations).

A *point estimate* of a population parameter is a single value of a statistic being used to estimate the population parameter. For example, the sample mean $\bar{x}$ is a point estimate of the population mean μ and the sample proportion p is a point estimate of the population proportion P. Denoted by P.E. Point estimate for one group is a single mean value statistic (e.g. $P.E. = \bar{x}$), while for two groups it is the difference in the two statistics (e.g. $P.E. = \bar{x}_1 - \bar{x}_2$).

An *interval estimate* is defined by two numbers, between which a population parameter is said to lie. For example, $a < \bar{X} < b$ is an interval estimate of the population mean μ. Also known as *Confidence Interval Estimation*. Denoted I.E.

8.2.1 Relating P.E. and I.E.

The interval estimate of a confidence interval is defined by the sample statistic ± margin of error.

A general expression relating P.E. and I.E. is given by:

$$I.E. = P.E. \pm M.E.$$
$$Lower\ Bound = P.E. - M.E.$$
$$Upper\ Bound = P.E. + M.E.$$

DOI: 10.1201/9781003292654-11

where
  I.E. is the Interval Estimate or Confidence Interval at a given level of confidence
  P.E. is the Point Estimate
  M.E. is the Margin of Error

8.2.2 Computing P.E.

If dealing with mean, the P.E. for one population will be given by $P.E. = \bar{x}$ and P.E. for two population will be given by $P.E. = \bar{x}_1 - \bar{x}_2$.

If dealing with proportion, the P.E. for one population will be given by $P.E. = \bar{p}$ and P.E. for two population will be given by $P.E. = \bar{p}_1 - \bar{p}_2$.

8.2.3 Computing M.E.

Margin of error is the range of values above and below the sample statistic in a confidence interval. Margin of error can be computed using *Z distribution* or *t distribution* and the standard error (S.E.) of mean or proportion.

Z distribution is used in two situations: for cases *when we are dealing with means and the standard deviation is given/known* and for cases *when we are dealing with proportions*. In this case, we use $Z_{\alpha/2}$ values from the Z distribution table.

t distribution is used for cases *when we are dealing with means if the standard deviation is not given but computed*. In this case, we use $t_{\alpha/2}$ values from the t distribution table with a given number of degree of freedom (df).

A general expression for computing M.E.

$$M.E. = Z_{\alpha/2} \times S.E.$$

or

$$M.E. = t_{\alpha/2} \times S.E.$$

8.2.4 Computing $Z_{\alpha/2}$ values

In this case, we use the Z distribution table and confidence levels or $\alpha/2$ values. The confidence level is the degree of belief (usually it is between 90% and 99%). Conventionally, Z distribution table has been used. Z alpha/2 values are also known as Z critical values.
Steps:

- Compute the $\frac{Confidence\ Level}{2}$ value
- Check in the Z table the corresponding Z value of the area equals to $\frac{Confidence\ Level}{2}$. The Z table value is the $Z_{\alpha/2}$ value.

Example: **Estimate the $Z_{\alpha/2}$ values for the following confidence levels in the table below**

Confidence Levels	$\alpha = (1 - CL)$	$\alpha/2$	$Z_{\alpha/2}$
90% = 0.90			
95% = 0.95			
99% = 0.99			

SOLUTION

First, we complete the columns for α and $\alpha/2$ values
  Using 90% confidence level to get the $Z_{\alpha/2}$ value

- Compute the $\frac{Confidence\ Level}{2}$ value, which is $\frac{0.90}{2} = 0.45$.
- Check in the Z table the corresponding Z value of the area equals to 0.45, we get = 1.64. The Z table value is the $Z_{\alpha/2}$ value.

The same process is repeated for 95% and 99% confidence levels to generate the $Z_{\alpha/2}$ values

Confidence Levels	$\alpha = (1 - CL)$	$\alpha/2$	$Z_{\alpha/2}$
90% = 0.90	0.10	0.05	1.64
95% = 0.95	0.05	0.025	1.96
99% = 0.99	0.01	0.005	2.57

In SPSS, we use IDF.NORMAL function to compute the Z critical values

8.2.5 Computing $t_{\alpha/2}$ Values

In this case, we use the t distribution table, degree of freedom (df) and confidence levels or $\alpha/2$ values. The df is always computed based on the sample size. t alpha/2 values are also known as t critical values.

Example: Estimate the $t_{\alpha/2}$ values for the following confidence levels for the given df values.

Confidence Levels	df	α	$\alpha/2$	$t_{\alpha/2}$
90% = 0.90	12			
95% = 0.95	8			
99% = 0.99	10			

SOLUTION:

First, we complete the columns for α and $\alpha/2$ values

To estimate the $t_{\alpha/2}$ values, we look up for the value that intersects a given confidence level with the given df from the t-distribution table

Confidence Levels	df	α	$\alpha/2$	$t_{\alpha/2}$
90% = 0.90	12	0.10	0.05	1.782
95% = 0.95	8	0.05	0.025	2.306
99% = 0.99	10	0.01	0.005	3.169

In SPSS, we use IDF.T function to compute the t critical values.

8.2.6 Computing Standard Errors (S.E.)

We illustrate the formulas for computing standard errors (S.E.) when dealing with one or two groups both for mean(s) and proportion(s).

1. *S.E. for One Population (one group)*

   When dealing with one population/group, we will consider one population mean and one population proportion.

   a. *S.E. for One Population Mean*

   When dealing with one population mean, we will consider situations when σ (standard deviation) is the given or known and when σ (standard deviation) is not known or not given.

   i. If σ (standard deviation) is known or given, S.E. will be computed as

   $$S.E. = \frac{\sigma}{\sqrt{n}}$$

where σ is the given or known standard deviation and n is the sample size.

ii. If σ (standard deviation) is not known or not given

In this case, we computed standard deviation and denote it with s, S.E. will be computed as

$$S.E. = \frac{s}{\sqrt{n}}$$

and $df = n - 1$

b. **S.E. for One Population Proportion**

In this case, we will compute S.E. as

$$S.E. = \sqrt{\frac{p(1-P)}{n}}$$

where p is the proportion.

2. **S.E. for Two Populations (Two Groups)**

When dealing with two groups, we make assumptions while computing S.E.s. If dealing with two means then we will make an assumption of equal variance or unequal variance. If dealing with two proportions we will make an assumption of equal proportions or different proportions. We will also consider two population means and two population proportions.

a. **S.E. for Two Population Means**

In this case, we will consider situation when σ_1 and σ_2 is known or given and when σ_1 and σ_2 is not known or not given.

i. *If σ_1 and σ_2 is known or given*

A. Assuming equal variance

$$S.E. = s_p\sqrt{\frac{1}{n_1} + \frac{1}{n_2}}$$

where

$$s_p = \sqrt{\frac{(n_1-1)\sigma_1^2 + (n_2-1)\sigma_2^2}{n_1+n_2-2}}$$

where σ_1 is the known/given standard deviation for population (group) 1, σ_2 is the known/given standard deviation for population (group) 2, n_1 is the size for population (group) 1, and n_2 is the size for population (group) 2.

B. Assuming unequal variance

$$S.E. = \sqrt{\left(\frac{\sigma_1^2}{n_1}\right) + \left(\frac{\sigma_2^2}{n_2}\right)}$$

ii. *If σ_1 and σ_2 is not known or not given*

A. Assuming equal variance

$$S.E. = s_p\sqrt{\frac{1}{n_1} + \frac{1}{n_2}}$$

where

$$s_p = \sqrt{\frac{(n_1-1)s_1^2 + (n_2-1)s_2^2}{n_1+n_2-2}}$$

with

$$df = n_1 + n_2 - 2$$

B. Assuming unequal variance

$$S.E. = \sqrt{\left(\frac{s_1^2}{n_1}\right) + \left(\frac{s_2^2}{n_2}\right)}$$

with

$$df = \frac{\left(\dfrac{s_1^2}{n_1} + \dfrac{s_2^2}{n_2}\right)^2}{\left(\dfrac{1}{n_1 - 1}\right)\left(\dfrac{s_1^2}{n_1}\right)^2 + \left(\dfrac{1}{n_2 - 1}\right)\left(\dfrac{s_2^2}{n_2}\right)^2}$$

b. **S.E. for Two Population Proportions**

In this case, we make an assumption of equal proportion or different proportions.

A. Assuming equal proportions

$$S.E. = \sqrt{p(1-p) \times \left[\frac{1}{n_1} + \frac{1}{n_2}\right]}$$

where

$$p = \frac{p_1 \times n_1 + p_2 \times n_2}{n_1 + n_2}$$

where p_1 is the sample proportion for population (group) 1, p_2 is the sample proportion for population (group) 2, n_1 is the size for population (group) 1 and n_2 is the size for population (group) 2.

B. Assume different proportions

$$S.E. = \sqrt{\frac{p_1(1-p_1)}{n_1} + \frac{p_2(1-p_2)}{n_2}}$$

3. **S.E. for Matched or Paired Samples**

Matched/paired samples are two observations from the same person. For example, marks in a given subject at mid semester (x_1) and end of semester (x_2). When we have matched samples, we first get the difference in the observations ($d = x_1 - x_2$), then use the data for the differences (d) to estimate the mean of the difference ($\bar{d}$) and standard deviation of the difference (s_d). S.E. is then computed as

$$S.E. = \frac{s_d}{\sqrt{n}}$$

with

$$df = n - 1$$

8.2.7 Computing Confidence Interval

Once you have computed S.E.s, the S.E. values together with either $Z_{\alpha/2}$ or $t_{\alpha/2}$ values are substituted into M.E. formula to estimate the M.E. value. The M.E. value together with the P.E. value are substituted into the I.E. formula to estimate the I.E. value which we can use to estimate the lower bound and the upper bound values.

1. **Computing Confidence Interval Estimate for One Population (one group)**
   a. *Confidence Interval for One Population Mean*
      i. *If σ (standard deviation) is known*

      Once you have computed S.E., the value of S.E. is substituted into M.E. formula to have the M.E. value. The M.E. value is then substituted into I.E. formula to have the I.E. value.

      Confidence Interval = I.E.

      But I.E. is computed as

      $$I.E. = P.E. \pm M.E.$$

   $P.E. = \overline{x}$ since we are dealing with mean for one group/one population

   M.E. will be computed as $M.E. = Z_{\alpha/2} \times S.E.$ since we are dealing with mean and standard deviation is given/known.

   To estimate $Z_{\alpha/2}$, we first divide CL by 2, e.g. for 94% confidence level

   $$\frac{CL}{2} = \frac{0.94}{2} = 0.47$$

   Then look for the Z value that is equals to the area of 0.47. So $Z_{\alpha/2} = 1.88$. Or use IDF. NORMAL (0.97, 0, 1) in SPSS to obtain 1.88.

   S.E. will be computed as $S.E. = \frac{\sigma}{\sqrt{n}}$ since we are dealing with one population mean standard deviation is given/known

   We then compute M.E. as

   $$M.E. = Z_{\alpha/2} \times S.E.$$

   Finally, compute I.E. as

   $$I.E. = P.E. \pm M.E.$$

   Substitute the values of P.E. and M.E. into the I.E. formula to get the lower bound and the upper bound

   $$Lower\ Bound = P.E. - M.E.$$

   $$Upper\ Bound = P.E. + M.E.$$

   In SPSS, we go to File > New > Syntax > SPSS syntax window will populate.

   Copy and paste the SPSS syntax but replace the values in the following order: sample, sample mean, standard deviation and the confidence level.
   **The first number is the sample size (n), the second number is the sample mean ($\overline{x}$),
   **and the third number is the population standard deviation (σ)
   **Replace the four values below with your own.
   data list list/n sample_mean population_sd.
   begin data
   35 105 15
   end data.

```
Compute mean = sample_mean.
Compute square_root_n =SQRT(n).
Compute standard_difference = population_sd/square_root_n.
Compute z_critical = IDF.NORMAL(0.975,0,1).
Compute margin_error = z_critical*standard_difference.
Compute lower_bound = mean-margin_error.
Compute upper_bound = mean+margin_error.
EXECUTE.
Formats mean lower_bound upper_bound (f8.5).
LIST z_critical mean lower_bound upper_bound.
```

Select the Syntax and click Run Selection

The SPSS output will have the P.E. and I.E. values

Interpret the SPSS results/output in terms of the *P.E.* (Mean) and *I.E.* (95% Confidence Interval of the Difference in terms of lower bound and upper bound).

Example

A report stated that the average university tuition fees per undergraduate student in private universities in Kenya was $1599. Suppose this average cost was based on a sample of 50 undergraduate university students studying in a private university in Kenya and that the population standard deviation is $\sigma = \$600$. Develop a 95% confidence interval estimate for the mean university tuition fees per undergraduate student in private universities in Kenya.

We need to compute I.E., which is also the confidence interval

We compute I.E. as

$$I.E. = P.E. \pm M.E.$$

$P.E. = \overline{x}$ since we have mean for one population, so $P.E. = \overline{x} = 1599$

We compute $M.E. = Z_{\alpha/2} \times S.E.$ since the standard deviation is given

To estimate $Z_{\alpha/2}$, we first divide CL by 2

$$\frac{CL}{2} = \frac{0.95}{2} = 0.475$$

We look for the Z value that is equals to the area of 0.475. So $Z_{\alpha/2} = 1.96$. Alternatively, using IDF. NORMAL(0.95, 0, 1) in SPSS gives Z critical value as 1.96.

S.E. is computed as $S.E. = \frac{\sigma}{\sqrt{n}}$ since we are dealing with one population mean

$$S.E. = \frac{\sigma}{\sqrt{n}} = \frac{600}{\sqrt{50}} = 84.8528$$

We can now compute M.E.

$$M.E. = Z_{\alpha/2} \times S.E. = 1.96 \times 84.8528 = 166.3115$$

We can now compute I.E.

$$I.E. = P.E. \pm M.E. = 1599 \pm 166.3115$$

$$Lower\ Bound = P.E. - M.E. = 1599 - 166.3115 = 1432.6885$$

$$Upper\ Bound = P.E. + M.E. = 1599 + 166.3115 = 1765.3115$$

In SPSS, to estimate Point and Interval Estimate for one population mean we follow the path: we will use SPSS syntax since Z test is not inbuilt within SPSS with drop down menu. What one need to do is to replace the values (sample, sample mean, population standard deviation) in the syntax.

File > New > Syntax > SPSS syntax window will populate (Figure 8.1)

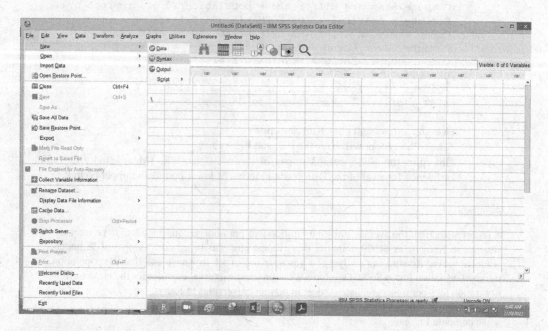

FIGURE 8.1   SPSS Menu for Syntax

Copy and paste the SPSS syntax but replace the values in the following order: sample, sample mean, standard deviation and the confidence level

**The first number is the sample size (n), the second number is the sample mean ($\overline{x}$),

**and the third number is the population standard deviation (σ)

**Replace the four values below with your own.

data list list/n sample_mean population_sd.

begin data

50   1599   600

end data.

```
Compute mean = sample_mean.
Compute square_root_n = SQRT(n).
Compute standard_difference = population_sd/square_root_n.
Compute z_critical = IDF.NORMAL(0.975,0,1).
Compute margin_error = z_critical*standard_difference.
Compute lower_bound = mean-margin_error.
Compute upper_bound = mean+margin_error.
EXECUTE.
Formats mean lower_bound upper_bound (f8.5).
LIST z_critical mean lower_bound upper_bound.
```

Select the Syntax and click Run Selection (Figure 8.2)

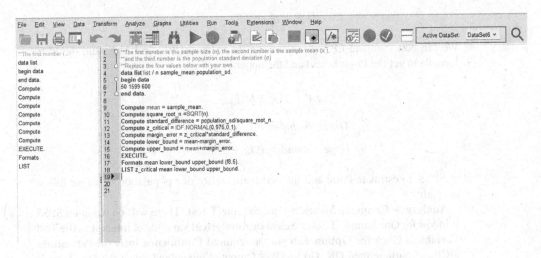

FIGURE 8.2   SPSS Syntax for computing Confidence Intervals for Single Population Mean

The SPSS output will have the P.E. and I.E. values (Figure 8.3)

List

```
z_critical    mean lower_bound upper_bound

  1.96  1599.000   1432.692   1765.308

Number of cases read:  1    Number of cases listed:  1
```

FIGURE 8.3   Confidence Intervals for Single Population Mean

P.E. = mean = 1599, lower bound = 1432.692 and upper bound = 1765.308.
From the above output, we are 95% confidence that the mean undergraduate tuition fees will range from 1432.692 to 1765.308.

    ii. *If σ (standard deviation) is not known or not given*
        Once you have computed S.E., the value of S.E. is substituted into M.E. formula to have the M.E. value. The M.E. value is then substituted into I.E. formula to have the I.E. value.
        Confidence Interval = I.E.
        But I.E. is computed as

$$I.E. = P.E. \pm M.E.$$

P.E. = $\bar{x}$ since we are dealing with mean for one group/one population.

M.E. will be computed as $M.E. = t_{\alpha/2} \times S.E.$ since we are dealing with mean and standard deviation is not given but will be computed.

To get $t_{\alpha/2}$ value, we need df and confidence level. The df is computed as $df = n-1$ since we are dealing with one population mean. $t_{\alpha/2}$ value will be the intersection of df and the confidence level. Or using SPSS, the t critical value will be obtained using IDF.T(prob, df).

S.E. is computed as $S.E. = \frac{s}{\sqrt{n}}$ since we are dealing with one population mean

We then compute M.E.

$$M.E. = t_{\alpha/2} \times S.E.$$

We can now compute I.E. by substituting the values of P.E. and M.E. into the M.E. formula to get the lower bound and the upper bound

$$I.E. = P.E. \pm M.E.$$

$$Lower\ Bound = P.E. - M.E.$$

$$Upper\ Bound = P.E. + M.E.$$

In SPSS, to estimate Point and Interval Estimate for one population mean we follow the path:
Analyze > Compare Means > One-Sample T test. There will be a pop up SPSS Window for One Sample T Test > Select **the numerical variable of interest** as **the Test Variable** > Click the **Option Tab** put the required **Confidence Interval Percentage** > Click Continue then OK. Go to SPSS Output Window and select the One-Sample Test table.

Example

The early morning trading volumes (millions of shares) at Nairobi Securities Exchange for 18 days in January are as shown below.

216, 143, 221, 169, 198, 202, 210, 122, 187, 212, 232, 181, 199, 220, 258, 233, 219, 221

Estimate the 98% confidence level for the mean early morning trading volumes at Nairobi Securities Exchange

Create the variable in the Variable View Window then key the data into SPSS using the data view (Figure 8.4)

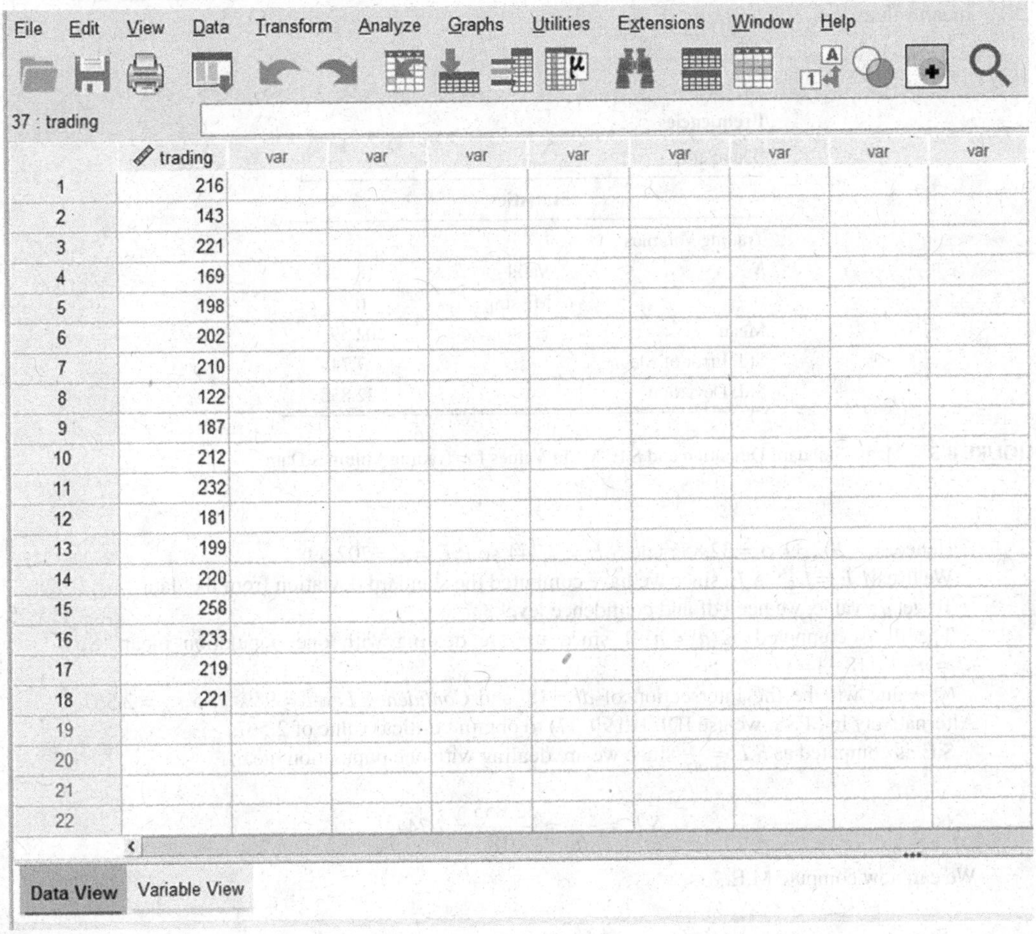

File	Edit	View	Data	Transform	Analyze	Graphs	Utilities	Extensions	Window	Help

37 : trading

	trading	var	var	var	var	var	var	var	var	var
1	216									
2	143									
3	221									
4	169									
5	198									
6	202									
7	210									
8	122									
9	187									
10	212									
11	232									
12	181									
13	199									
14	220									
15	258									
16	233									
17	219									
18	221									
19										
20										
21										
22										

Data View   Variable View

FIGURE 8.4   SPSS Data View Window for Trading Volumes

We need to compute I.E. which is also the confidence interval and I.E. is computed as

$$I.E. = P.E. \pm M.E.$$

$P.E. = \bar{x}$ since we have mean for one population,

We use SPSS to compute mean, standard deviation and S.E. mean: we go to Analyze > Descriptive Statistics > Frequencies > there will be SPSS pop up window for Frequencies > select trading as the Variable(s) > click Statistics and select Mean, Standard deviation, S.E. mean then Continue and Click OK. Form the SPSS output select mean, standard deviation and S.E. mean values.

Frequencies

[DataSet4]

Statistics		
Trading Volumes		
N	Valid	18
	Missing	0
Mean		202.39
Std. Error of Mean		7.744
Std. Deviation		32.855

FIGURE 8.5   Mean, Standard Deviation and S.E. Mean Values for Trading Volumes Data

Hence, $\bar{x} = 202.39$, $\sigma = 32.855$ and $S.E. = 7.744$ so $P.E. = \bar{x} = 202.39$

We use $M.E. = t_{\alpha/2} \times S.E.$ since we have computed the standard deviation from the data

To get $t_{\alpha/2}$ value, we need df and confidence level.

The df is computed as $df = n - 1$ since we are dealing with one population mean. So $df = n - 1 = 18 - 1 = 17$

$t_{\alpha/2}$ value will be the intersection of $df = 18$ and *Confidence Level* = 98%. So $t_{\alpha/2} = 2.567$. Alternatively in SPSS, we use IDF.T(0.99, 17) to obtain t critical value of 2.567.

S.E. is computed as $S.E. = \frac{s}{\sqrt{n}}$ since we are dealing with one population mean

$$S.E. = \frac{s}{\sqrt{n}} = \frac{32.855}{\sqrt{18}} = 7.744$$

We can now compute M.E.

$$M.E. = t_{\alpha/2} \times S.E. = 2.567 \times 7.744 = 19.8788$$

We can now compute I.E.

$$I.E. = P.E. \pm M.E. = 202.39 \pm 19.8788$$

$$Lower\ Bound = P.E. - M.E. = 202.39 - 19.8788 = 182.5112$$

$$Upper\ Bound = P.E. + M.E. = 202.39 + 19.8788 = 222.2688$$

In SPSS, we go to **Analyze > Compare Means > One-Sample T test**. There will be a pop up SPSS Window for One Sample T Test > Select **trading** as **the Test Variable** > Click the **Option Tab** put the required **Confidence Interval Percentage (98%)**> Click Continue then OK (Figure 8.6)

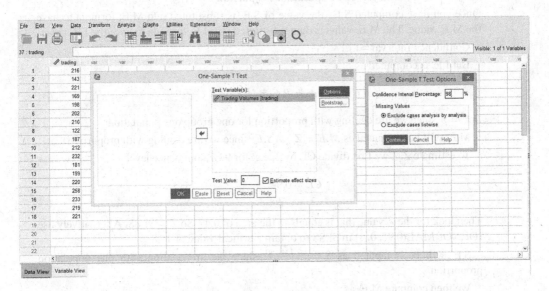

FIGURE 8.6   SPSS One-Sample t-Test Pop Up Window for generating Confidence Intervals for Single Mean

The SPSS output will have the P.E. and I.E. values

T-Test

	One-Sample Statistics			
	N	**Mean**	**Std. Deviation**	**Std. Error Mean**
Trading Volumes	18	202.39	32.855	7.744

	One-Sample Test					
	Test value = 0					
					98% Confidence Interval of the Difference	
	T	**df**	**Sig. (2-tailed)**	**Mean Difference**	**Lower**	**Upper**
Trading Volumes	26.135	17	<.001	202.389	182.51	222.27

FIGURE 8.7   Confidence Interval for Mean Trading Volumes

P.E. = Mean Difference = 1599, lower = 182.51 and upper = 222.27.

From the above output, we are 98% confidence that the mean trading volumes will be between 182.51 and 222.27.

b. *Confidence Interval for One Population Proportion*

Once you have computed S.E., the value of S.E. is substituted into M.E. formula to have the M.E. value. The M.E. value is then substituted into I.E. formula to have the I.E. value.

Confidence Interval = I.E.

But I.E. is computed as

$$I.E. = P.E. \pm M.E.$$

$P.E. = \bar{p}$ since we are dealing with proportion for one group/one population

M.E. will be computed as $M.E. = Z_{\alpha/2} \times S.E.$ since we are dealing with proportion.

To estimate $Z_{\alpha/2}$, we first divide CL by 2, e.g. for 94% confidence level

$$\frac{CL}{2} = \frac{0.94}{2} = 0.47$$

Then look for the Z value that is equals to the area of 0.47. So $Z_{\alpha/2} = 1.88$. Alternatively use IDF.NORMAL(0.97, 0, 1) in SPSS to obtain z critical value of 1.88

S.E. will be computed as $S.E. = \sqrt{\frac{p(1-p)}{n}}$ since we are dealing with one population proportion

We then compute M.E. as

$$M.E. = Z_{\alpha/2} \times S.E.$$

Finally, compute I.E. as

$$I.E. = P.E. \pm M.E.$$

Substitute the values of P.E. and M.E. into the I.E. formula to get the lower bound and the upper bound

$$Lower\ Bound = P.E. - M.E.$$

$$Upper\ Bound = P.E. + M.E.$$

In SPSS, we go to **Analyze > Compare Means > One Sample Proportions**. There will be a pop up SPSS Window for One Sample Proportions > Select **the categorical variable of interest** as **the Test Variable** > Click **Define Success and select the proportion of interest** > Click the **Confidence Level Tab** put the required **Confidence Interval Percentage** > Click Continue then OK. Interpret the SPSS results/output in One Sample Proportions Confidence Intervals table in terms of the *P.E.* (Proportions) and *I.E.* (95% Confidence Interval of the Difference).

Example

Using car_sales.sav data, determine the 95% confidence interval estimate for the proportion of Automobile vehicles

Locate the car_sales.sav data, then generate a frequency table for vehicles using Analyze > Descriptive Statistics > Frequencies > there will be a pop up window for Frequencies > select vehicle type as Variable(s) > Ensure that Display frequency table is Checked then click OK (Figure 8.8).

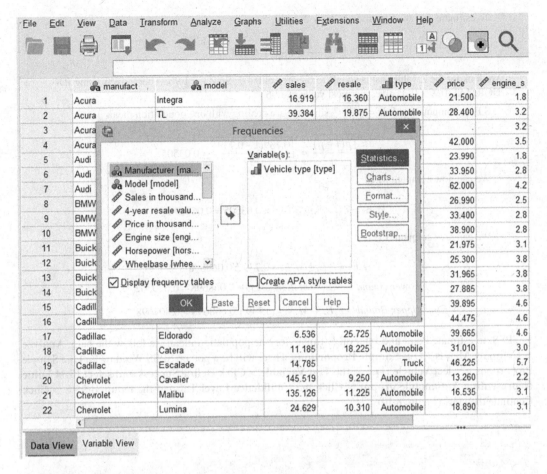

FIGURE 8.8 SPSS Menu for generating Frequencies for Vehicle Type

The SPSS Output will display frequency table showing the proportions

		Frequency	Percent	Valid Percent	Cumulative Percent
Vehicle Type					
Valid	Automobile	116	73.9	73.9	73.9
	Truck	41	26.1	26.1	100.0
	Total	157	100.0	100.00	

FIGURE 8.9 Frequency Table for Vehicle Type

From the above SPSS output, the proportions of Automobile vehicle is $p = \frac{x}{n} = \frac{116}{157} = 0.739$
We need to compute I.E., which is also the confidence interval
We compute I.E. as

$$I.E. = P.E. \pm M.E.$$

$P.E. = \bar{p}$ since we have proportion for one population, so $P.E. = \bar{p} = 0.739$
We compute $M.E. = Z_{\alpha/2} \times S.E.$ since the standard deviation is given

To estimate $Z_{\alpha/2}$, we first divide CL by 2

$$\frac{CL}{2} = \frac{0.95}{2} = 0.475$$

We look for the Z value that is equals to the area of 0.475. So $Z_{\alpha/2} = 1.96$. Alternatively using IDF. NORMAL(0.975, 0, 1) gives z critical value of 1.96

S.E. is computed as $S.E. = \sqrt{\frac{p(1-p)}{n}}$ since we are dealing with one population mean

$$S.E. = \sqrt{\frac{p(1-p)}{n}} = \sqrt{\frac{0.739(1-0.739)}{157}} = 0.03505$$

We can now compute M.E.

$$M.E. = Z_{\alpha/2} \times S.E. = 1.96 \times 0.03505 = 0.068698$$

We can now compute I.E.

$$I.E. = P.E. \pm M.E. = 0.739 \pm 0.068698$$

$$Lower\ Bound = P.E. - M.E. = 0.739 - 0.068698 = 0.670302$$

$$Upper\ Bound = P.E. + M.E. = 0.739 + 0.068698 = 0.807698$$

In SPSS, to estimate Point and Interval Estimate for one population mean we follow the path: **Analyze > Compare Means > One Sample Proportions**. There will be a pop up SPSS Window for One Sample Proportions > Select **Vehicle type** as **the Test Variable** > Click **Define Success and select First Value (representing Automobile group) >** Click the **Confidence Level Tab** put the required **95%** > Click Continue then OK (Figure 8.10)

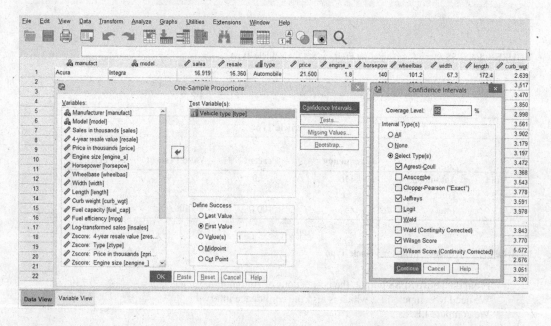

FIGURE 8.10   SPSS One-Sample Proportion Pop Up Window for generating Confidence Intervals for Single Proportion

The SPSS output will display the P.E. and I.E. for the one sample proportion

Proportions

		Observed			Asymptotic	95% Confidence Interval	
	Interval Type	Successe	Trials	Proportion	Standard Error	Lower	Upper
Vehicle type = Automobile	Agresti-Coull	116	157	.739	.035	.665	.802
	Jeffreys	116	157	.739	.035	.666	.803
	Wilson Score	116	157	.739	.035	.665	.801

One-Sample Proportions Confidence Intervals

FIGURE 8.11 Confidence Intervals for Proportions of Automobile Vehicle Types

From the SPSS output we can see that (using Jeffreys) the *P.E.* = *0.739* (the value under Proportions) and *I.E.* (95% Confidence Interval of the Difference) is given by lower bound = 0.666 and upper bound = 0.803. We are 95% confident that the proportions of Automobile vehicles type will be between 66.6% and 80.3%.

8.2.8 Computing Confidence Interval Estimate for Two Populations (Two Groups)

a. *Confidence Interval for Two Population Means*

i. *If σ_1 and σ_2 is known or given*

Once you have computed S.E., the value of S.E. is substituted into M.E. formula to have the M.E. value. The M.E. value is then substituted into I.E. formula to have the I.E. value.

Confidence Interval for the difference in mean = I.E. for the difference in mean

But I.E. is computed as

$$I.E. = P.E. \pm M.E.$$

$P.E. = \bar{x}_1 - \bar{x}_2$ since we means for the two population,

We compute M.E. using $M.E. = Z_{\alpha/2} \times S.E.$ since the std deviations are given and we are dealing with two population means

To estimate $Z_{\alpha/2}$, we first divide CL by 2, e.g. for 98% confidence level

$$\frac{CL}{2} = \frac{0.98}{2} = 0.49$$

We look for the Z value that is equals to the area of 0.49. So $Z_{\alpha/2} = 2.32$. Alternatively use IDF.NORMAL(0.99,0,1) in SPSS to obtain z critical value of 2.32

F-test is usually used to determine the assumption for computing the S.E.

S.E. will be computed as follows with the assumption is of equal variances as

$$S.E. = s_p \sqrt{\frac{1}{n_1} + \frac{1}{n_2}}$$

where

$$s_p = \sqrt{\frac{(n_1-1)\sigma_1^2 + (n_2-1)\sigma_2^2}{n_1+n_2-2}}$$

$$S.E. = s_p \sqrt{\frac{1}{n_1} + \frac{1}{n_2}}$$

And if assuming unequal variance, S.E. will be computed as

$$S.E. = \sqrt{\left(\frac{\sigma_1^2}{n_1}\right) + \left(\frac{\sigma_2^2}{n_2}\right)}$$

We can now compute M.E.

$$M.E. = Z_{\alpha/2} \times S.E.$$

We can now compute I.E.

$$I.E. = P.E. \pm M.E.$$

$$Lower\ Bound = P.E. - M.E.$$

$$Upper\ Bound = P.E. + M.E.$$

In SPSS, we go to File > New > Syntax > SPSS syntax window will populate > Copy and paste the SPSS syntax below but replace the values in the following order: sample size for group 1, sample size for group 2, sample mean fro group 1, sample mean fro group 2, variance for group 1 and variance for group 2. Also put the confidence level of interest.

**The first number is the sample size for group 1(n1=100), the second number is the sample size for group 2(n2=100),

**the third number is the sample mean for group 1 (200), the fourth number is the sample mean for group 2 (190)

** the fifth number is the population variance for group 1 (1600) and the sixth number is the population variance for group 2 (400)

**Replace the four values below with your own.

**The p value that is reported is based on a two-tailed test.

**To obtain the one-tailed p value, simply divide the two-tailed p value by 2.

data list list/n1 n2 sample_mean1 sample_mean2 population_var1 population_var2.
begin data
100   100   200   190   1600   400
end data.

```
Compute mean_difference = (sample_mean1 - sample_mean2).
Compute standard_difference = SQRT((population_var1/n1)+(population_
var2/n2)).
Compute z_critical = IDF.NORMAL(0.98,0,1).
Compute margin_error = z_critical*standard_difference.
Compute lower_bound = mean_difference -margin_error.
Compute upper_bound = mean_difference +margin_error.
EXECUTE.
Formats mean_difference lower_bound upper_bound (f8.5).
LIST z_critical mean_difference lower_bound upper_bound.
**Select the Syntax and click Run Selection
```

The SPSS output will have P.E. and I.E. values.

Interpret the SPSS results/output in terms of the *P.E.* (Mean) and *I.E.* (95% Confidence Interval of the Difference in terms of lower bound and upper bound).

Example

The average expenditure in a sample survey of 40 male consumers when they go partying was found to be \$135.67, and the average expenditure in a sample survey of 30 female consumers when they go partying was found to be \$68.64. Based on past surveys, the standard deviation for male consumers is assumed to be \$35, while the standard deviation for female consumers is assumed to be \$20.

a. What is the point estimate of the difference between the population mean expenditure for males and the population mean expenditure for females?

$$n_1 = 40, \bar{x}_1 = 135.67, \sigma_1 = 35, n_2 = 30, \bar{x}_2 = 68.64, \sigma_2 = 20$$

P.E. for two groups or population is given by $P.E. = \bar{x}_1 - \bar{x}_2$ since we are dealing with two population means

$$P.E. = \bar{x}_1 - \bar{x}_2 = 135.67 - 68.64 = 67.03$$

b. Develop a 96% confidence interval for the difference between the two population means (Assume unequal variance)

Confidence interval for the difference between the two population means = I.E. for the difference between the two population means

$$I.E. = P.E. \pm M.E.$$

$$Lower\ Bound = P.E. - M.E.$$

$$Upper\ Bound = P.E. + M.E.$$

$$P.E. = \bar{x}_1 - \bar{x}_2 = 135.67 - 68.64 = 67.03$$

To compute M.E. we use $M.E = Z_{\alpha/2} \times S.E$ since we are dealing with two population means and standard deviation are given

To compute $Z_{\alpha/2}$ value
- Compute the $\frac{Confidence\ Level}{2}$ value, i.e. $\frac{0.96}{2} = 0.48$
- Check in the Z table the corresponding Z value of the area equals to 0.48. The Z table value is the $Z_{\alpha/2}$ value. Hence, $Z_{\alpha/2} = 2.05$. Alternatively, using IDF.NORMAL(0.98, 0, 1) in SPSS gives a z critical value of 2.05

To compute S.E., we use

$$S.E. = \sqrt{\frac{\sigma_1^2}{n_1} + \frac{\sigma_2^2}{n_2}} = \sqrt{\frac{35^2}{40} + \frac{20^2}{30}} = 6.6301$$

since we are dealing with two population means and standard deviations have been given and we are *assuming unequal variance* (note that F-test can be used to determine the assumptions to be made)

We can now compute M.E.

$$M.E. = Z_{\alpha/2} \times S.E. = 2.05 \times 6.6301 = 13.5917$$

We can now compute I.E.

$$I.E. = P.E. \pm M.E. = 67.03 \pm 13.5917$$

$$Lower\ Bound = P.E. - M.E. = 67.03 - 13.5917 = 53.4383$$

$$Upper\ Bound = P.E. + M.E. = 67.03 + 13.5917 = 80.6217$$

In SPSS, we go to File > New > Syntax > SPSS syntax window will populate > Copy and paste the SPSS syntax below but replace the values in the following order: sample size for group 1, sample size for group 2, sample mean fro group 1, sample mean fro group 2, variance for group 1 and variance for group 2. Also put the confidence level of interest.
**The first number is the sample size for group 1(n1=40), the second number is the sample size for group 2(n2=30),

**the third number is the sample mean for group 1 (135.67), the fourth number is the sample mean for group 2 (68.64)

** the fifth number is the population variance for group 1 (35*35) and the sixth number is the population variance for group 2 (20*20)

**Replace the four values below with your own.

**The p value that is reported is based on a two-tailed test.

**To obtain the one-tailed p value, simply divide the two-tailed p value by 2.

data list list/n1 n2 sample_mean1 sample_mean2 population_var1 population_var2.

begin data

40   30   135.67   68.64   1225   400

end data.

```
Compute mean_difference = (sample_mean1 - sample_mean2).
Compute standard_difference =SQRT((population_var1/n1)+(population_
var2/n2)).
Compute z_critical = IDF.NORMAL(0.98,0,1).
Compute margin_error = z_critical*standard_difference.
Compute lower_bound = mean_difference -margin_error.
Compute upper_bound = mean_difference +margin_error.
EXECUTE.
Formats mean_difference lower_bound upper_bound (f8.5).
LIST z_critical mean_difference lower_bound upper_bound.
**Select the Syntax and click Run Selection
```

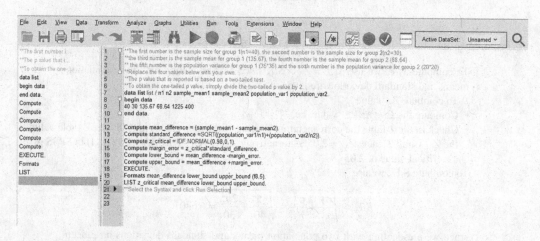

FIGURE 8.12   SPSS Syntax for computing Confidence Intervals for Two Population Mean

Select the Syntax and click Run Selection
   The SPSS output will have P.E. and I.E. values.

```
z_critical mean_difference lower_bound upper_bound

   2.05      67.03000     53.41342   80.64658


Number of cases read:  1    Number of cases listed:  1
```

FIGURE 8.13   Confidence Intervals for the difference between Two Population Mean

We are 98% confident that the difference between the mean expenditure for male and female consumers will be between $53.413 and $80.646.

ii. *If σ_1 and σ_2 is not known or not given*

Once you have computed S.E., the value of S.E. is substituted into M.E. formula to have the M.E. value. The M.E. value is then substituted into I.E. formula to have the I.E. value.

Confidence Interval for the difference in two means = I.E. for the difference in two means

But I.E. is computed as

$$I.E. = P.E. \pm M.E.$$

$P.E. = \bar{x}_1 - \bar{x}_2$ since we means for the two population,

M.E. will be computed as $M.E. = t_{\alpha/2} \times S.E.$ since we are dealing with two means and standard deviation values are not given but will be computed.

To get $t_{\alpha/2}$ value we need df and confidence level. The df is computed as $df = n_1 + n_2 - 2$

if assuming equal variance and $df = \dfrac{\left(\frac{s_1^2}{n_1} + \frac{s_2^2}{n_2}\right)^2}{\left(\frac{1}{n_1-1}\right)\left(\frac{s_1^2}{n_1}\right)^2 + \left(\frac{1}{n_2-1}\right)\left(\frac{s_2^2}{n_2}\right)^2}$ if assuming unequal variance

Hence, $t_{\alpha/2}$ value will be the intersection of *df* and the confidence level. Alternatively, we can use IDF.T(prob, df) in SPSS to obtain the t critical value.

S.E. is computed as $S.E. = s_p \sqrt{\frac{1}{n_1} + \frac{1}{n_2}}$ if assuming equal variance and $S.E. = \sqrt{\left(\frac{s_1^2}{n_1}\right) + \left(\frac{s_2^2}{n_2}\right)}$ if assuming unequal variance

We then compute M.E.

$$M.E. = t_{\alpha/2} \times S.E.$$

We can now compute I.E. by substituting the values of P.E. and M.E. into the M.E. formula to get the lower bound and the upper bound

$$I.E. = P.E. \pm M.E.$$

$$Lower\ Bound = P.E. - M.E.$$

$$Upper\ Bound = P.E. + M.E.$$

To estimate Point and Interval Estimate for the two population mean using t test in SPSS we follow the path:

Analyze > Compare Means > Independent Samples T Test. There will be a pop up SPSS Window for Independent Samples T test > Select **the numerical variable of interest** as **the Test Variable** > Select the *categorical variable of interest* as the **Grouping Variable** and **Define Groups** as per the coded values and click Continue > Click the **Option Tab** put the required **Confidence Interval Percentage** > Click Continue then OK

Interpret the SPSS results/output in terms of the *P.E.* (Mean Difference) and *I.E.* (95% Confidence Interval of the Difference) with the required assumptions of *Equal Variance* or *Unequal Variance*.

a. If the p-value of the Levene's Test for Equality of Variance is greater than 0.05 we Assume Equal Variance.

b. If the p-value of the Levene's Test for Equality of Variance is less than 0.05 we Assume Unequal Variance.

Example: Consider the following data for tonnes of cargo being transported by two airlines

Swiss

9.1 15.1 8.8 10.0 7.5 10.5 8.3 9.1 6.0 5.8 12.1 9.3

Atlantic

$$4.7 \quad 5.0 \quad 4.2 \quad 3.3 \quad 5.5 \quad 2.2 \quad 4.1 \quad 2.6 \quad 3.4 \quad 7.0 \quad 6.3$$

a. What is the *point estimate of the difference between the two population means?*
b. Develop a *96% confidence interval of the difference between* the daily population means for the two airports (Assume equal variance)

We create the variables in the Variable View Window (Figure 8.14)

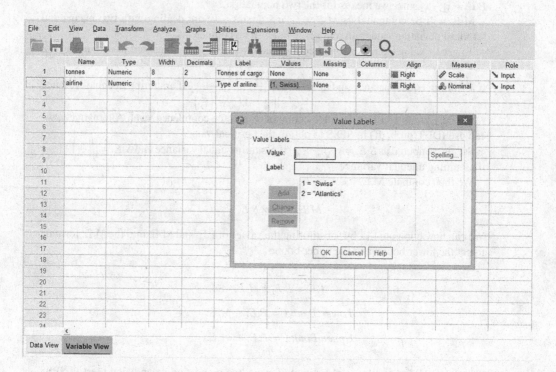

FIGURE 8.14   SPSS Variable View Window displaying Tonnes of Cargo and Type of Airline

Then key the data into SPSS using the Data View Window (Figure 8.15)

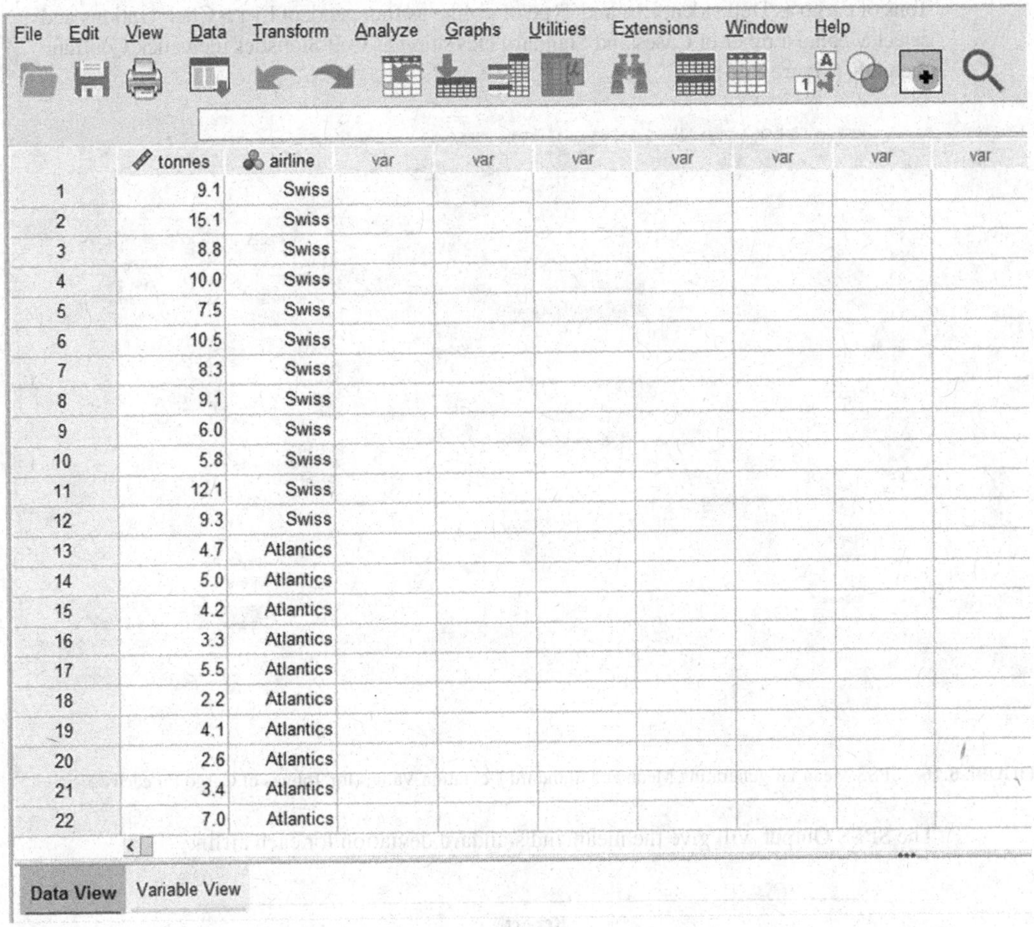

	tonnes	airline	var	var	var	var	var	var	var
1	9.1	Swiss							
2	15.1	Swiss							
3	8.8	Swiss							
4	10.0	Swiss							
5	7.5	Swiss							
6	10.5	Swiss							
7	8.3	Swiss							
8	9.1	Swiss							
9	6.0	Swiss							
10	5.8	Swiss							
11	12.1	Swiss							
12	9.3	Swiss							
13	4.7	Atlantics							
14	5.0	Atlantics							
15	4.2	Atlantics							
16	3.3	Atlantics							
17	5.5	Atlantics							
18	2.2	Atlantics							
19	4.1	Atlantics							
20	2.6	Atlantics							
21	3.4	Atlantics							
22	7.0	Atlantics							

Data View Variable View

FIGURE 8.15 SPSS Data View Window displaying Tonnes of Cargo and Type of Airline

Then we compute the mean and standard deviation values for the two groups using Analyze > Compare Means > Means > there will be a pop up window for Means > select Tons of cargo as Dependent List and Type of airline as Independent List > Click Options and select Mean, Number of Cases and Standard Deviation as Cell Statistics then click Continue then OK (Figure 8.16).

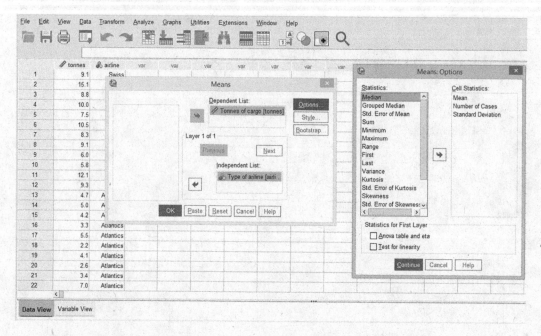

FIGURE 8.16   SPSS Menu for generating Mean and Standard Deviation Values for Tonnes of Cargo for each Airline

The SPSS Output will give the mean and standard deviation for each airline

Report			
Tons of cargo			
Type of Ariline	**Mean**	**N**	**Std. Deviation**
Swiss	9.300	12	2.5427
Atlantics	4.391	11	1.4983
Total	6.952	23	3.2465

FIGURE 8.17   Mean and Standard Deviation Values for Tonnes of Cargo for each Airline

We have $n_1 = 12$, $\bar{x}_1 = 9.3$, $s_1 = 2.5427$, $n_2 = 11$, $\bar{x}_2 = 4.391$, $s_2 = 1.4983$

P.E. for two groups or population is given by $P.E. = \bar{x}_1 - \bar{x}_2$ since we are dealing with two population means

$$P.E. = \bar{x}_1 - \bar{x}_2 = 9.3 - 4.391 = 4.909$$

We use $M.E. = t_{\alpha/2} \times S.E.$ since we have computed the standard deviation from the data

To get $t_{\alpha/2}$ value we need df and confidence level

The df is computed as $df = n_1 + n_2 - 2$ since we are assuming equal variance. So $df = n_1 + n_2 - 2 = 12 + 11 - 2 = 21$

$t_{\alpha/2}$ value will be the intersection of $df = 21$ and *Confidence Level* = 96%. So $t_{\alpha/2} = 2.189$. Alternatively, using IDF.T(0.98, 21) in SPSS gives t critical value of 2.189

S.E. is computed as $S.E. = s_p\sqrt{\frac{1}{n_1} + \frac{1}{n_2}}$ since we are assuming equal variance (note that F-test can be used to determine the assumption)

$$S.E. = s_p\sqrt{\frac{1}{n_1} + \frac{1}{n_2}}$$

where

$$s_p = \sqrt{\frac{(n_1-1)s_1^2 + (n_2-1)s_2^2}{n_1 + n_2 - 2}} = \sqrt{\frac{(12-1)2.5427^2 + (11-1)1.4983^2}{12+11-2}} = 2.1108$$

$$S.E. = s_p\sqrt{\frac{1}{n_1} + \frac{1}{n_2}} = 2.1108 \times \sqrt{\frac{1}{12} + \frac{1}{11}} = 0.8811$$

We can now compute M.E.

$$M.E. = t_{\alpha/2} \times S.E. = 2.189 \times 0.8811 = 1.9287$$

We can now compute I.E.

$$I.E. = P.E. \pm M.E. = 4.909 \pm 1.9287$$

$$Lower\ Bound = P.E. - M.E. = 4.909 - 1.9287 = 2.9803$$

$$Upper\ Bound = P.E. + M.E. = 4.909 + 1.9287 = 6.8377$$

Analyze > Compare Means > Independent Samples T Test. There will be a pop up SPSS Window for Independent Samples T test > Select **Tons of cargo** as **the Test Variable** > Select the **Type of airline** as the **Grouping Variable** and click **Define Groups** then put the code (as per the coded values) for Group 1 as 1 and Group 2 as 2 and click Continue > Click the **Option Tab** put the **98% Confidence Interval Percentage** > Click Continue then OK (Figure 8.18)

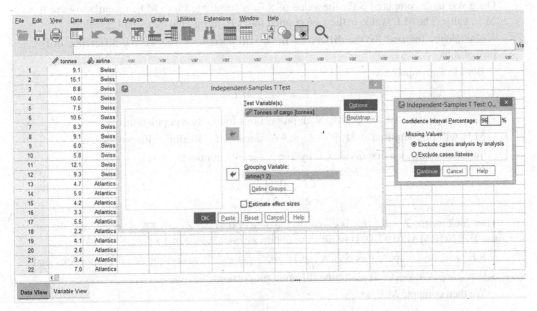

FIGURE 8.18 SPSS Independent-Samples t-Test Pop Up Window for generating Confidence Intervals for Two Means

SPSS Output

T-Test

Group Statistics

	Type of Ariline	N	Mean	Std. Deviation	Std. Error Mean
Tons of cargo	Swiss	12	9.300	2.5427	.7340
	Atlantics	11	4.391	1.4983	.4518

Independent Samples Test

		Levene's Test for Equality of Variances		t-test for Equality of Means					96% Confidence Interval of the Difference	
		F	Sig.	T	df	Sig. (2-tailed)	Mean Difference	Std. Error Difference	Lower	Upper
Tons of cargo	Equal variances assumed	.916	.349	5.571	21w	<.001	4.9091	.8811	2.9799	6.8382
	Equal variances not assumed			5.696	18.061	<.001	4.9091	.8619	3.0016	6.8166

FIGURE 8.19   Confidence Interval for the difference between Two Population Means

The *P.E.* = Mean Difference value=4.9091

Since the p-value of the Levene's Test for Equality of Variance is less than 0.05 we Assume Unequal Variance. Hence, lower bound will be 3.0016 and upper bound will be 6.8166

We are 98% confident that the mean difference in tons of cargo for the two airlines will be between 3.0016 and 6.8166.

If we have assumed equal variance then the lower bound will be 2.9799 and the upper bound will be 6.8382.

b. ***Confidence Interval for Two Population Proportions***

Once you have computed S.E., the value of S.E. is substituted into M.E. formula to have the M.E. value. The M.E. value is then substituted into I.E. formula to have the I.E. value.

Confidence Interval for the difference in two proportions = I.E. for the difference in two proportions

But I.E. is computed as

$$I.E. = P.E. \pm M.E.$$

$P.E. = \bar{p}_1 - \bar{p}_2$ since we are dealing with proportions for the two population,

M.E. will be computed as $M.E. = Z_{\alpha/2} \times S.E.$ since we are dealing with proportion.

To estimate $Z_{\alpha/2}$, we first divide CL by 2, e.g. for 94% confidence level

$$\frac{CL}{2} = \frac{0.94}{2} = 0.47$$

Then look for the Z value that is equals to the area of 0.47. So $Z_{\alpha/2} = 1.88$. Alternatively, we can use IDF.NORMAL(0.97, 0, 1) in SPSS to obtain z critical value of 1.88

S.E. will be computed as $S.E. = \sqrt{p(1-p) \times \left[\frac{1}{n_1} + \frac{1}{n_2}\right]}$ if assuming equal proportions and $S.E. = \sqrt{\frac{p_1(1-p_1)}{n_1} + \frac{p_2(1-p_2)}{n_2}}$ if assuming different proportions

We then compute M.E. as

$$M.E. = Z_{\alpha/2} \times S.E.$$

Finally, compute I.E. as

$$I.E. = P.E. \pm M.E.$$

Substitute the values of P.E. and M.E. into the I.E. formula to get the lower bound and the upper bound

$$Lower \ Bound = P.E. - M.E.$$

$$Upper \ Bound = P.E. + M.E.$$

To estimate Point and Interval Estimate for two population proportions in SPSS we follow the path:

Analyze > Compare Means > Independent-Samples Proportions. There will be a pop up SPSS Window for Independent-Sample Proportions > Select **the numerical variable of interest** as **the Test Variable >** then select **the categorical variable of interest** as **the Grouping Variable >** Under **Define Groups** put the coded values for the two categories/proportions of interest > Click the **Confidence Level Tab** put the required **Confidence Interval Percentage** > Click Continue then OK. Interpret the SPSS results/output in One Sample Proportions Confidence Intervals table in terms of the *P.E.* (Proportions) and *I.E.* (95% Confidence Interval of the Difference).

Example

Using customer_subset.sav data, determine the 95% confidence interval estimate for the difference between proportion of male customers who defaulted and proportion of male customers who didn't default on a bank.

First, we generate the number of male who defaulted and those who did not default using the path:

Analyze > Descriptive Statistics > Crosstabs > SPSS window for crosstabs will populate > Select any of the categorical variable of interest (Gender) and put it in Row(s), then select the other categorical variable of interest (Ever defaulted on a bank) and put it in Column(s) > Click Cells and a pop up window for Cell Display will populate then select Observed and Column percentages > Click Continue then OK (Figure 8.20).

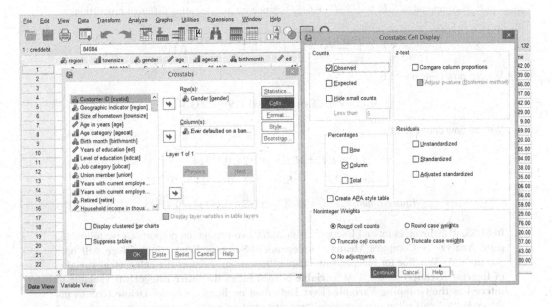

FIGURE 8.20   SPSS Crosstabs Pop Up Window displaying Cells Options

SPSS output

Gender * Ever defaulted on a bank loan Crosstabulation					
			Ever defaulted on a bank loan		
			No	**Yes**	**Total**
Gender	Male	Count	29	7	36
		% within Ever defaulted on a bank loan	50.0%	31.8%	45.0%
	Female	Count	29	15	44
		% within Ever defaulted on a bank loan	50.0%	68.2%	55.0%
Total		Count	58	22	80
		% within Ever defaulted on a bank loan	100.0%	100.0%	100.0%

FIGURE 8.21   Cross tabulation for Gender and Default Status

From the SPSS output, we can see that $p_1 = \frac{29}{58} = 0.50$, $p_2 = \frac{7}{22} = 0.318$, $n_1 = 58$, $n_2 = 22$
    We need to compute I.E., which is also the confidence interval
    We compute I.E. as

$$I.E. = P.E. \pm M.E.$$

$P.E. = \bar{p}_1 - \bar{p}_2$ since we are dealing with proportions for the two population, so
$P.E. = 0.50 - 0.318 = 0.182$
    We compute $M.E. = Z_{\alpha/2} \times S.E.$ since the standard deviation is given
    To estimate $Z_{\alpha/2}$, we first divide CL by 2

$$\frac{CL}{2} = \frac{0.95}{2} = 0.475$$

We look for the Z value that is equals to the area of 0.475. So $Z_{\alpha/2} = 1.96$. Alternatively, we can
use IDF.NORMAL(0.975, 0, 1) in SPSS to obtain the z critical value of 1.96
    S.E. is computed as $S.E. = \sqrt{\frac{p_1(1-p_1)}{n_1} + \frac{p_2(1-p_2)}{n_2}}$ since we are assuming different proportions

$$S.E. = \sqrt{\frac{0.50(1-0.50)}{58} + \frac{0.318(1-0.318)}{22}} = 0.119$$

We can now compute M.E.

$$M.E. = Z_{\alpha/2} \times S.E. = 1.96 \times 0.119 = 0.23324$$

We can now compute I.E.

$$I.E. = P.E. \pm M.E. = 0.182 \pm 0.23324$$

$$Lower\ Bound = P.E. - M.E. = 0.182 - 0.23324 = -0.05124$$

$$Upper\ Bound = P.E. + M.E. = 0.182 + 0.23324 = 0.41524$$

In SPSS, to estimate Point and Interval Estimate for two population proportions we follow the
path: **Analyze > Compare Means > Independent-Samples Proportions**. There will be a
pop up SPSS, Window for Independent-Sample Proportions > Select **the categorical variable
of interest (Gender)** as **the Test Variable** > then select **the other categorical variable of
interest** as **the Grouping Variable (Ever Defaulted on Bank)** > Under **Define Groups** put
the coded values for the two categories/proportions of interest – Grouping variable (0 and 1

for Ever defaulted) > Under Define Success select the First Value (representing the Gender of interest = Male) > Click the **Confidence Level Tab** put the required **95% Confidence Interval Percentage** > Click Continue then OK (Figure 8.22). Interpret the SPSS results/output in Two Sample Proportions Confidence Intervals table in terms of the *P.E.* (Proportions) and *I.E.* (95% Confidence Interval of the Difference).

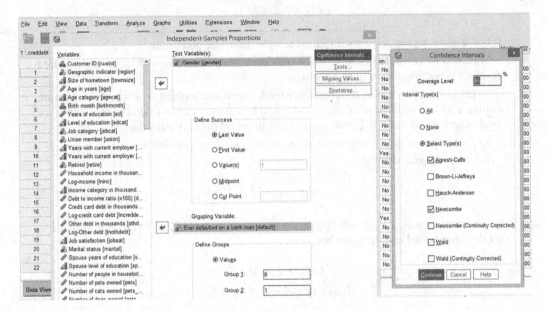

FIGURE 8.22 SPSS Independent-Samples Proportions Pop Up Window for generating Confidence Intervals for Two Proportions

The SPSS output will display the P.E. and I.E. for the one sample proportion

				95% Confidence Interval of the Difference	
	Interval Type	Difference in Proportions	Asymptotic Standard Error	Lower	Upper
Gender = Male	Agresti-Caffo	.182	.119	−.060	.394
	Newcombe	.182	.119	−.061	.380

Independent-Samples Proportions Confidence Intervals

FIGURE 8.23 Confidence Interval for the difference between Two Population Proportions

From the SPSS output we can see that (using Agresti or Newcombe) the *P.E. = Difference in Proportions = 0.182* and *I.E.* (95% Confidence Interval of the Difference) is given by lower bound = −0.60 and upper bound = 0.394 (Figure 8.23).

We are 95% confident that the difference between the proportion of male customers who defaulted and proportion of male customers who didn't default is between −0.6% and 39.4%.

c. *Confidence Interval for Matched or Paired Samples*

Once you have computed S.E., the value of S.E. is substituted into M.E. formula to have the M.E. value. The M.E. value is then substituted into I.E. formula to have the I.E. value.

We first get the difference in the observations ($d = x_1 - x_2$), then use the data for the differences (d) to estimate the mean of the difference ($\bar{d}$) and standard deviation of the difference (s_d).

Confidence Interval = I.E.

But I.E. is computed as

$$I.E. = P.E. \pm M.E.$$

$P.E. = \bar{d}$ since we are dealing with paired samples from one population

M.E. will be computed as $M.E. = t_{\alpha/2} \times S.E.$ since we are dealing with mean and standard deviation is not given but will be computed.

To get $t_{\alpha/2}$ value we need df and confidence level. The df is computed as $df = n - 1$ since we are dealing with one population mean. $t_{\alpha/2}$ value will be the intersection of df and the confidence level. Alternatively use IDF.T(prob, df) in SPSS to obtain the t critical value.

S.E. is computed as $S.E. = \frac{s_d}{\sqrt{n}}$ since we are dealing with one population mean

We then compute M.E.

$$M.E. = t_{\alpha/2} \times S.E.$$

We can now compute I.E. by substituting the values of P.E. and M.E. into the M.E. formula to get the lower bound and the upper bound

$$I.E. = P.E. \pm M.E.$$

$$Lower\ Bound = P.E. - M.E.$$

$$Upper\ Bound = P.E. + M.E.$$

To estimate Point and Interval Estimate for matched/paired means (mean while considering two dependent groups) in SPSS we follow the path:

Analyze > Compare Means > Paired Samples T-test. There will be SPSS pop up Window for Paired Samples T–Test > Select *the paired numerical variable of interest* for **Variable 1** and **Variable 2** > Click the **Option Tab** put the required **Confidence Interval Percentage** > Click Continue then OK. The SPSS Output will give the mean and confidence interval

Example

Consider the following data for 12 students that were used to compare students placement math score and the first math quiz score.

Student	Placement	First Quiz
1	540	474
2	432	380
3	528	463
4	574	612
5	448	420
6	502	526
7	480	430
8	499	459
9	610	615
10	572	541
11	390	335
12	593	613

a. What is the point estimate for the population mean difference in the scores?
b. Develop a 98% confidence interval for the population mean difference in the scores?

We first get the difference in the observations ($d = x_1 - x_2$), then use the data for the differences (d) to estimate the mean of the difference (d) and standard deviation of the difference (s_d).

Student	Placement Math Score	First Quiz Math Score	d = Placement Math – First Quiz Math
1	540	474	66
2	432	380	52
3	528	463	65
4	574	612	−38
5	448	420	28
6	502	526	−24
7	480	430	50
8	499	459	40
9	610	615	−5
10	572	541	31
11	390	335	55
12	593	613	−20

We then create the variable for the difference and key the difference data (d) into SPSS to compute mean and standard deviation values using SPSS via: Analyze > Descriptive Statistics > Frequencies > there will be SPSS pop up window for Frequencies > Select difference data as the Variable(s) > Click Statistics and select Mean, Standard Deviation and S.E. mean values > click Continue then OK (Figure 8.24).

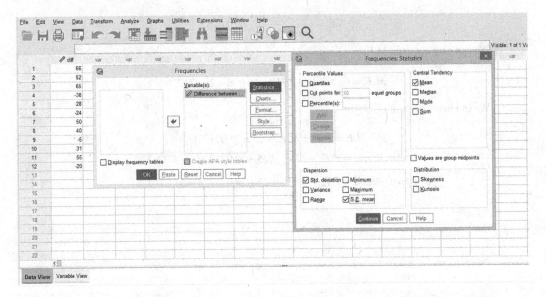

FIGURE 8.24 SPSS Frequency Menu for generating Mean, Standard Deviation and S.E. Mean Values for Differences in Scores

The SPSS output will give Mean, Standard Deviation and S.E. mean values

Statistics		
Difference between math placement and math first quiz score		
N	Valid	12
	Missing	0
Mean		25.00
Std. Error of Mean		10.696
Std. Deviation		37.050

FIGURE 8.25   Mean, Standard Deviation and S.E. Mean Values for Differences in Scores

From the above SPSS output $\bar{d} = 25$, $s_d = 37.050$, $n = 12$, and $S.E. = \frac{s_d}{\sqrt{n}} = \frac{37.050}{\sqrt{12}} = 10.696$
Confidence Interval = I.E.
But I.E. is computed as

$$I.E. = P.E. \pm M.E.$$

$P.E. = \bar{d} = 25$ since we are dealing with paired data for one group/one population
We use $M.E. = t_{\alpha/2} \times S.E.$ since we have computed the standard deviation from the data
To get $t_{\alpha/2}$ value we need df and confidence level
The df is computed as $df = n - 1$ since we are dealing with one population mean. So $df = n - 1 = 12 - 1 = 11$

$t_{\alpha/2}$ value will be the intersection of $df = 11$ and *Confidence Level* = 98%. So $t_{\alpha/2} = 2.718$. Using IDF.T(0.99, 11) in SPSS we obtain t critical value of 2.718
S.E. is computed as $S.E. = \frac{s_d}{\sqrt{n}}$ since we are dealing with matched/paired sample

$$S.E. = \frac{s_d}{\sqrt{n}} = \frac{37.050}{\sqrt{12}} = 10.696$$

We can now compute M.E.

$$M.E. = t_{\alpha/2} \times S.E. = 2.718 \times 10.696 = 29.071728$$

We can now compute I.E.

$$I.E. = P.E. \pm M.E. = 25 \pm 29.071728$$

$$Lower\ Bound = P.E. - M.E. = 25 - 29.071728 = -4.071728$$

$$Upper\ Bound = P.E. + M.E. = 25 + 29.071728 = 54.071728$$

Using SPSS, we first key the paired data into SPSS after creating the variables in the Variable View Window (Figure 8.26)

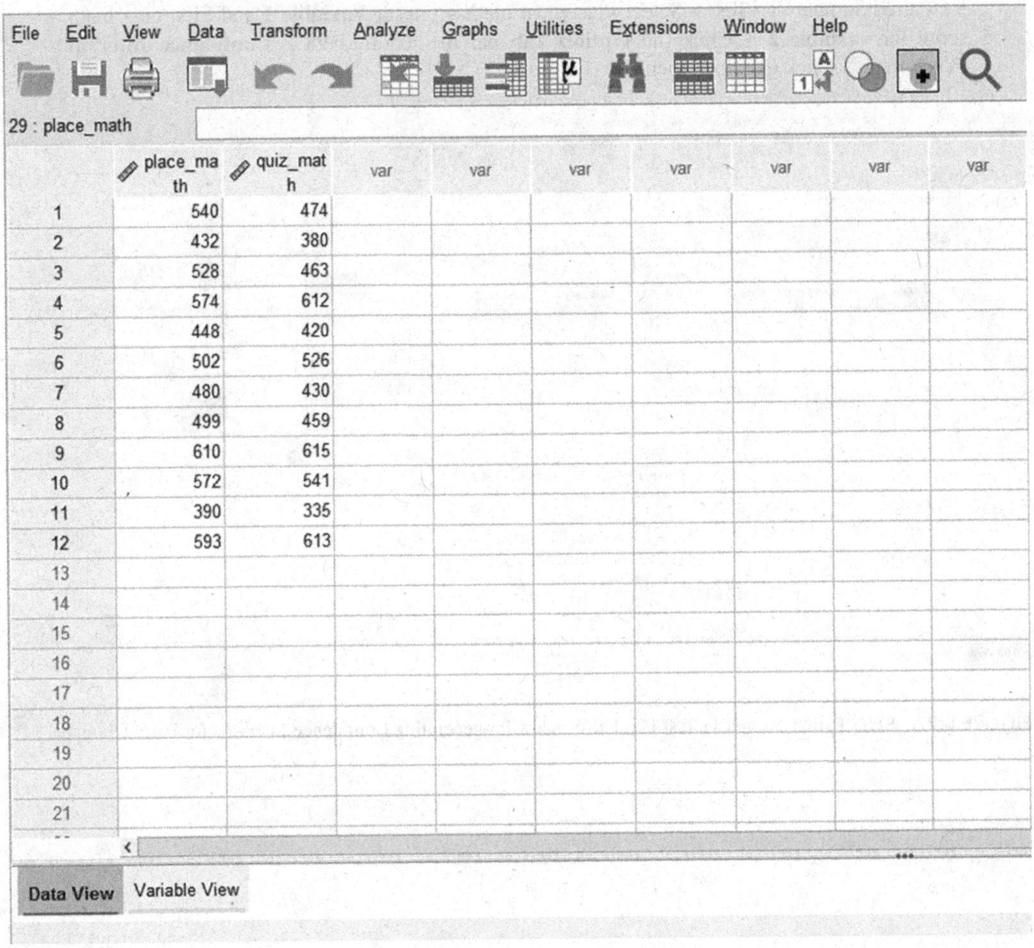

| File | Edit | View | Data | Transform | Analyze | Graphs | Utilities | Extensions | Window | Help |

29 : place_math

	place_math	quiz_math	var	var	var	var	var	var	var
1	540	474							
2	432	380							
3	528	463							
4	574	612							
5	448	420							
6	502	526							
7	480	430							
8	499	459							
9	610	615							
10	572	541							
11	390	335							
12	593	613							
13									
14									
15									
16									
17									
18									
19									
20									
21									

Data View   Variable View

FIGURE 8.26   SPSS Data View displaying the Two Scores

To estimate Point and Interval Estimate for matched/paired means (mean while considering two dependent groups) in SPSS we follow the path:

Analyze > Compare Means > Paired Samples T test. There will be SPSS pop up Window for Paired Samples T –Test > Select Placement math score for **Variable 1** and First quiz math score for **variable 2** > Click the **Option Tab** put the required **98% Confidence Interval Percentage** > Click Continue then OK (Figure 8.27).

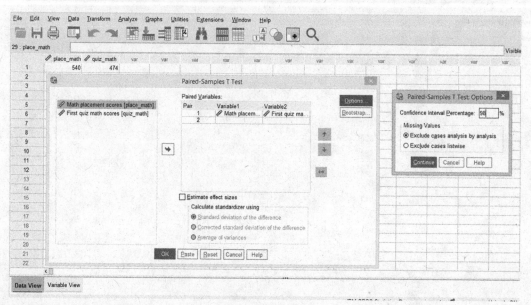

FIGURE 8.27   SPSS Paired-Samples t-Test Pop Up Window for generating Confidence Intervals for Paired Samples

The SPSS Output will give the mean and confidence interval

T-Test

[DataSet4]

		Mean	N	Std. Deviation	Std. Error Mean
		Paired Samples Statistics			
Pair 1	Math placement scores	514.00	12	68.274	19.709
	First quiz math scores	489.00	12	93.466	26.981

		N	Correlation	Sig.
		Paired Samples Correlations		
Pair 1	Math placement scores and first quiz math scores	12	.942	<.001

Paired Samples Test

		Paired Differences							
				Std. Error Mean	98% Confidence Interval of the Difference		t	df	Sig. (2-tailed)
		Mean	Std. Deviation		Lower	Upper			
Pair 1	Math placement scores – First quiz math scores	25.000	37.050	10.696	–4.071	54.071	2.337	11	.039

FIGURE 8.28 Confidence Interval for the Mean difference for Two Population

From the SPSS results/output in we can see that the *P.E.* = Mean = 25.0 and lower bound =–4.071, while the upper bound = 54.071.

We are 98% confident that the difference between placement math score and the first quiz math score is between –4.071 and 54.071 (Figure 8.28)

8.3 Sample Size Determination

To compute the required sample size, we always use the *M.E. formula*. Then substitute the given S.E. of interest and make *n* the subject to estimate the sample size. In SPSS, these formulas can be typed in the syntax window or using Transform > Compute Variable > type in the formula in the Numeric Expression Window.

Sample Size when Estimating μ

$$n = \frac{Z_{\alpha/2}^2 \sigma^2}{E^2}$$

where

E = error of estimation

σ = population standard deviation

n = sample size

Example

A certain stock exchange survey showed that the mean trading volume was while standard deviation for daily trading volume was 4 minutes. Based on these findings another study wants to estimate the population mean trading volumes with a margin of error of 75 seconds, what sample size should be used? Assume 96% confidence.

To compute the sample size we use M.E. formula. We then substitute the formula for S.E. and make n the subject

We use $M.E. = Z_{\alpha/2} \times S.E.$ since the standard deviation is given and we are dealing with mean.

We use S.E. as $S.E. = \frac{\sigma}{\sqrt{n}}$ since we the standard deviation has been given and are dealing with one population mean

We substitute the S.E. formula into the M.E. formula and make n the subject

$$M.E. = Z_{\alpha/2} \times S.E.$$

$$M.E. = Z_{\alpha/2} \times \frac{\sigma}{\sqrt{n}}$$

Making n the subject

$$\sqrt{n} = Z_{\alpha/2} \times \frac{\sigma}{M.E.}$$

Which can be re-written as

$$n = \left(\frac{Z_{\alpha/2} \times \sigma}{M.E.} \right)^2$$

To get the $Z_{\alpha/2}$ value, we divide the confidence level by 2

$$\frac{CL}{2} = \frac{0.96}{2} = 0.48$$

So the Z value equivalent to the area of 0.48 gives us the $Z_{\alpha/2}$ value. So $Z_{\alpha/2} = 2.05$. Alternatively, we use IDF.NORMAL(0.98, 0, 1) in SPSS to obtain z critical value of 2.05

Substituting into the formula for n

$$n = \left(\frac{2.05 \times 4}{1.25} \right)^2 = 43.0336 \approx 43$$

Sample size when Estimating p

$$n = \frac{Z^2 pq}{E^2}$$

where

p = population proportion
$q = 1 - p$
E = error of estimation
n = sample size

Example

A certain research center conducted a telephone survey of 2500 residents to learn about the major issues among the residents. The survey results showed that 1550 of the respondents think that security is a major concern. What sample size would you recommend if the research centers' goal is to estimate the current proportion of residents who think that security is a major concern with a margin of error of 0.04? Use a 98% confidence level.

Surveyed 2500

1550 think it's a major concern

Proportion of those who think security is a major concern: $\bar{p} = \frac{x}{n} = \frac{1550}{2500} = 0.62$

To compute the required sample size, we use the *M.E. formula*. Then substitute the given S.E. of interest and make n the subject to estimate the sample size.

$$M.E. = Z_{\alpha/2} \times S.E.$$

$$S.E. = \sqrt{\frac{p(1-P)}{n}}$$

$$M.E. = Z_{\alpha/2} \times \sqrt{\frac{p(1-P)}{n}}$$

Making n the subject we square both sides to remove the square root

$$M.E.^2 = Z_{\alpha/2}^2 \times \frac{p(1-p)}{n}$$

Therefore, n becomes

$$n = Z_{\alpha/2}^2 \times \frac{p(1-p)}{M.E.^2}$$

To compute $Z_{\alpha/2}$ value

- Compute the $\frac{Confidence\ Level}{2}$ value, i.e. $\frac{0.98}{2} = 0.49$
- Check in the Z table the corresponding Z value of the area equals to 0.49. The Z table value is the $Z_{\alpha/2}$ value. Hence, $Z_{\alpha/2} = 2.32$. Alternatively, we use IDF.NORMAL(0.99, 0, 1) in SPSS to obtain z critical value of 2.32

We know that $p = 0.62$, $M.E. = 0.04$

We substitute into the formula for n

$$n = Z_{\alpha/2}^2 \times \frac{p(1-p)}{M.E.^2} = 2.32^2 \times \frac{0.62(1-0.62)}{0.04^2} = 792.56 \approx 793$$

Practice Exercise

1. Using customer_subset.sav data,
   a. Determine the point and the 98% confidence interval estimate for household income
   b. What sample size would you recommend if another study's goal is to estimate the current mean household income with a margin of error of 0.01? Use a 98% confidence level.

2. Using customer_subset.sav data, determine the point and 93% confidence interval estimate for the difference in household income for some college level of education and post-undergraduate degree education level.

3. Using bankloan.sav data, determine the 95% confidence interval estimate for the proportion of respondents who previously defaulted

4. Using customer_subset.sav data, determine the 95% confidence interval estimate for the difference between proportion of college degree holders customers who defaulted and proportion of college degree holders customers who didn't default on a bank.

5. Consider the following data on the scores on a 10-poit scale for individuals before training and after training

	Scores	
Individual	**After Training**	**Before Training**
1	6	5
2	6	4
3	7	7
4	4	3
5	3	5
6	9	8
7	7	5
8	6	6

a. What is the point estimate for the population mean difference in the scores?

b. Develop a 96% confidence interval for the population mean difference in the scores.

9

Statistical Inference: Hypothesis Testing

9.1 Testing of Hypotheses

Testing Hypothesis (also known as Hypothesis Testing) is part of statistical inference that uses statistic to make decisions and conclusions about hypothesized values of a population.

9.1.1 The Nature of Hypothesis Testing

In hypothesis testing we begin by making a tentative assumption about a population parameter. This tentative assumption is called *the null hypothesis* and is denoted by H_0. We then define another hypothesis, called *the alternative hypothesis*, which is the opposite of what is stated in the null hypothesis. The alternative hypothesis is denoted by H_a. We then take a random sample from the population and subject it to test statistics, and on the basis of the corresponding sample characteristic (test statistics values), we either reject or fail to reject the hypothesis with a particular level of confidence. The hypothesis testing procedure uses data from a sample to test the two competing statements indicated by H_0 and H_a.

In hypothesis testing we will indicate the statistical tests (parametric or nonparametric) to be used to test/perform the analysis of a given hypothesis. The following table provides the comparisons between the parametric and nonparametric tests.

Population of Interest	Parametric Test (Means)	Nonparametric Test (Medians)
One population	One sample t-test	Sign test
		One sample Wilcoxon test
Two population	Independent samples t-test	Mann-Whitney test
Paired samples	Paired samples t-test	Wilcoxon Rank sum test
More than two population	One-way ANOVA	Kruskal Wallis test
Paired samples	Pearson coefficient of correlation	Spearman rank coefficient of correlation

9.1.2 Hypothesis

A hypothesis is a tentative assumption about the distribution of a phenomenon that a research intends to establish. There are three types of hypotheses: *Research hypotheses, Statistical Hypotheses* and *Substantive hypotheses*. A *research hypothesis* is a statement of what the researcher believes will be the outcome of an experiment or a study. Before studies are undertaken, business researchers often have some idea or theory based on experience or previous work as to how the study will turn out. These ideas, theories, or notions established before an experiment or study is conducted are research hypotheses. In order to scientifically test research hypotheses, a more formal hypothesis structure needs to be set up using *statistical hypotheses*. All statistical hypotheses consist of two parts, a null hypothesis and an alternative hypothesis. *Substantive* result is when the outcome of a statistical study produces results that are important to the decision maker.

In order to test hypotheses, business researchers formulate their research hypotheses into statistical hypotheses. There are two types of statistical hypotheses: *Null hypothesis* and *alternative hypothesis*. The null hypothesis, denoted by H_0, is usually the hypothesis that sample observations result purely from chance. Generally, the null hypothesis states that the *"null" condition exists; that is, there is nothing new happening, the old theory is still true, the old standard is correct and the system is in control.* The alternative hypothesis, denoted by H_1 or H_a, is the hypothesis that sample observations are influenced by some nonrandom cause. The alternative hypothesis states that *the new theory is true, there are new standards, the system is out of control and/or something is happening.*

9.2 Forms for Null and Alternative Hypothesis

Depending on the situation, hypothesis tests about a population parameter may take one of three forms: Two use inequalities in the null hypothesis; the third uses an equality in the null hypothesis. For hypothesis tests involving a population mean, we let μ_0 denote the hypothesized value and we must choose one of the following three forms for the hypothesis test.

$$H_0: \mu \geq \mu_0$$
$$H_a: \mu < \mu_0 \tag{i}$$

$$H_0: \mu \leq \mu_0$$
$$H_a: \mu > \mu_0 \tag{ii}$$

$$H_0: \mu = \mu_0$$
$$H_a: \mu \neq \mu_0 \tag{iii}$$

The first two forms are called *one-tailed tests*. The third form is called *a two-tailed test*. A test of a statistical hypothesis, where the region of rejection is on only one side of the sampling distribution, is called a *one-tailed test*. This is a *directional hypothesis*. A test of a statistical hypothesis, where the region of rejection is on both sides of the sampling distribution, is called a *two-tailed test*. This is a *nondirectional hypothesis.*

9.3 Types of Errors in Hypothesis Testing

Any decision we make based on a hypothesis test may be incorrect because we have used partial information obtained from a sample to draw conclusions about the entire population. There are two types of incorrect decisions: *Type I error* and *Type II error. Type I error* occurs when the researcher rejects the null hypothesis when it is true. The probability of committing a Type I error is called the *significance level.* This probability is also called *alpha,* and is often denoted by α. *Type II error* occurs when the researcher fails to reject a null hypothesis that is false. The probability of committing a Type II error is called *Beta,* and is often denoted by β. The probability of *not* committing a Type II error is called the *Power of the test.*

9.4 Decision Rules for Rejecting Null Hypothesis

There are two approaches can be used to make decision about the Null hypothesis: *Critical value approach and p-value approach.*

9.4.1 Critical-Value Approach to Hypothesis Testing

We choose the critical value "cutoff point" based on the significance level (α) of the hypothesis test. These are the $Z_{\alpha/2}$ or $t_{\alpha/2}$ or F_α or χ^2_α values. We then compare the critical value with the corresponding test statistics value (Z-test, t-test, F-test and chi-square test).

How to make decision:

- If the test statistics value (e.g. Z-test) is greater than (>) the critical value (e.g. $Z_{\alpha/2}$) – we reject the null hypothesis (H_0)
- If the test statistics value (e.g. Z-test) is less than (<) the critical value (e.g. $Z_{\alpha/2}$) – we fail to reject the null hypothesis (H_0)

This is the conventional approach for decision making in hypothesis testing.

9.4.2 p-Value Approach to Hypothesis Testing

We estimate the p-value based on the test statistics value (e.g. z-test, t-test, F-test and chi-square test). p-Value is the strength that supports the null hypothesis. *p-Value is usually computed using statistical software like Statistical Package for Social Science (SPSS).* We then compare the p-value with the significance level (α – level) of the test. The conventional method for estimating p-value for Z-test, t-test, F-test and chi-square test has always been done using the respective tables, i.e. using z-test table, using t-test table, using F-test table and using chi-square test table.

> *Using Z – table to obtain the p-values.* p-Value is the extreme area under the curve using z-test value. We estimate the area from the center of the normal distribution to the z-test value, then subtract it from 0.5. If it's one tailed the p-value will be what has been estimated. If its two tailed, then you multiply p-value by 2.
>
> *Using t – table to obtain p-value:* we check along the $df = n - 1$ where the $t-test$ value will fall. It will fall between some two values. Then check up the t-table for the corresponding probability values. These are the p-values. If it's a one tailed test (i.e. a directional hypothesis), then the p-value will be the given range of the probability values. If it's a two tailed test (i.e. a nondirectional hypothesis), then the p-value will be the given range of the probability values multiplied by 2.
>
> *Using F – table to estimate the p-value:* Check along the $df = n - 1, m - 1$ where the F-test value will fall. The corresponding range for the α values where F-test will fall will be the p-value. The p-value is always two tailed (for F-test).
>
> *To obtain p-value:* using the χ^2-table, we check along the df where the $\chi^2 - test$ value will fall. It will fall between some two values. Then check up the χ^2-table for the corresponding probability values. These are the p-values. The p-value will be the given range of the probability values.

How to make decision:

- If the p-value is less than (<) α – level – we reject the null hypothesis (H_0)
- If the p-value is greater than (>) α – level – we fail to reject the null hypothesis (H_0)

We will use this approach to make decision in hypothesis testing.

9.4.3 Steps in Hypothesis Testing

There will be six steps in hypothesis testing both for critical value approach and p-value approach.

9.4.3.1 Steps in the Critical-Value Approach to Hypothesis Testing

Step 1: State the null and alternative hypotheses

Step 2: Decide on/state the significance level, α. If not given, the default $\alpha = 0.05$

Step 3: Compute the value of the test statistic (e.g. Z-test)

Step 4: Determine the critical value (e.g. $Z_{\alpha/2}$)

Step 5: Decision on H_0 using critical value – compare the Z-test with $Z_{\alpha/2}$ value. If the value of the test statistic (Z-test) is greater than the critical value ($Z_{\alpha/2}$), reject H_0; otherwise, do not reject H_0

Step 6: Make conclusions at a given level of confidence based on the decision in Step 5

9.4.3.2 Steps in the p-Value Approach to Hypothesis Testing

Step 1: State the null and alternative hypotheses

Step 2: Decide on/state the significance level, α. If not given, the default $\alpha = 0.05$

Step 3: Compute the value of the test statistic (Z-test, t-test, F-test and Chi-square test)

Step 4: Determine the p-value (output of the analysis from SPSS)

Step 5: Decision on H_0 using p-value – compare the p-value with α value. If p-value $< \alpha$, reject H_0; otherwise, do not reject H_0

Step 6: Make conclusions at a given level of confidence based on the decision in Step 5

Practice Exercise

1. The manager of the Jacaranda Resort Hotel stated that the mean guest bill for a weekend is $600 or less. A member of the hotel's accounting staff noticed that the total charges for guest bills have been increasing in recent months. The accountant will use a sample of future weekend guest bills to test the manager's claim.

   a. Formulate the null and alternative hypotheses that can be used to test the manager's claim.

   b. State the Type I and the Type II error in this situation.

2. A report from KenGen showed that the cost of electricity for an efficient home in a particular neighborhood of Nairobi was Ksh. 1050/= per month. A researcher believes that the cost of electricity for a comparable neighborhood in Mombasa is lower. A sample of homes in this Mombasa neighborhood will be taken and the sample mean monthly cost of electricity will be used to test the following null and alternative hypotheses.

$$H_0: \mu \geq 1050$$

$$H_a: \mu < 1050$$

   a. Assume the sample data led to rejection of the null hypothesis. What would be your conclusion about the cost of electricity in the Mombasa neighborhood?

   b. What is the Type I error in this situation?

   c. What is the Type II error in this situation?

10

Testing Hypothesis about One Population

10.1 Application of Hypothesis Testing

In this section, we will conduct a hypothesis testing procedure by applying the necessary statistical tests in the appropriate steps.

10.1.1 Testing of Hypothesis about One Population

In this case, we will consider one population mean and one population proportion.

a. *Testing Hypothesis about One Population Mean*

In this case we are testing to determine if the population mean is significantly different/greater/ less than the hypothesized population value.

i. If σ is known/given, then we will use one-sample Z-test

ii. If σ is unknown/not given, then we will use one-sample t-test

One-Sample t-test is used to determine if a given quantitative/numerical variable is statistically significantly different from a given hypothesized population mean value if σ is unknown/not given. For example, if the mean weight of school going age is significantly different from 45 kg. *One-Sample Z-test* is used to determine if a given quantitative/numerical variable is statistically significantly different from a given hypothesized population mean value if σ is known/ given. For example, if the mean weight of school going age is significantly different from 45 kg but σ is known/given.

Step 1: State the null and alternative hypotheses.

The Null Hypothesis is H_0: $\mu = \mu_0$ and the alternative hypothesis is H_a: $\mu \neq \mu_0$ (two tailed)

Or

The Null Hypothesis is H_0: $\mu \geq \mu_0$ and the alternative hypothesis is H_a: $\mu < \mu_0$ (left/one tailed)

Or

The Null Hypothesis is H_0: $\mu \leq \mu_0$ and the alternative hypothesis is H_a: $\mu > \mu_0$ (right/one tailed)

Step 2: Decide on/State the significance level, α. If not given, the default $\alpha = 0.05$.

Step 3: Compute the value of the test statistic – *One-Sample z–test (if σ is given/known).*

$$z - test = \frac{\bar{x} - \mu_0}{S.E.}$$

$$S.E. = \frac{\sigma}{\sqrt{n}}$$

where $\bar{x}$ is the mean from the sample, μ_0 is the hypothesized mean (mean value in the hypothesis), σ is the given/known standard deviation, n is the sample.

DOI: 10.1201/9781003292654-13

In Statistical Package for Social Science (SPSS), we compute the one-sample Z-test using syntax since Z-test is not inbuilt within SPSS with drop down menu. What one need to do is to replace the values in the syntax.

File > New > Syntax > SPSS syntax window will populate > Copy and paste the SPSS syntax but replace the values in the following order: sample size, sample mean, population mean and standard deviation.

**The first number is the sample size (n = 35), the second number is the sample mean (105),

**the third number is the population mean (100)

**and the fourth number is the population standard deviation (15)

**Replace the four values below with your own.

**The p-value that is reported is based on a two-tailed test.

**To obtain the one-tailed p-value, simply divide the two-tailed p-value by two.

data list list/n sample_mean population_mean population_sd.

begin data

35   105   100   15

end data.

```
Compute mean_difference = sample_mean - population_mean.
Compute square_root_n =SQRT(n).
Compute standard_difference = population_sd/square_root_n.
Compute z_statistic = mean_difference/standard_difference.
Compute chi_square = z_statistic*z_statistic.
Compute p_value = SIG.CHISQ(chi_square, 1).
Compute cohens_d = mean_difference/population_sd.
EXECUTE.
Formats z_statistic p_value cohens_d (f8.5).
LIST z_statistic p_value cohens_d.
```

Select the Syntax and click Run Selection

The SPSS output will have the Z-test value for Step 3 and p-value for Step 4.

Step 4: Estimate the p-value, p or z critical value. Z critical value can be estimated in SPSS using IDF.NORMAL(prob, 0, 1).

State the p-value from the SPSS output.

Step 5: Decision on H_0: we use p-value to make decision on H_0.

If p-value < α, reject H_0; otherwise, do not reject H_0.

We can also use the critical value to make decision on H_0 i.e. if z-test > z critical value, reject H_0; otherwise fail to reject H_0. .

Step 6: Make conclusions at a given level of confidence based on the decision in Step 5.

Incase σ *is unknown/not given*, we then use *One-Sample t-test* for Step 3 given by

$$t - test = \frac{\bar{x} - \mu_0}{S.E.}$$

$$df = n - 1$$

The t–test can be generated via SPSS.

In SPSS, we compute the one-sample t-test using the path:

Analyze > Compare Means > One-Sample T test > There will be SPSS pop up window for One-Sample T test > Select **the variable of interest** as **the Test Variable** > Put the **Test Value** (i.e. the hypothesized population mean values) > Click the **Option Tab** put the required **Confidence Interval Percentage (e.g. 96% since $\alpha = 0.04$)** that matches the **given α – level** > then click Continue then OK

This will give use the t-test value, df, and the p-value for Steps 3 and 4. We can also estimate the t critical value for step 4 in SPSS using IDF.T(prob,df)

Step 5: Decision on H_0 using p-value:

If $P \leq \alpha$, reject H_0; otherwise, do not reject H_0.

We can also use the critical value to make decision on H_0 i.e. if t-test > t critical value, reject H_0; otherwise fail to reject H_0.

Step 6: Make conclusions at a given level of confidence based on the decision in Step 5.

Example 1: Two-Tailed Test

A car inventor developed a new fuel-economical car. The car inventor claims that the new fuel-economical car will can travel continuously for 5 hours (300 minutes) on a 20 liters of petrol. The inventor selects a sample of 50 newly developed fuel-economical car for testing. The new fuel-economical car run for an average of 295 minutes, with a standard deviation of 20 minutes. Test the null hypothesis that the mean run time is 300 minutes against the alternative hypothesis that the mean run time is not 300 minutes. Use a 0.05 level of significance.

SOLUTION:

Step 1: State the Null and Alternative Hypothesis

Null hypothesis (H_0): $\mu = 300$

Alternative hypothesis (H_1): $\mu \neq 300$

This is a two-tailed test.

Step 2: State the significance level for the test, $\alpha = 0.05$.

Step 3: Compute the value of the test statistic: we will use a *one-sample Z-test* since σ is known or given and the mean for one group has also been given

$$\bar{x} = 295, \ \mu_0 = 300, \ \sigma = 20, \ n = 50,$$

$$Z - test = \frac{(\bar{x} - \mu_0)}{S.E.}$$

$$S.E. = \frac{\sigma}{\sqrt{n}}$$

$$S.E. = \frac{20}{\sqrt{50}} = 2.8284$$

$$Z - test = \frac{(295 - 300)}{2.8284} = -1.77$$

We ignore the sign of the Z-test value, $Z - test = 1.77$.

In SPSS, we compute the one-sample Z-test using syntax since Z-test is not inbuilt within SPSS with drop down menu. What one need to do is to replace the values of the sample size, sample mean, population mean and standard deviation in the syntax (Figure 10.1).

File > New > Syntax > SPSS syntax window will populate > (Figure 10.1).

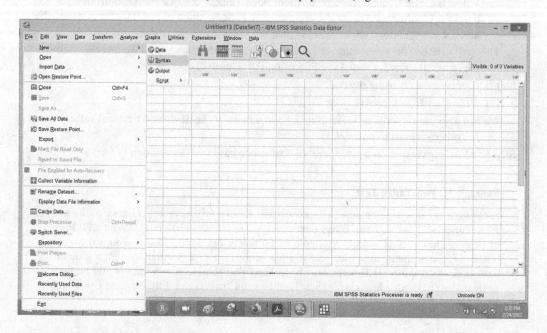

FIGURE 10.1   SPSS Menu for Syntax Window

Copy and paste the SPSS syntax but replace the values in the following order: sample size, sample mean, population mean and standard deviation (Figure 10.2).

**The first number is the sample size ($n = 35$), the second number is the sample mean (105),

**the third number is the population mean (100)

**and the fourth number is the population standard deviation (15)

**Replace the four values below with your own.

**The p value that is reported is based on a two-tailed test.

**To obtain the one-tailed p-value, simply divide the two-tailed p-value by two.

```
data list list/n sample_mean population_mean population_sd.
begin data
50   295   300   20
end data.

Compute mean_difference = sample_mean - population_mean.
Compute square_root_n =SQRT(n).
Compute standard_difference = population_sd/square_root_n.
Compute z_statistic = mean_difference/standard_difference.
Compute chi_square = z_statistic*z_statistic.
Compute p_value = SIG.CHISQ(chi_square, 1).
Compute cohens_d = mean_difference/population_sd.
EXECUTE.
Formats z_statistic p_value cohens_d (f8.5).
LIST z_statistic p_value cohens_d.
```

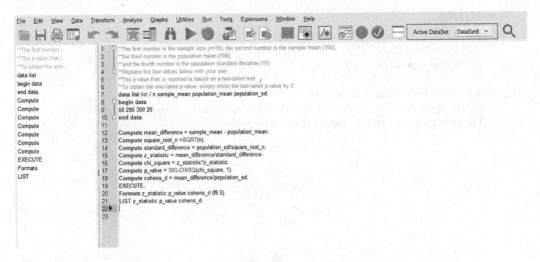

FIGURE 10.2 SPSS Syntax for One Sample Z-test for One Population Mean

Select the Syntax and click Run Selection
The SPSS output will have the Z-test value for Step 3 and p-value for Step 4 (Figure 10.3).

```
z_statistic  p_value cohens_d

 -1.76777   .07710 -.25000


Number of cases read:  1    Number of cases listed:  1
```

FIGURE 10.3 Z-test Value for One Population Mean

We ignore the sign of the Z-test value, $Z-test = 1.77$
Step 4: Determine the p-value
From the above SPSS output the p-value = 0.07710. Since it's a two-tailed test, we take the
p-value as it is
Step 5: Decision on H_0 using p-value: We compare p-value with alpha level for the test, $\alpha = 0.05$
To make decision: If the p-value is less than α value, reject H_0; otherwise, do not reject H_0.
Since p-value is greater than (>) α value, we fail to reject H_0.
Step 6: Make conclusions at a given level of confidence based on the decision in Step 5.
At 95% level of confidence, the car inventor's claim was true.

Example 2: One-Tailed Test

The statistics lecturer for Business Statistics class thinks that the average Quiz One test score is
at least 110. To prove the point, the statistics lecturer administers Quiz One test to 20 randomly
selected students. Among the sampled students, the average Quiz One test score is 108 with a
standard deviation of 10. Based on these results, should the statistics lecturer reject or fail to
reject her original hypothesis? Assume a significance level of 0.01.

SOLUTION:

Step 1: State the Null and Alternative Hypothesis
  Null hypothesis: $H_0: \mu \geq 110$
  Alternative hypothesis: $H_a: \mu < 110$
  This is a one-tailed test.
Step 2: State the significance level: $\alpha = 0.01$
Step 3: Compute the value of the test statistic: we will use a *one-sample Z-test* since σ is
  known or given and we have mean value for one group

$$\bar{x} = 108, \; \mu_0 = 110, \; \sigma = 10, \; n = 20,$$

$$Z - test = \frac{(\bar{x} - \mu_0)}{S.E.}$$

$$S.E. = \frac{\sigma}{\sqrt{n}}$$

$$S.E. = \frac{10}{\sqrt{20}} = 2.236$$

$$Z - test = \frac{(108 - 110)}{2.236} = -0.894$$

We ignore the sign of the Z-test value, $Z - test = 0.894$

In SPSS, we compute the one-sample Z-test using syntax since Z-test is not inbuilt within
  SPSS with drop down menu. What one need to do is to replace the values in the syntax
  (Figure 10.4).

File > New > Syntax > SPSS syntax window will populate >

Copy and paste the SPSS syntax but replace the values in the following order: sample size,
  sample mean, population mean and standard deviation

**The first number is the sample size ($n = 10$), the second number is the sample mean
  (108),

**the third number is the population mean (110)

**and the fourth number is the population standard deviation (10)

**Replace the four values below with your own.

**The p value that is reported is based on a two-tailed test.

**To obtain the one-tailed p-value, simply divide the two-tailed p-value by two.

data list list/n sample_mean population_mean population_sd.
begin data
20   108   110   10
end data.

```
Compute mean_difference = sample_mean - population_mean.
Compute square_root_n =SQRT(n).
Compute standard_difference = population_sd/square_root_n.
Compute z_statistic = mean_difference/standard_difference.
Compute chi_square = z_statistic*z_statistic.
Compute p_value = SIG.CHISQ(chi_square, 1).
Compute cohens_d = mean_difference/population_sd.
EXECUTE.
Formats z_statistic p_value cohens_d (f8.5).
LIST z_statistic p_value cohens_d.
```

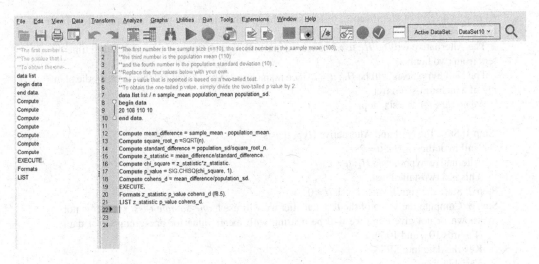

FIGURE 10.4 SPSS Syntax for One Sample Z-test for One Population Mean

Select the Syntax and click Run Selection

The SPSS output will have the Z-test value for Step 3 and p-value for Step 4 (Figure 10.5).

```
z_statistic  p_value cohens_d

  -.89443    .37109  -.20000


Number of cases read:  1    Number of cases listed:  1
```

FIGURE 10.5 Z-test Value for One Population Mean

We ignore the sign of the Z-test value, $Z-test = 0.89443$

Step 4: Determine the p-value

From the above SPSS output the p-value = 0.37109. Since it's a one-tailed test, we divide the p-value by two, hence p-value = 0.185545.

Step 5: Decision on H_0 using p-value: We compare p-value with alpha level for the test, $\alpha = 0.01$

To make decision: If the p-value is less than α value, reject H_0; otherwise, do not reject H_0.

Since p-value is greater than (>) α value, we fail to reject H_0.

Step 6: Make conclusions at a given level of confidence based on the decision in Step 5.

At 99% level of confidence, the statistics lecturer's claim that the average Quiz One test score for Business Statistics class is at least 110 was true.

Example 3: Two-tailed test

Consider the following data on tree planting labor time (times in hours) for trees designed for landscaping residential homes.

1.7   1.5   2.6   2.2   2.4   2.3   2.6   3.0   1.4   2.3   1.9   2.1

At $\alpha = 0.04$ level of significance, test to see whether the mean tree-planting time differs from two hours.

The Alternative will be $H_1: \mu \neq 2$, since we are finding out if the mean tree-planting time differs from two hours.

The Null hypothesis will be $H_0: \mu = 2$ since managers use two hours of labor time for the planting of a medium-sized tree.

We go through the six steps

Step 1: State the Null and Alternative Hypothesis
   Null hypothesis: $H_0: \mu = 2$
   Alternative hypothesis: $H_a: \mu \neq 2$
   This is a two-tailed test.
Step 2: State the significance level: $\alpha = 0.04$
Step 3: Compute the value of the test statistic: we will use a *one-sample t-test* since σ is not known or not given and we will be dealing with mean value for one group/set of data (Figures 10.6 and 10.7).
   Key the data into SPSS
   Variable View

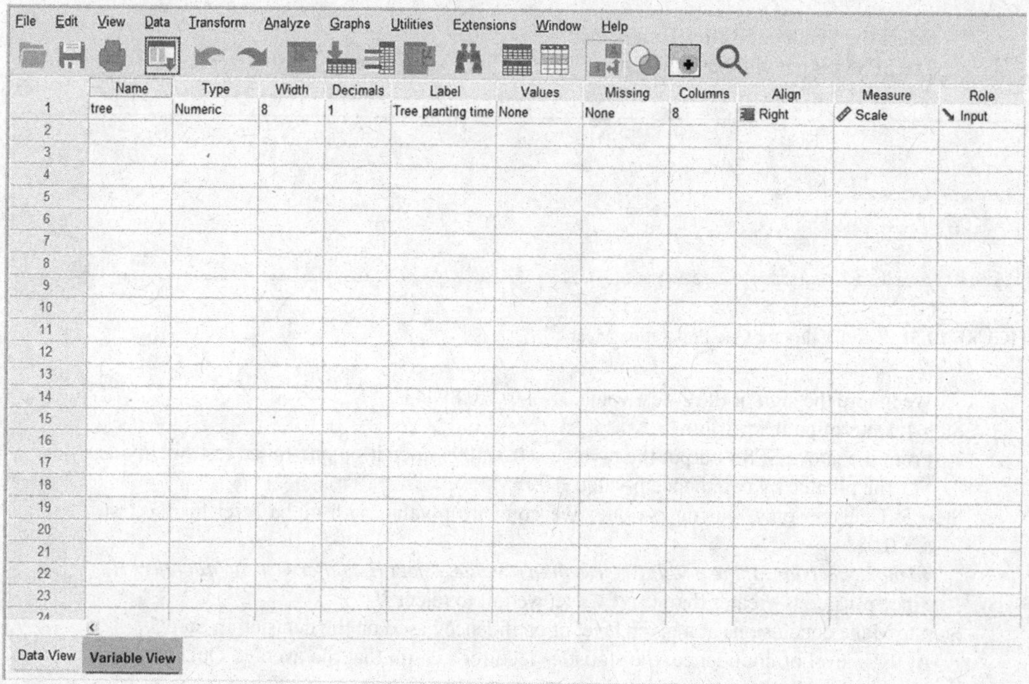

	Name	Type	Width	Decimals	Label	Values	Missing	Columns	Align	Measure	Role
1	tree	Numeric	8	1	Tree planting time	None	None	8	≣ Right	⌀ Scale	↘ Input
2											
3											
4											
5											
6											
7											
8											
9											
10											
11											
12											
13											
14											
15											
16											
17											
18											
19											
20											
21											
22											
23											
24											

Data View   **Variable View**

FIGURE 10.6   SPSS Variable View Window for Tree Planting Time

Data View

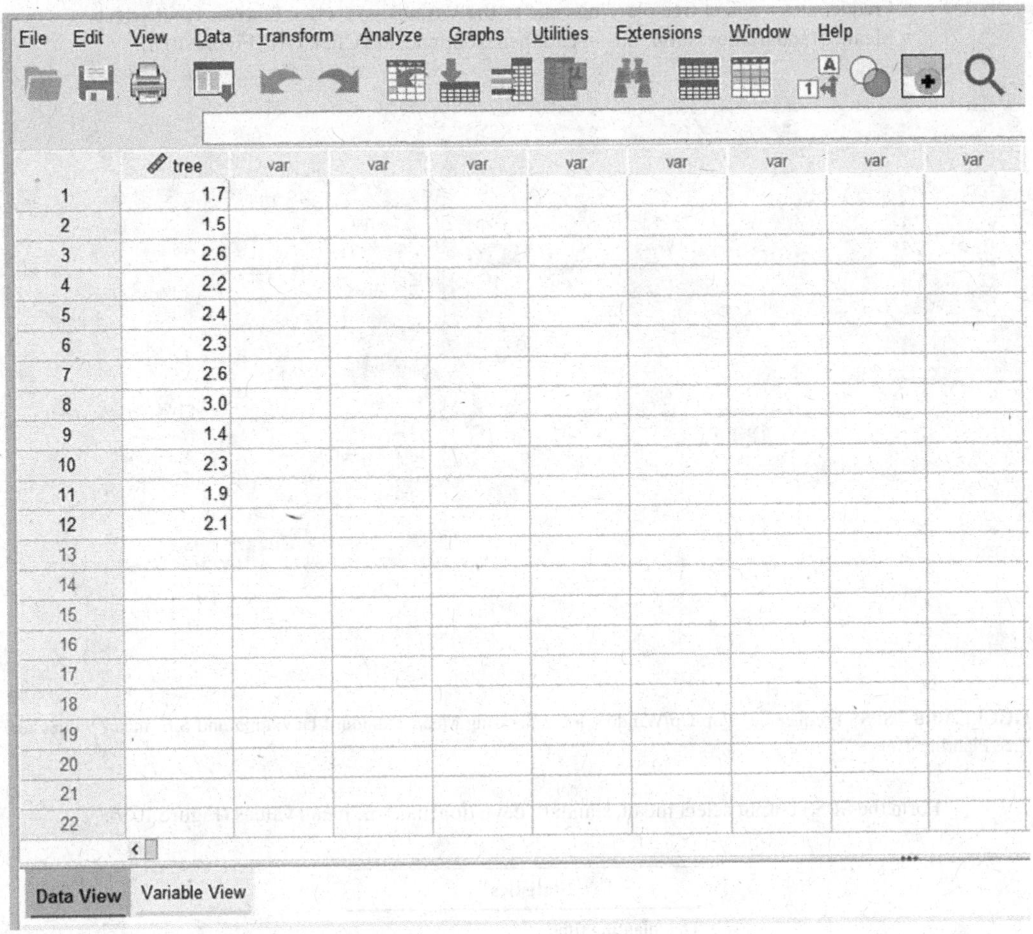
File Edit View Data Transform Analyze Graphs Utilities Extensions Window Help
tree var var var var var var var var
1 1.7
2 1.5
3 2.6
4 2.2
5 2.4
6 2.3
7 2.6
8 3.0
9 1.4
10 2.3
11 1.9
12 2.1
13
14
15
16
17
18
19
20
21
22
Data View Variable View

FIGURE 10.7 SPSS Data View Window for Tree Planting Time

We use SPSS to compute mean, standard deviation and S.E. mean: we go to Analyze > Descriptive Statistics > Frequencies > there will be SPSS pop up window for Frequencies > select tree planting time as the Variable(s) > click Statistics and select Mean, Standard deviation, S.E. mean then Continue and Click OK (Figure 10.8).

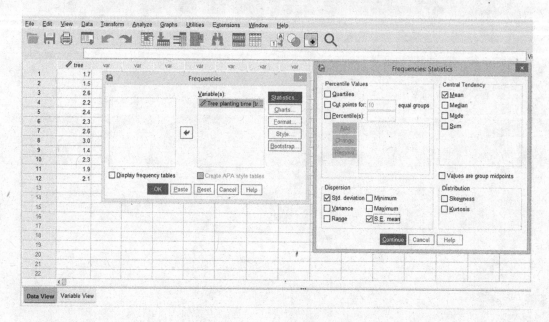

FIGURE 10.8 SPSS Frequencies Pop Up Window for generating Mean, Standard Deviation and S.E. mean Values for Tree Planting Time

Form the SPSS output select mean, standard deviation and S.E. mean values (Figure 10.9).

Statistics		
Tree planting time		
N	Valid	12
	Missing	0
Mean		2.167
Std. Error of Mean		.1373
Std. Deviation		.4755

FIGURE 10.9 Mean, Standard Deviation and S.E. mean Values for Tree Planting Time

Hence, $\bar{x} = 2.167$, $\mu_0 = 2$, $s = 0.4755$, $S.E. = 0.1373$ and $n = 12$.
We compute t-test as:

$$t-test = \frac{(\bar{x} - \mu_0)}{S.E.}$$

$$S.E. = \frac{s}{\sqrt{n}} = \frac{0.4755}{\sqrt{12}} = 0.1373$$

$$t-test = \frac{(2.167 - 2)}{0.1373} = 1.2163$$

$$df = n - 1 = 12 - 1 = 11$$

In SPSS, we compute the one-sample t-test using the path (Figures 10.10 and 10.11):
Analyze > Compare Means > One-Sample T test > There will be SPSS pop up window
for One-Sample T test > Select **the variable of interest (Tree planting time)** as **the
Test Variable** > Put the **Test Value** (what the hypothesis is testing = 2) > Click the
Option Tab put the required **Confidence Interval Percentage (96% since α = 0.04)**
that matches the **given α – level** > then click Continue then OK.

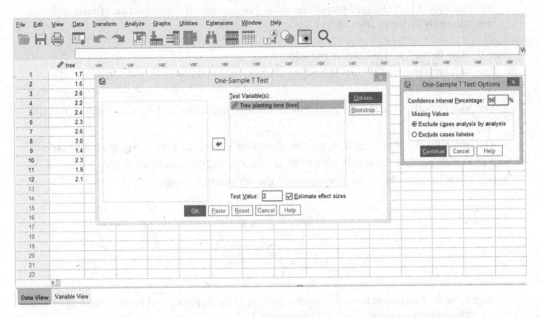

FIGURE 10.10   SPSS One-Sample t-Test Pop Up Window for generating t-Test Value for One Population Mean

The SPSS Output.

One-Sample Test

Test Value = 2

	t	df	Sig. (2-tailed)	Mean Difference	96% Confidence Interval of the Difference	
					Lower	Upper
Tree planting time	1.214	11	.250	.1667	−.153	.486

FIGURE 10.11   t-Test Value for One Sample Mean

Select the *t-test value* (t), *degree of freedom* (df) and *p-value* [Sig. (2-tailed)] from the
SPSS Output.

$$t-test = 1.214, \ df = 11$$

Step 4: Determine the p-value
Using t distribution table, we check along df = 11 where t-test value = 1.2163 will fall and
look up for the probability values.
Using the SPSS output above, the p-value is the value of Sig. Hence, p-value = 0.250
Step 5: Decision on H_0 using p-value: We compare p-value = 0.250 with $\alpha = 0.04$
To make decision: If the p-value is less than α value, reject H_0; otherwise, do not reject H_0.
Since p-value is greater than (>) α value, we fail to reject H_0.

Step 6: Make conclusions at a given level of confidence based on the decision in Step 5.
   At 96% level of confidence, the mean tree planting time is not different from 2 hours that
      the managers estimated.

b. ***Testing Hypothesis about One Population Proportion***

   One-Sample Z-test is used to determine if a given proportion is statistically significantly different from a given hypothesized proportion value. For example, if the proportion of male students
   is significantly different from 35%.

Step 1: State the null and alternative hypotheses.

   The Null Hypothesis is H_0: $p = p_0$ and the alternative hypothesis is H_a:$p \neq p_0$ (two tailed).
   Or

   The Null Hypothesis is H_0: $p \geq p_0$ and the alternative hypothesis is H_a:$p < p_0$ (left/one tailed).
   Or

   The Null Hypothesis is H_0: $p \leq p_0$ and the alternative hypothesis is H_a:$p > p_0$ (right/one tailed).

Step 2: Decide on/state the significance level, α. If not given, the default $\alpha = 0.05$.

Step 3: Compute the value of the test statistic – *One Proportion Z-test.*

$$Z - test = \frac{p - p_0}{S.E.}$$

$$S.E. = \sqrt{\frac{p_0(1 - p_0)}{n}}$$

   where p is the proportion from the sample data, p_0 is the hypothesized proportion (proportion
      value in the hypothesis), and n is the sample.

   To estimate Z-test for one population proportion in SPSS we follow the path:

   First we determine the proportion of interest from the categorical variable using Analyze >
      Descriptive Statistics > there will be a pop up SPSS window for Frequencies > Select
      the Categorical Variable of Interest and put it in the Variable(s) window > Tick Display
      Frequency Table then OK. The SPSS output will give a frequency table with the proportions/
      percentages.

   Analyze > Compare Means > One-Sample Proportions. There will be a pop up SPSS
      Window for One-Sample Proportions > Select **the categorical variable of interest** as **the
      Test Variable** > Click **Define Success and select the proportion of interest** > Click the
      Confidence Level Tab put the required **Confidence Interval Percentage** > Click Tests >
      Select the different tests and change the Test Value to have the hypothesized value (0.8) >
      Click Continue then OK.

   From the SPSS Output you select *Z-test value* and *p-value.*

Step 4: Determine the p-value.

   From the SPSS output you select the p-value or z critical value. Z critical value can be estimated
      in SPSS using IDF.NORMAL(prob, 0, 1).

Step 5: Decision on H_0 using p-value: We compare p-value = 0.250 with $\alpha = 0.04$

   To make decision: If the p-value is less than α value, reject H_0; otherwise, do not reject H_0.

   We can also use the critical value to make decision on H_0 i.e. if z-test > z critical value, reject
      H_0; otherwise fail to reject H_0.

Step 6: Make conclusions at a given level of confidence based on the decision in Step 5.

Example 1: Two-Tailed Test

The Chief Executive Officer (CEO) of a car sales company claimed that 80% of the vehicles types are automobiles. A local research firm conducted a survey and the survey results are in SPSS data (car_sales.sav). Based on these findings, can we reject the CEO's hypothesis that 80% of the vehicles are automobiles? Use a 0.05 level of significance.

SOLUTION:

The six steps of hypothesis testing is also followed.
Step 1: State the null and alternative hypotheses.
  Null hypothesis (H_0): $p = 0.80$
  Alternative hypothesis (H_1): $p \neq 0.80$
  This is a two-tailed test.
Step 2: Decide on/State the significance level: $\alpha = 0.05$.
Step 3: Compute the value of the test statistic – *One-Sample Proportion Z-test*.
  Using SPSS, we first generate a frequency table using (Figure 10.12).
  Analyze > Descriptive Statistics > there will be a pop up SPSS window for Frequencies >
    Select the Categorical Variable of Interest (**Vehicles Type**) and put it in the Variable(s)
    window > Tick Display Frequency Table then OK.

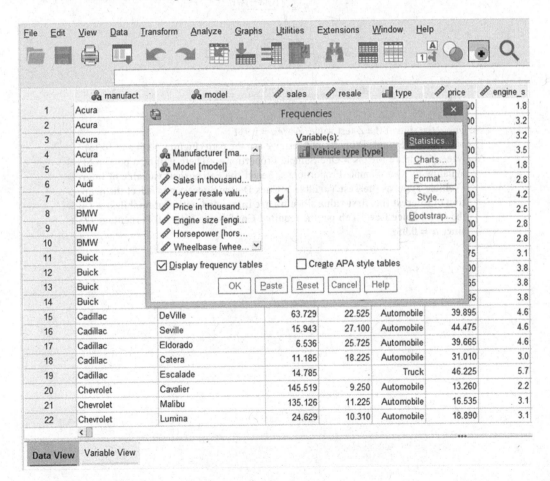

FIGURE 10.12   SPSS Pop Up Window for Frequencies

The SPSS output will give a frequency table with the proportions/percentages (Figure 10.13).

Vehicle Type

		Frequency	Percent	Valid Percent	Cumulative Percent
Valid	Automobile	116	73.9	73.9	73.9
	Truck	41	26.1	26.1	100.0
	Total	157	100.0	100.0	

FIGURE 10.13 Frequency Table for Vehicle Type

From the SPSS output (for Automobile), $p = 0.739$, $p_0 = 0.80$, $n = 157$
These values can be substituted into the formula below to get Z-test value

$$Z - test = \frac{p - p_0}{S.E.}$$

$$S.E. = \sqrt{\frac{p_0(1-p_0)}{n}}$$

$$S.E. = \sqrt{\frac{0.80 \times 0.20)}{157}} = 0.03192$$

$$Z - test = \frac{0.739 - 0.80}{0.03192} = -1.911$$

We ignore the sign of the Z-test value, $Z - test = 1.911$
To compute One-Sample Proportion Z-test, we use the path (Figures 10.14 and 10.15)
Analyze > Compare Means > One-Sample Proportions > There will be a pop up SPSS
   Window for One-Sample Proportions > Select **the categorical variable of interest
   (Vehicle Type)** as **the Test Variable** > Click **Define Success and select the propor-
   tion of interest (i.e. first value since we are interested with Automobiles)** > Click
   the **Confidence Level Tab** put the required **Confidence Interval Percentage (95%**
   since $\alpha = 0.05$)

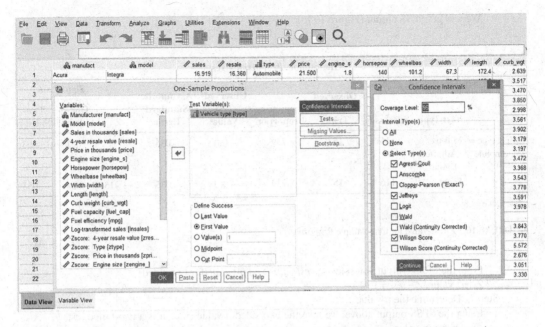

FIGURE 10.14 SPSS One-Sample Proportion Pop Up Window for generating Z-test Value for Single Proportion

Click Tests > Select the different tests and change the Test Value to have the hypothesized value (0.8) > Click Continue then OK

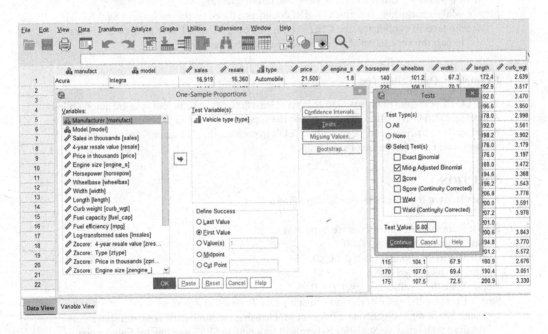

FIGURE 10.15 SPSS One-Sample Proportion Pop Up Window displaying the Tests for Single Proportion

We will get SPSS output (Figure 10.16).

	Test Type	Successes	Observed Trials	Proportion	Observed-Test Value[a]	Asymptotic Standard Error	Z	Significance One-Sided p	Significance Two-Sided p
Vehicle type = Automobile	Mid-p Adjusted Binomial	116	157	.739	−.061	−.035		−.031	−.063
	Score	116	157	.739	−.061	−.035	−1.915	−.028	−.055

One-Sample Proportions Tests

[a] Test value = .8

FIGURE 10.16   Z-test Value for One Sample Proportion

From the SPSS Output you select *Z-test value* and *p-value* (for two sided)
Z-test value, $Z-test = 1.915$

Step 4: Determine the p-value
Using the SPSS output above, we take the two sided p-value since it is a two-tailed test.
Hence, p-value = 0.055.

Step 5: Decision on H_0 using p-value: We compare p-value = 0.0550 with $\alpha = 0.05$
To make decision: If the p-value is less than α value, reject H_0; otherwise, do not reject H_0.
Since p-value is greater than (>) α value, we fail to reject H_0.

Step 6: Make conclusions at a given level of confidence based on the decision in Step 5.
At 95% level of confidence, the CEO's claim was true, i.e. the claim that 80% of his customers are very satisfied with the service they receive is true.

Example 2: One-Tailed Test

The CEO of a car sales company claimed that at least 80% of the vehicles types are automobiles. A local research firm conducted a survey and the survey results are in SPSS data (car_sales.sav). Based on these findings, can we reject the CEO's hypothesis that at least 80% of the vehicles are automobiles? Use a 0.05 level of significance.

SOLUTION:

The six steps of hypothesis testing is also followed

Step 1: State the null and alternative hypotheses.
Null hypothesis (H_0): $p \geq 0.80$
Alternative hypothesis (H_1): $p < 0.80$
This is a one-tailed test.

Step 2: State the significance level: $\alpha = 0.05$.

Step 3: Compute the value of the test statistic – *One Proportion Z-test*.
Using SPSS, we first generate a frequency table using the same path as shown above
Analyze > Descriptive Statistics > there will be a pop up SPSS window for Frequencies >
Select the Categorical Variable of Interest (Vehicles Type) and put it in the Variable(s) window > Tick Display Frequency Table then OK.

The SPSS output will give a frequency table with the proportions/percentages.

		Frequency	Percent	Valid Percent	Cumulative Percent
	Vehicle Type				
Valid	Automobile	116	73.9	73.9	73.9
	Truck	41	26.1	26.1	100.0
	Total	157	100.0	100.0	

Frequency Table for Vehicle Type

From the SPSS output (for Automobile), $p = 0.739$, $p_0 = 0.80$, $n = 157$
These values can be substituted into the formula below to get Z-test value.

$$Z - test = \frac{p - p_0}{S.E.}$$

$$S.E. = \sqrt{\frac{p_0 (1 - p_0)}{n}}$$

$$S.E. = \sqrt{\frac{0.80 \times 0.20)}{157}} = 0.03192$$

$$Z - test = \frac{0.739 - 0.80}{0.03192} = -1.911$$

We ignore the sign of the Z-test value, $Z - test = 1.911$
To compute One-Sample Proportion Z-test we use the path as above
Analyze > Compare Means > One-Sample Proportions > There will be a pop up SPSS
Window for One-Sample Proportions > Select **the categorical variable of interest (Vehicle
Type) as the Test Variable >** Click **Define Success and select the proportion of interest
(i.e. first value since we are interested with automobiles) >** Click the **Confidence Level
Tab** put the required **Confidence Interval Percentage (95% since $\alpha = 0.05$)**

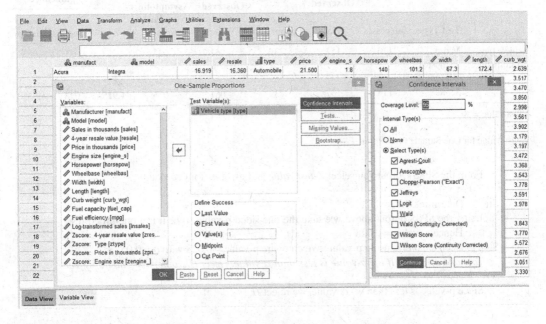

SPSS One-Sample Proportion Pop Up Window for generating Z-test Value for Single Proportion

Click Tests > Select the different tests and change the Test Value to have the hypothesized value (0.8) > Click Continue then OK

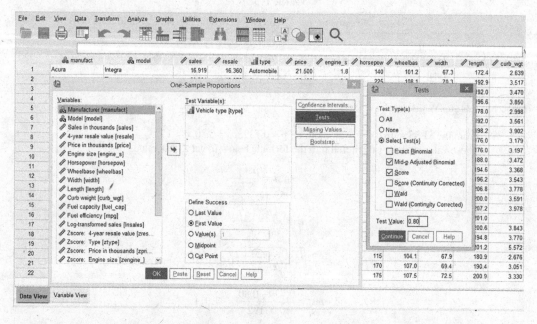

SPSS One-Sample Proportion Pop Up Window displaying the Tests for Single Proportion

We will get SPSS output

One-Sample Proportions Tests

	Test Type	Successes	Observed Trials	Proportion	Observed-Test Value[a]	Asymptotic Standard Error	Z	Significance One-Sided p	Two-Sided p
Vehicle type = Automobile	Mid-p Adjusted Binomial	116	157	.739	−.061	−.035		−.031	−.063
	Score	116	157	.739	−.061	−.035	−1.915	−.028	−.055

[a] Test value = .8

Z-test Value for One Sample Proportion

From the SPSS Output you select *Z-test value* and *p-value* (for one sided)
Z-test value, $Z-test = 1.915$
Step 4: Determine the p-value
Using the SPSS output above, we take the one sided p-value since it is a one-tailed test.
Hence, p-value = 0.028
Step 5: Decision on H_0 using p-value: We compare p-value = 0.028 with $\alpha = 0.05$
To make decision: If the p-value is less than α value, reject H_0; otherwise, do not reject H_0.
Since p-value is less than (<) α value, we reject H_0.

Step 6: Make conclusions at a given level of confidence based on the decision in Step 5.
   At 95% level of confidence, the CEO's claim was not true, i.e. the claim that at least 80% of the vehicles are automobiles is not true.
   We can also use SPSS Syntax to compute z-test if the proportion value and sample value have been given.

10.2 Practice Exercise

1. EPZ Company is considering a promotion that involves mailing discount coupons to all their credit card customers. This promotion will be considered a success if more than 18% of those receiving the coupons use them. Before going national with the promotion, coupons were sent to a sample of 100 credit card customers found that 12% of those receiving the coupons use them. Use $\alpha = 0.02$ to conduct your hypothesis test. Should EPZ go national with the promotion? (Use p-value and critical value approach).

2. Using SPSS data (*bankloan.sav*), determine if the mean credit debt is significantly different from 2.25 at $\alpha = 0.05$ level of significance. (Use SPSS and p-value approach).

11

Testing Hypothesis about Two Populations

11.1 Application of Hypothesis Testing

In this section, we will conduct a hypothesis testing procedure by applying the necessary statistical tests in the appropriate steps.

11.1.1 Testing of Hypothesis about Two Populations

In this section, we will conduct a hypothesis testing procedure for two populations by applying the necessary statistical tests in the six steps of hypothesis testing.

a. *Hypothesis Test for Two Population Means*

In this case, we will consider two population means and two population proportions.

In this case, we are testing to determine if the population mean is significantly different/greater/less than the hypothesized population value.

i. If σ_1 and σ_2 are given/known, then we use a two-sample Z-test.

ii. If σ_1 and σ_2 are not given but computed, then we use a two-sample independent t-test.

Two-sample independent t-test is used to determine if there is a statistically significant difference in mean for a given quantitative/numerical variable between two independent groups when σ_1 and σ_2 are not given/not known. For example, to determine if salary is statistically significant different between males and females if σ_1 and σ_2 are not given/not known.

Two-sample Z t-test is used to determine if there is a statistically significant difference in mean for a given quantitative/numerical variable between two independent groups when σ_1 and σ_2 are given/known. For example, to determine if salary is statistically significant different between males and females if σ_1 and σ_2 are given/known.

Step 1: State the null and alternative hypotheses.

The Null Hypothesis is H_0: $\mu_1 - \mu_2 = \mu_0$ and the alternative hypothesis is H_a: $\mu_1 - \mu_2 \neq \mu_0$ (two tailed).

Or

The Null Hypothesis is H_0: $\mu_1 - \mu_2 \geq \mu_0$ and the alternative hypothesis is H_a: $\mu_1 - \mu_2 < \mu_0$ (left/one tailed).

Or

The Null Hypothesis is H_0: $\mu_1 - \mu_2 \leq \mu_0$ and the alternative hypothesis is H_a: $\mu_1 - \mu_2 > \mu_0$ (right/one tailed).

Note: $\mu_0 = 0$ unless stated

Step 2: Decide on/state the significance level, α. If not given, the default $\alpha = 0.05$.

DOI: 10.1201/9781003292654-14

Step 3: We test for the assumption the compute Z-test or t-test.
a. We first test for the assumptions of equal variance or unequal variance

H_0: $\sigma_1^2 = \sigma_2^2$ (Equal variances Assumed)

H_a: $\sigma_1^2 \neq \sigma_2^2$ (Equal variances not Assumed)

State the alpha/significance level: $\alpha = 0.05$

Compute the F-test: $F - test = \frac{\sigma_1^2}{\sigma_2^2}$ (SPSS will compute F-test)

Determine the p-value or F critical value in SPSS using IDF.F(pro, df1, df2)

Make decision on H_0 using p-value: If p-value is less than α, then reject H_0. We can also use the critical value to make decision on H_0 i.e. if F-test > F critical value, reject H_0; otherwise fail to reject H_0.

Make the assumption at 95% level of confidence based on the decision on H_0

b. Compute the value of the test statistic based on the Assumptions of Equal Variance or Unequal Variance (F-test is used to make decisions on which assumption is appropriate).

  i.   If σ_1 and σ_2 are given/known then we use two sample Z-test

$$z - test = \frac{(\bar{x}_1 - \bar{x}_2) - \mu_0}{S.E.}$$

Assuming equal variances, S.E. is computed as

$$S.E. = s_p \sqrt{\frac{1}{n_1} + \frac{1}{n_2}}$$

$$s_p = \sqrt{\frac{(n_1 - 1)\sigma_1^2 + (n_2 - 1)\sigma_2^2}{n_1 + n_2 - 2}}$$

Assuming unequal variance, S.E. is computed as

$$S.E. = \sqrt{\left(\frac{\sigma_1^2}{n_1}\right) + \left(\frac{\sigma_2^2}{n_2}\right)}$$

where $\bar{x}_1$ is the sample mean for group 1, $\bar{x}_2$ is the sample mean for group 2, σ_1 is the standard deviation for group 1, σ_2 is the standard deviation for group 2, n_1 is the sample for group 1, and n_2 is the sample for group 2.

In SPSS, we compute the two sample Z-test using syntax since Z-test for two means is not inbuilt within SPSS with drop down menu. What one need to do is to replace the values in the syntax

File > New > Syntax > SPSS syntax window will populate > Copy and paste the SPSS syntax but replace the values in the following order: sample size, sample mean, population mean and standard deviation

**The first number is the sample size for group 1 (n1 = 23), the second number is the sample size for group 2 (n2 = 30),

**the third number is the sample mean for group 1 (78), the fourth number is the sample mean for group 2 (85)

** the fifth number is the population variance for group 1 (225) and the sixth number is the population variance for group 2 (100)

**Replace the four values below with your own.

**The p value that is reported is based on a two-tailed test.

**To obtain the one-tailed p value, simply divide the two-tailed p value by 2.

data list list/n1 n2 sample_mean1 sample_mean2 population_var1 population_var2.

begin data

25   30   78   85   225   100

end data.

```
Compute f_statistic= population_var1/population_var2.
Compute p_ftest=1-CDF.F(2.25,24,29).
Compute mean_difference = sample_mean1 - sample_mean2.
Compute standard_difference =SQRT((population_var1/n1)+(population_var2/
n2)).
Compute z_statistic = mean_difference/standard_difference.
Compute chi_square = z_statistic*z_statistic.
Compute p_value = SIG.CHISQ(chi_square, 1).
EXECUTE.
Formats f_statistic p_ftest z_statistic p_value (f8.5).
LIST f_statistic p_ftest z_statistic p_value.
```

**Select the Syntax and click Run Selection

The SPSS output will have the F-test, p-value for F-test, Z-test value for step 3 and p-value for Z-test for Step 4.

Step 4: Estimate the p-value.

State the p-value from the SPSS output. We can also estimate the Z critical value using IDF. NORMAL(prob, 0, 1) in SPSS.

Step 5: Decision on H_0: we use p-value to make decision on H_0.

If $P < \alpha$, reject H_0; otherwise, do not reject H_0. We can also use the critical value to make decision on H_0 i.e. if Z-test > Z critical value, reject H_0; otherwise fail to reject H_0.

ii. If σ_1 and σ_2 are not given but computed then we use two sample independent t-test as follows

$$t - test = \frac{(\bar{x}_1 - \bar{x}_2) - \mu_0}{S.E.}$$

Assuming equal variances, S.E. is computed as

$$S.E. = s_p \sqrt{\frac{1}{n_1} + \frac{1}{n_2}}$$

where

$$s_p = \sqrt{\frac{(n_1 - 1) s_1^2 + (n_2 - 1) s_2^2}{n_1 + n_2 - 2}}$$

with

$$df = n_1 + n_2 - 2$$

Assuming unequal variance, S.E. is computed as

$$S.E. = \sqrt{\left(\frac{s_1^2}{n_1}\right) + \left(\frac{s_2^2}{n_2}\right)}$$

with

$$df = \frac{\left(\frac{s_1^2}{n_1} + \frac{s_2^2}{n_2}\right)^2}{\left(\frac{1}{n_1 - 1}\right)\left(\frac{s_1^2}{n_1}\right)^2 + \left(\frac{1}{n_2 - 1}\right)\left(\frac{s_2^2}{n_2}\right)^2}$$

where $\bar{x}_1$ is the sample mean for group 1, $\bar{x}_2$ is the sample mean for group 2, s_1 is the computed standard deviation for group 1, s_2 is the computed standard deviation for group 2, n_1 is the sample for group 1, and n_2 is the sample for group 2.

Two-sample independent t-test in SPSS drop down menu. In SPSS, we compute the independent samples t-test using the path:

Analyze > Compare Means > Independent-Sample T test > There will be SPSS pop up window for Independent-Samples T test > Select **the numerical variable of interest as the Test Variable** > Select the **Grouping Variable** and Define Groups using the **Coded Values** of the variable (these are the two different groups that the hypothesis is testing) > Click the **Option Tab** put the required **Confidence Interval Percentage** that matches the **given α – level** > Click Continue then OK.

From the SPSS Output, decide on which assumptions of *Equal Variance* or *Unequal Variance* based on the F-test for Equality of Variance (**Levene's Test for Equality of Variances**) and p-value (**sig**).

a. *If the p-value of the Levene's Test for Equality of Variance is greater than 0.05 we Assume Equal Variance and select the t-test value (t), degree of freedom (df) and p-value (Sig. (2-tailed)) with the assumption of equal variance*

b. *If the p-value of the Levene's Test for Equality of Variance is less than 0.05 we Assume Unequal Variance and select the t-test value (t), degree of freedom (df) and p-value (Sig. (2-tailed)) with the assumption of unequal variance*

Step 4: Estimate the p-value.

State the p-value from the SPSS output.

Step 5: Decision on H_0: we use p-value to make decision on H_0.

If p-value $< \alpha$, reject H_0; otherwise, do not reject H_0.

Step 6: Make conclusions at a given level of confidence based on the decision in Step 5.

Example 1: Two-Tailed Test

Every semester, Business Statistics course is usually allocated two sections: Section A and Section B. Students are usually randomly assigned to one of the two sections. After the assignment, Section A had 30 students while Section B had 25 students. At the end of the semester, each class took the same standardized test. Section A students had an average test score of 78, with a standard deviation of 10; while Section B students had an average test score of 85, with a standard deviation of 15. Test the hypothesis that the performance of students in Section A and Section B are not different. Use a 0.10 level of significance.

SOLUTION:

Step 1: State the null and alternative hypotheses.
   Null hypothesis: H_0: $\mu_1 - \mu_2 = 0$ (no differences in the mean scores)
   Alternative hypothesis: H_1: $\mu_1 - \mu_2 \neq 0$ (there are differences in the mean scores)
   This is a two-tailed test.
Step 2: State the significance level: $\alpha = 0.10$
Step 3:
a. We first test for the assumptions of equal variance or unequal variance using F-test
   H_0: $\sigma_1^2 = \sigma_2^2$ (Equal variances Assumed)
   H_a: $\sigma_1^2 \neq \sigma_2^2$ (Equal variances not Assumed)
   State the alpha/significance level: $\alpha = 0.05$
   Compute the F-test: $F - test = \frac{\sigma_1^2}{\sigma_2^2}$ (SPSS will compute F-test)

$$F - test = \frac{\sigma_1^2}{\sigma_2^2} = \frac{15^2}{10^2} = 2.25$$

$$df = \begin{cases} numerator = n - 1 = 25 - 1 = 24 \\ denominator = m - 1 = 30 - 1 = 29 \end{cases}$$

Determine the p-value: check along df = 24,29 where F-test value = 2.25 will fall

To compute p-value: go to SPSS and use: 1-CDF.F(2.25,24,29) = 0.0193

Make decision on H_0 using p-value: Compare p-value = 0.0193 and $\alpha = 0.05$. Since p-value is less than α, then reject H_0

Make the assumption at 95% level of confidence we assume unequal variance. We can then compute S.E. with assumption of unequal variance

We can assume unequal variance since F-test value $\neq 1$

We compute S.E. with an assumption of unequal variance

$$S.E. = \sqrt{\left(\frac{10^2}{30}\right) + \left(\frac{15^2}{25}\right)} = 3.51$$

b. Compute the value of the test statistic (Z-test) based on the assumptions of unequal variance from the F-test.

Since σ_1 and σ_2 are given/known then we use two sample Z-test

$$z-test = \frac{(\bar{x}_1 - \bar{x}_2) - \mu_0}{S.E.}$$

$$z = \frac{[78-85] - 0}{3.51} = -1.99$$

In SPSS, we compute the two sample Z-test using the syntax below (Figure 11.1)

File > New > Syntax > SPSS syntax window will populate > Copy and paste the SPSS syntax but replace the values in the following order: sample size for group 1, sample size for group 2, sample mean fro group 1, sample mean fro group 2, variance for group 1 and variance for group 2

**The first number is the sample size for group 1(n1 = 23), the second number is the sample size for group 2(n2 = 30),

**the third number is the sample mean for group 1 (78), the fourth number is the sample mean for group 2 (85)

** the fifth number is the population variance for group 1 (225) and the sixth number is the population variance for group 2 (100)

**Replace the four values below with your own.

**The p value that is reported is based on a two-tailed test.

**To obtain the one-tailed p value, simply divide the two-tailed p value by 2.

```
data list list/n1 n2 sample_mean1 sample_mean2 population_var1 population_var2.
begin data
25  30  78  85  225  100
end data.

Compute f_statistic= population_var1/population_var2.
Compute p_ftest=1-CDF.F(2.25,24,29).
Compute mean_difference = sample_mean1 - sample_mean2.
Compute standard_difference =SQRT((population_var1/n1)+(population_var2/n2)).
Compute z_statistic = mean_difference/standard_difference.
Compute chi_square = z_statistic*z_statistic.
Compute p_value = SIG.CHISQ(chi_square, 1).
EXECUTE.
Formats f_statistic p_ftest z_statistic p_value (f8.5).
LIST f_statistic p_ftest z_statistic p_value.
```

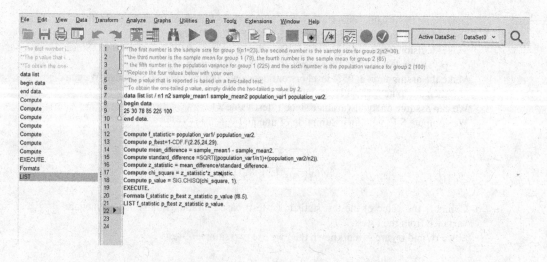

FIGURE 11.1   SPSS Syntax for Two Sample Z-test for Two Population Means

　　　　　**Select the Syntax and click Run Selection
　　　　　The SPSS output (Figure 11.2) will have the F-test, p-value for F-test, Z-test value for
　　　　　　　Step 3 and p-value for Z-test for Step 4.

```
f_statistic  p_ftest z_statistic  p_value

  2.25000    .01927   -1.99323    .04624

Number of cases read:  1    Number of cases listed:  1
```

FIGURE 11.2   Z Value for the difference between Two Population Means

　　　　　$F - test = 2.25$, p-value = 0.01927
　　　　　We ignore the sign of the Z-test value, $Z - test = 1.99$
　　　Step 4: Determine the p-value
　　　　　From the SPSS output, the p-value = 0.04624
　　　Step 5: Decision on H_0 using p-value: We compare p-value = 0.04624 with $\alpha = 0.10$
　　　　　*To make decision: If p-value is less than α value, reject H_0; otherwise, do not reject
　　　　　　H_0.*
　　　　　Since p-value = 0.04624 is less than (<) α value, we reject H_0.
　　　Step 6: Make conclusions at a given level of confidence based on the decision in Step 5.
　　　　　At 90% level of confidence, the claim that performance of students in Section A and
　　　　　　Section B are not different is rejected.

Example 2: One-Tailed Test

A solar engineer in charge claims that the new solar battery will operate continuously for *at
least* 7 minutes longer than the old solar battery. To test the claim, the company selects a simple
random sample of 100 new solar batteries and 100 old solar batteries. The old solar batteries run
continuously for 190 minutes with a standard deviation of 20 minutes while the new solar batteries run for 200 minutes with a standard deviation of 40 minutes. Test the solar engineer's claim
that the new solar batteries run at least 7 minutes longer than the old solar batteries. Use a 0.05
level of significance.

SOLUTION:

Step 1: State the null and alternative hypotheses.
Null hypothesis: $H_0: \mu_1 - \mu_2 \geq 7$
Alternative hypothesis: $H_a: \mu_1 - \mu_2 < 7$
This is a one-tailed test.
Step 2: State the significance level: $\alpha = 0.05$
Step 3:
a. We first test for the assumptions of equal variance or unequal variance using F-test
$H_0: \sigma_1^2 = \sigma_2^2$ (Equal variances Assumed)
  $H_a: \sigma_1^2 \neq \sigma_2^2$ (Equal variances not Assumed)
State the alpha/significance level: $\alpha = 0.05$
Compute the F-test:

$$F-test = \frac{\sigma_1^2}{\sigma_2^2} = \frac{40^2}{20^2} = 4$$

$$df = \begin{cases} numerator = n-1 = 100-1 = 99 \\ denominator = m-1 = 100-1 = 99 \end{cases}$$

We can assume unequal variance since F-test value $\neq 1$
Determine the p-value: check along df = 99,99 where F-test value = 4 will fall
To compute p-value: go to SPSS and use: 1-CDF.F(4,99,99) = 0.000
Make decision on H_0 using p-value: Compare p-value = 0.000 and $\alpha = 0.05$. Since p-value is less than α, then reject H_0
Make the assumption at 95% level of confidence we assume unequal variance. We can then compute S.E. with assumption of unequal variance
We can assume unequal variance since F-test value $\neq 1$
We compute S.E. with an assumption of unequal variance

$$S.E. = \sqrt{\left(\frac{40^2}{100}\right) + \left(\frac{20^2}{100}\right)} = 4.472$$

b. Compute the value of the test statistic (Z-test) based on the assumptions of unequal variance from the F-test.
Since σ_1 and σ_2 are given/known then we use two sample Z-test

$$Z-test = \frac{(\bar{x}_1 - \bar{x}_2) - \mu_0}{S.E.}$$

$$Z-test = \frac{[200-190]-7}{4.472} = 0.67$$

In SPSS, we compute the two sample Z-test using the syntax below (Figure 11.3).
File > New > Syntax > SPSS syntax window will populate > Copy and paste the SPSS syntax but replace the values in the following order: sample size for group 1, sample size for group 2, sample mean for group 1, sample mean for group 2, variance for group 1 and variance for group 2
**The first number is the sample size for group 1(n1 = 100), the second number is the sample size for group 2(n2 = 100),
**the third number is the sample mean for group 1 (200), the fourth number is the sample mean for group 2 (190)
** the fifth number is the population variance for group 1 (1600) and the sixth number is the population variance for group 2 (400)
**Replace the four values below with your own.
**The p value that is reported is based on a two-tailed test.
**To obtain the one-tailed p value, simply divide the two-tailed p value by 2.

```
data list list/n1 n2 sample_mean1 sample_mean2 population_var1 population_var2
population_mean.
begin data
100  100  200  190  1600  400  7
end data.
```

```
Compute f_statistic= population_var1/population_var2.
Compute p_ftest=1-CDF.F(4,99,99).
Compute mean_difference = (sample_mean1 - sample_mean2)-7.
Compute standard_difference =SQRT((population_var1/n1)+(population_
var2/n2)).
Compute z_statistic = mean_difference/standard_difference.
Compute chi_square = z_statistic*z_statistic.
Compute p_value = SIG.CHISQ(chi_square, 1).
EXECUTE.
Formats f_statistic p_ftest z_statistic p_value (f8.5).
LIST f_statistic p_ftest z_statistic p_value.
```

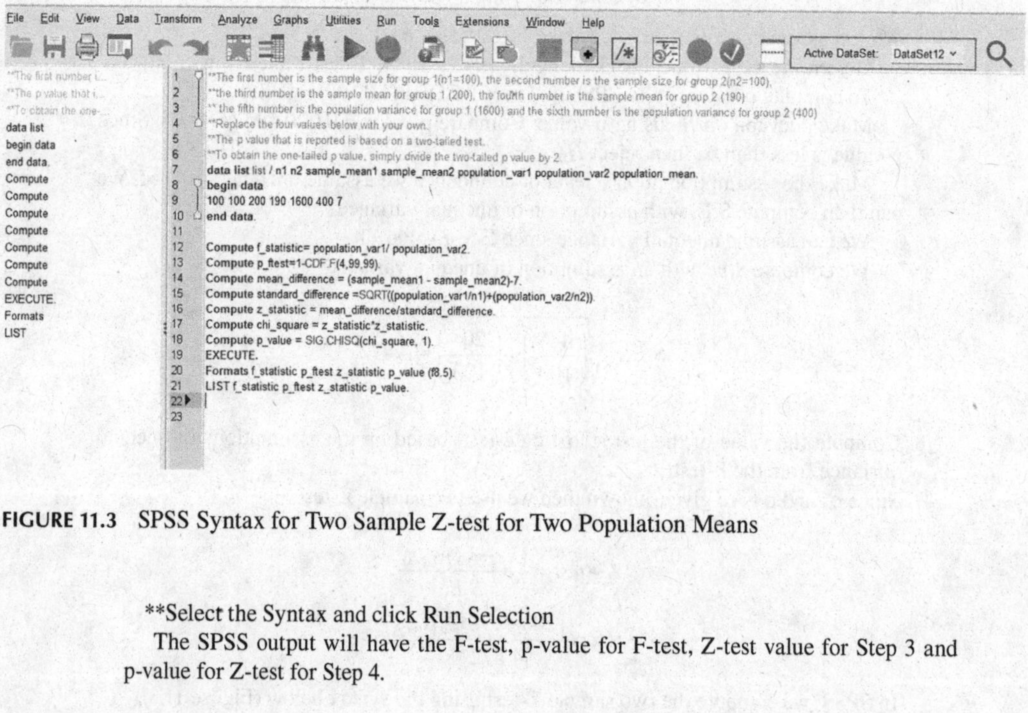

FIGURE 11.3 SPSS Syntax for Two Sample Z-test for Two Population Means

**Select the Syntax and click Run Selection
The SPSS output will have the F-test, p-value for F-test, Z-test value for Step 3 and
p-value for Z-test for Step 4.

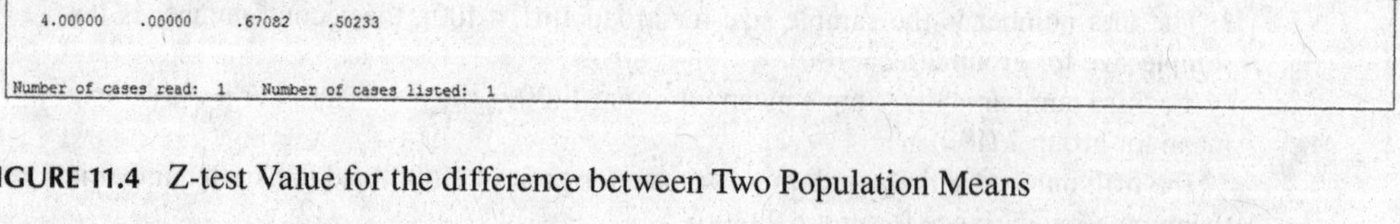

```
f_statistic  p_ftest z_statistic  p_value

  4.00000   .00000   .67082   .50233


Number of cases read:  1    Number of cases listed:  1
```

FIGURE 11.4 Z-test Value for the difference between Two Population Means

$F-test = 4$, p-value = 0.000

We ignore the sign of the Z-test value, $Z-test = 0.67$

Step 4: Determine the p-value

From the SPSS output, the p-value = 0.50233. Since its one tailed test we divide p-value by 2. Hence, p-value = 0.25117

Step 5: Decision on H_0 using p-value: We compare p-value = 0.25117 with $\alpha = 0.05$

To make decision: If p-value is less than α value, reject H_0; otherwise, do not reject H_0.

Since p-value = 0.25117 is greater than (>) α value, we fail to reject H_0.

Step 6: Make conclusions at a given level of confidence based on the decision in Step 5.

At 95% level of confidence, the solar engineer's claim that the new solar batteries run at least 7 minutes longer than the old solar batteries is true.

Example: Two sample independent t-test

The National Association of Home Builders provided data on the cost of the most popular home remodeling projects. Sample data on cost in thousands of dollars for two types of remodeling projects are as follows.

Kitchen	Master Bedroom
25.2	18.0
17.4	22.9
22.8	26.4
21.9	24.8
19.7	26.9
23.0	17.8
16.9	21.0
21.8	25.2
23.6	
22.5	

At $\alpha = 0.02$, are the population mean remodeling costs for the two types of projects significantly different?

SOLUTION:

Step 1: State the null and alternative hypotheses.

Null hypothesis: H_0: $\mu_k - \mu_m = 0$ (no differences in the mean remodeling cost)

Alternative hypothesis: H_1: $\mu_k - \mu_m \neq 0$ (there are differences in the mean remodeling cost)

This is a two-tailed test.

Step 2: State the significance level: $\alpha = 0.02$

Step 3:

i. We first test for the assumptions of equal variance or unequal variance using F-test

We key the data into SPSS

Variable View

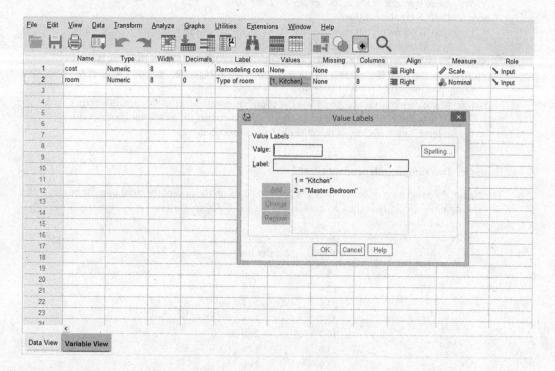

FIGURE 11.5   SPSS Variable View Window displaying Remodeling Cost and Type of Room

Data View

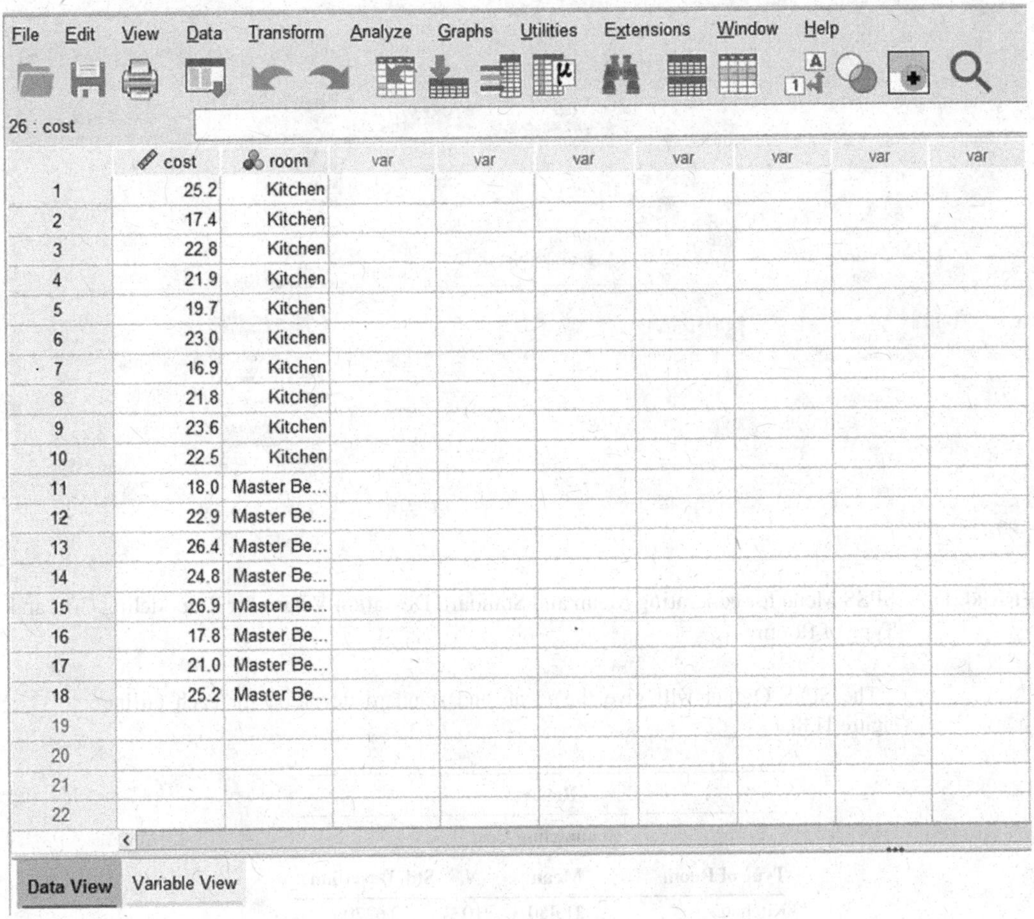
File Edit View Data Transform Analyze Graphs Utilities Extensions Window Help
26 : cost
cost room var var var var var var var
1 25.2 Kitchen
2 17.4 Kitchen
3 22.8 Kitchen
4 21.9 Kitchen
5 19.7 Kitchen
6 23.0 Kitchen
7 16.9 Kitchen
8 21.8 Kitchen
9 23.6 Kitchen
10 22.5 Kitchen
11 18.0 Master Be...
12 22.9 Master Be...
13 26.4 Master Be...
14 24.8 Master Be...
15 26.9 Master Be...
16 17.8 Master Be...
17 21.0 Master Be...
18 25.2 Master Be...
19
20
21
22
Data View Variable View

FIGURE 11.6 SPSS Data View Window displaying Remodeling Cost and Type of Room

Then we compute the mean and standard deviation values for the two groups using Analyze > Compare Means > Means > there will be a pop up window for Means > select Remodeling cost as Dependent List and Type of room as Independent List > Click Options and select Mean, Number of Cases and Standard Deviation as Cell Statistics then click Continue then OK (Figure 11.7).

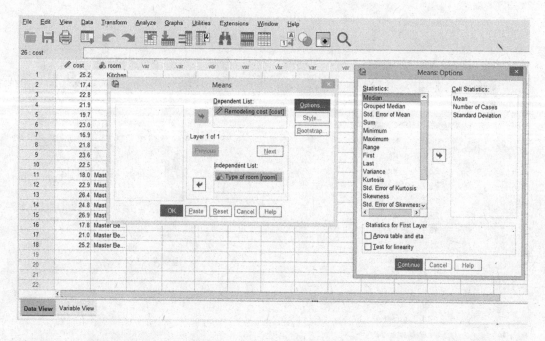

FIGURE 11.7 SPSS Menu for generating Mean and Standard Deviation Values for Remodeling Cost and Type of Room

The SPSS Output will give the mean and standard deviation for each airline (Figure 11.8).

Report			
Remodeling Cost			
Type of Room	**Mean**	**N**	**Std. Deviation**
Kitchen	21.480	10	2.6770
Master Bedroom	22.875	8	3.6011
Total	22.100	18	3.1052

FIGURE 11.8 Mean and Standard Deviation Values for Remodeling Cost and Type of Room

We have $n_1 = 10$, $\bar{x}_1 = 21.480$, $s_1 = 2.6770$, $n_2 = 8$, $\bar{x}_2 = 22.875$, $s_2 = 3.6011$

Testing for the Assumptions

H_0: $s_1^2 = s_2^2$ (Equal variances Assumed)

H_a: $s_1^2 \neq s_2^2$ (Equal variances not Assumed)

State the alpha/significance.level: $\alpha = 0.05$

Compute the F-test:

$$F-test = \frac{s_1^2}{s_2^2} = \frac{3.6011^2}{2.6770^2} = 1.8096$$

$$df = \begin{cases} numerator = n-1 = 8-1 = 7 \\ denominator = m-1 = 10-1 = 9 \end{cases}$$

We can assume unequal variance since F-test value $\neq 1$

Determine the p-value: check along df = 7,9 where F-test value = 1.8096 will fall

To compute p-value: go to SPSS and use: 1-CDF.F(1.8096,7,9) = 0.2002

Make decision on H_0 using p-value: Compare p-value = 0.2002 and $\alpha = 0.05$. Since p-value is greater than α, then fail to reject H_0

At 95% level of confidence we assume equal variance. We can then compute S.E. with assumption of equal variance

$$S.E. = s_p \sqrt{\frac{1}{n_1} + \frac{1}{n_2}}$$

where

$$s_p = \sqrt{\frac{(n_1 - 1)s_1^2 + (n_2 - 1)s_2^2}{n_1 + n_2 - 2}}$$

with

$$df = n_1 + n_2 - 2$$

$$s_p = \sqrt{\frac{(n_1 - 1)s_1^2 + (n_2 - 1)s_2^2}{n_1 + n_2 - 2}} = \sqrt{\frac{(10 - 1)2.677^2 + (8 - 1)3.6011^2}{10 + 8 - 2}} = 3.1152$$

$$S.E. = s_p \sqrt{\frac{1}{n_1} + \frac{1}{n_2}} = 3.1152 \sqrt{\frac{1}{10} + \frac{1}{8}} = 1.4777$$

$$df = n_1 + n_2 - 2 = 10 + 8 - 2 = 16$$

ii. Compute the value of the test statistic (t-test) based on the assumptions of equal variance from the F-test

We have $\mu_0 = 0$, $\bar{x}_1 = 21.480$, $\bar{x}_2 = 22.875$

$$t-test = \frac{(\bar{x}_1 - \bar{x}_2) - \mu_0}{S.E}$$

$$t = \frac{(22.875 - 21.480) - 0}{1.4777} = 0.9440$$

We ignore the sign of the t-test value, $t - test = 0.944$

In SPSS, we compute two sample independent t-test using the path:

Analyze > Compare Means > Independent-Sample T test > There will be SPSS popup window for Independent Samples t-test >Select **the variable of interest (Remodeling Cost)** as **the Test Variable** > Select the **Grouping Variable (Type of Room)** and Define Groups using the **Coded Values (Group 1 = 1 and Group 2 = 2)** of the variable (these are the two different groups that the hypothesis is testing) > Click the **Option Tab** put the required **Confidence Interval Percentage (98%)** that matches the **given α – level** (Figure 11.9)

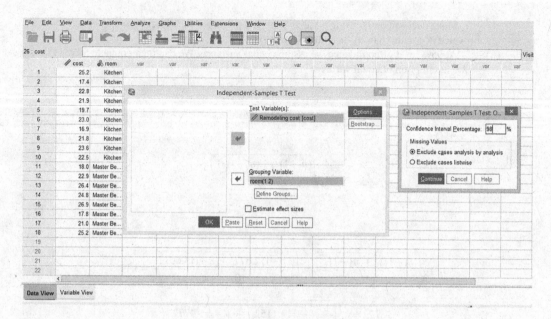

FIGURE 11.9   SPSS Independent-Samples t-Test Pop Up Window for generating t-Test Value for the difference between Two Population Means

The SPSS output

Independent Sample Test

		Levene's Test for Equality of Variances		t-Test for Equality of Means					98% Confidence Interval of the Difference	
		F	Sig.	t	Df	Sig. (2-tailed)	Mean Difference	Std. Error Difference	Lower	Upper
Remodeling cost	Equal variances assumed	1.284	.274	−.944	16	.359	−1.3950	1.4777	−5.2125	2.4225
	Equal variances not assumed			−.912	12.636	.379	−1.3950	1.5289	−5.4632	2.6732

FIGURE 11.10   t-Test Value for the difference between Two Population Means

From the SPSS output F-test = 1.284, p-value for F-test = 0.274, t-test = −0.944, df = 16, p-value for t-test = 0.359

Test for the assumptions using F-test

F-test = 1.284

p-value = 0.274

Since p-value is greater than $\alpha = 0.05$, we fail to reject the H_0, and *assume equal variances* at 95% level of confidence.

From the SPSS output based on the assumption of equal variance, the t-test = −0.944, df = 16

Step 4: Determine the p-value (for t-test):

To estimate p-value: we check along the $df = 16$ where the t-test value = 1.3403 will fall.

From the SPSS output, the p-value for the t-test = 0.359

Step 5: Decision on H_0 using p-value: We compare p-value = 0.359 with $\alpha = 0.02$

To make decision: If p-value is less than α value, reject H_0; otherwise, do not reject H_0.

Since p-value = 0.359 is greater than (>) α value, we fail reject H_0.

Step 6: Make conclusions at a given level of confidence based on the decision in Step 5.
At 95% level of confidence, the remodeling cost for kitchen and master bedroom is not significantly different.

b. *Hypothesis Test for Two Population Proportions*

The aim is to conduct a hypothesis test to determine whether the difference between two proportions is significant. The test procedure, called the **two sample proportion Z-test**, is appropriate when the samples are independent.

Step 1: State the null and alternative hypotheses.

The Null Hypothesis is H_0: $p_1 - p_2 = p_0$

and the alternative hypothesis is H_a: $p_1 - p_2 \neq p_0$ (two tailed).

Or

The Null Hypothesis is H_0: $p_1 - p_2 \geq p_0$

and the alternative hypothesis is H_a: $p_1 - p_2 < p_0$ (left/one tailed).

Or

The Null Hypothesis is H_0: $p_1 - p_2 \leq p_0$

and the alternative hypothesis is H_a: $p_1 - p_2 > p_0$ (right/one tailed).

Note: $p_0 = 0$ unless stated

Step 2: Decide on/State the significance level, α. If not given, the default $\alpha = 0.05$.

Step 3:

a. We first state the assumptions that we will use: either equal proportions or different proportions

b. We first compute S.E. with the stated assumption in 3 (a), then compute the value of the test statistic (Z-test)

Assuming equal population proportions, then

$$S.E. = \sqrt{p(1-p) \times \left[\frac{1}{n_1} + \frac{1}{n_2}\right]}$$

$$p = \frac{p_1 \times n_1 + p_2 \times n_2}{n_1 + n_2}$$

Assuming different population proportions, then

$$S.E. = \sqrt{\frac{p_1(1-p_1)}{n_1} + \frac{p_2(1-p_2)}{n_2}}$$

The test statistic is **Two Sample Proportions z - test**.

$$Z - test = \frac{(p_1 - p_2) - p_0}{S.E.}$$

where p_1 is the sample proportion for group 1, p_1 is the sample proportion for group 2, n_1 is the sample for group 1, and n_2 is the sample for group 2.

To compute Z-test for two population proportions in SPSS we follow the path:

Analyze > Compare Means > Independent-Samples Proportions. There will be a pop up SPSS Window for Independent-Sample Proportions > Select **the numerical variable of interest** as **the Test Variable** > then select **the categorical variable of interest** as **the Grouping Variable** > Under **Define Groups** put the coded values for the two categories/proportions of interest > Click the **Confidence Level Tab** put the required **Confidence Interval Percentage** > Click Continue then OK

We can also use SPSS syntax for one tailed Z-test for two proportions where we substitute the values in the syntax

To compute one-tailed Z-test for proportions in SPSS we use Snytax and substitute the values for p_1, p_2, n_1, and n_2

File > New > Syntax > SPSS syntax window will populate > Copy and paste the SPSS syntax but replace the values in the following order: sample size for group 1, sample size for group 2, sample proportion group 1 and sample proportion for group 2

**The first number is the sample size for group 1($n1 = 100$), the second number is the sample size for group 2($n2 = 200$),

**the third number is the sample proportion for group 1 (0.38) and the fourth number is the sample proportion for group 2 (0.51)

**The p value that is reported is based on a two-tailed test.

**To obtain the one-tailed p value, simply divide the two-tailed p value by 2.

data list list/n1 n2 sample_proportion1 sample_proportion2.

begin data

100   200   0.38   0.51

end data.

```
Compute prop_difference = sample_proportion1 - sample_proportion2.
Compute p_1=((sample_proportion1)*(1-sample_proportion1))/n1.
Compute p_2=((sample_proportion2)*(1-sample_proportion2))/n2.
Compute standard_difference =SQRT(p_1+p_2).
Compute z_statistic = prop_difference/standard_difference.
Compute chi_square = z_statistic*z_statistic.
Compute p_value = SIG.CHISQ(chi_square, 1).
EXECUTE.
Formats z_statistic p_value (f8.5).
LIST z_statistic p_value.
```

**Select the Syntax and click Run Selection

From the SPSS Output you select **Z-test value** and **p-value**

Step 4:   Determine the p-value.

From the SPSS output you select the p-value.

Step 5:   Decision on H_0 using p-value: We compare p-value with α

To make decision: If the p-value is less than α value, reject H_0; otherwise, do not reject H_0.

Step 6:   Make conclusions at a given level of confidence based on the decision in Step 5.

Example 1: Two-Tailed Test

The company states that the defaulting or not defaulting is equally effective among the men customers. A survey was conducted among to determine if the significant difference among those who had defaulted and those who have not defaulted among the different genders. Using customer_subset.sav data, is there a statistical significant difference among the proportion of male customers who defaulted and those who did not default on a bank? Use a 0.05 level of significance.

SOLUTION:

Step 1:   State the null and alternative hypotheses.

Null hypothesis (H_0): $p_1 - p_2 = 0$ (there is no difference between the proportion of male customers who have defaulted and the proportion of male customers who have not defaulted)

Alternative hypothesis (H_1): $p_1 - p_2 \neq 0$ (there is a difference between the proportion of male customers who have defaulted and the proportion of male customers who have not defaulted)

This is a two-tailed test.

Step 2:   State the significance level, $\alpha = 0.05$.

Step 3:

a. We first state the assumptions that we will use: either equal proportions or different proportions

   Assuming different proportions

b. Then we compute S.E. with the stated assumption in 3 (a) – equal proportion, then compute the value of the test statistic (Z-test)

First, we locate customer_subset.sav data then generate the number of male who defaulted and those who did not default using the path:

Analyze > Descriptive Statistics > Crosstabs > SPSS window for crosstabs will populate > Select any of the categorical variable of interest (Gender) and put it in Row(s), then select the other categorical variable of interest (Ever defaulted on a bank) and put it in Column(s) > Click Cells and a pop up window for Cell Display will populate then select Observed and Column percentages > Click Continue then OK (Figure 11.11).

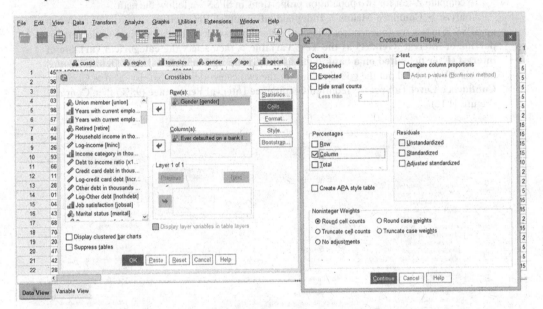

FIGURE 11.11   SPSS Crosstabs Pop Up Window displaying Cells Options

SPSS Output

Gender ×* Ever Defaulted on a Bank Loan Crosstabulation					
			Ever defaulted on a bank loan		
			No	Yes	Total
Gender	Male	Count	29	7	36
		% within Ever defaulted on a bank loan	50.0%	31.8%	45.0%
	Female	Count	29	15	44
		% within Ever defaulted on a bank loan	50.0%	68.2%	55.0%
Total		Count	58	22	80
		% within Ever defaulted on a bank loan	100.0%	100.0%	100.0%

FIGURE 11.12   Cross tabulation for Gender and Default Status

From the SPSS output, we can see that $p_1 = \frac{29}{58} = 0.50$, $p_2 = \frac{7}{22} = 0.318$, $n_1 = 58$, $n_2 = 22$
These values can be used to estimate S.E. and Z-test value

$$S.E. = \sqrt{\frac{p_1(1-p_1)}{n_1} + \frac{p_2(1-p_2)}{n_2}}$$

$$S.E. = \sqrt{\frac{0.50(1-0.50)}{58} + \frac{0.318(1-0.318)}{22}} = 0.119$$

$$z-test = \frac{(p_1-p_2)-0}{S.E.}$$

$$z-test = \frac{(0.50-0.318)}{0.119} = 1.529$$

To compute Z-test for two population proportions in SPSS we follow the path:
Analyze > Compare Means > Independent-Samples Proportions. There will be a pop up SPSS Window for Independent-Sample Proportions > Select **the categorical variable/proportion of interest (Gender)** as **the Test Variable** > then select **the categorical variable of interest (Ever defaulted on a bank loan)** as **the Grouping Variable** > Under **Define Groups** the select Values and put the coded values for the two categories (0 = No, 1 = Yes) > Click the **Confidence Level Tab** put the required **Confidence Interval Percentage (95%)** then Continue (Figure 11.12)

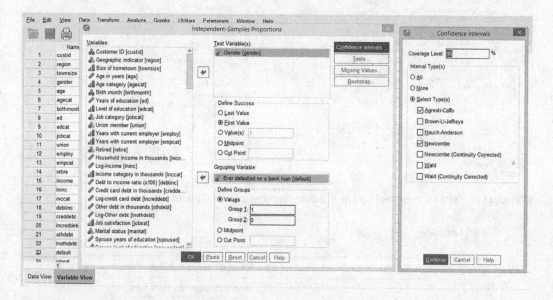

FIGURE 11.13   SPSS Independent-Samples Proportions Pop Up Window for generating Z-test Value for the difference between Two Population Proportions

Click Test and Select Wald > Click Continue then OK (Figure 11.14)

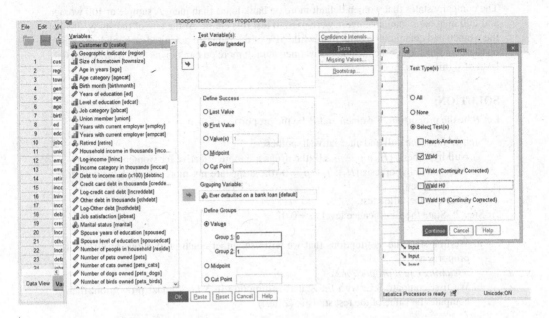

FIGURE 11.14   SPSS Independent-Samples Proportions Pop Up Window displaying the Tests for Two Population Proportions

SPSS output

					Significance	
	Test Type	**Difference in Proportions**	**Asymptotic Standard Error**	**Z**	**One-Sided p**	**Two-Sided p**
Gender = Male	Wald	.182	.119	1.527	.063	.127

Independent-Samples Proportions Tests

FIGURE 11.15   Z-test Value for the difference between Two Population Proportions

From the SPSS output we select the Z-test value and p-value (two sided)
We ignore the sign of the Z-test value, so $Z - test = 1.527$
Step 4: Determine the p-value (for t-test):
   p-Value: from the Z table estimate the extreme area under the curve using Z-test value = 1.527. We estimate the area from the center to Z value = 1.527. We then subtract the estimated area from 0.5 to have the p-value. If it's a two tailed test we multiply p-value by 2.
   From the SPSS output, the p-value = 0.127 (two sided)
Step 5: Decision on H_0 using p-value: We compare p-value = 0.127 with $\alpha = 0.05$
   To make decision: If p-value is less than α value, reject H_0; otherwise, do not reject H_0.
   Since p-value = 0.127 is greater than (>) α value, we fail reject H_0.
Step 6: Make conclusions at a given level of confidence based on the decision in Step 5.
   At 95% level of confidence, there is no difference between the proportion of male customers who have defaulted and the proportion of male customers who have not defaulted.

Example 2: One-Tailed Test

The company states that women default more on bank loans than men. A sample of 100 women and 200 men bank loan customers were used to test this claim. At the end of the study, 38% of the women had defaulted on bank loan while 51% of the men had defaulted on bank loan. Based on these findings, can we conclude that women default more on bank loans than men? Use a 0.01 level of significance.

SOLUTION:

Let P_1 be the proportion of women and P_2 be the proportion of men

   Step 1: State the null and alternative hypotheses.
       Null hypothesis (H_0): $p_1 - p_2 \geq 0$ (the drugs are more effective for women than men)
       Alternative hypothesis (H_1): $p_1 - p_2 \neq 0$ (the drugs are not more effective for women than men)
       This is a one-tailed test.
   Step 2: State the significance level, $\alpha = 0.01$.
   Step 3:
   a. We first state the assumptions that we will use: either equal proportions or different proportions
       Assuming equal proportions
   b. We first compute S.E. with the stated assumption in 3 (a) – equal proportion, then compute the value of the test statistic (Z-test)

$$p_1 = 0.38, \ p_2 = 0.51, \ n_1 = 100, \ n_2 = 200$$

$$S.E. = \sqrt{p(1-p) \times \left[\frac{1}{n_1} + \frac{1}{n_2} \right]}$$

$$p = \frac{(p_1 \times n_1) + (p_2 \times n_2)}{n_1 + n_2}$$

$$p = \frac{(0.38 \times 100) + (0.51 \times 200)}{100 + 200} = 0.467$$

$$S.E. = \sqrt{0.467 \times 0.533 \times \left[\frac{1}{100} + \frac{1}{200} \right]} = 0.061$$

$$z - test = \frac{(p_1 - p_2) - 0}{S.E.}$$

$$z - test = \frac{(0.38 - 0.51)}{0.061} = -2.13$$

We ignore the sign of the Z-test value, $Z - test = 2.13$
   To compute one-tailed Z-test for proportions in SPSS we use Snytax and substitute the values for p_1, p_2, n_1, and n_2
   File > New > Syntax > SPSS syntax window will populate > Copy and paste the SPSS syntax but replace the values in the following order: sample size for group 1, sample size for group 2, sample proportion group 1 and sample proportion for group 2
   **The first number is the sample size for group 1($n1 = 100$), the second number is the sample size for group 2($n2 = 200$),
   **the third number is the sample proportion for group 1 (0.38) and the fourth number is the sample mean for group 2 (0.51)
   **The p value that is reported is based on a two-tailed test.
   **To obtain the one-tailed p value, simply divide the two-tailed p value by 2.

```
data list list/n1 n2 sample_proportion1 sample_proportion2.
begin data
100   200   0.38   0.51
end data.

Compute prop_difference = sample_proportion1 - sample_proportion2.
Compute p_1=((sample_proportion1)*(1-sample_proportion1))/n1.
Compute p_2=((sample_proportion2)*(1-sample_proportion2))/n2.
Compute standard_difference =SQRT(p_1+p_2).
Compute z_statistic = prop_difference/standard_difference.
Compute chi_square = z_statistic*z_statistic.
Compute p_value = SIG.CHISQ(chi_square, 1).
EXECUTE.
Formats z_statistic p_value (f8.5).
LIST z_statistic p_value.
```

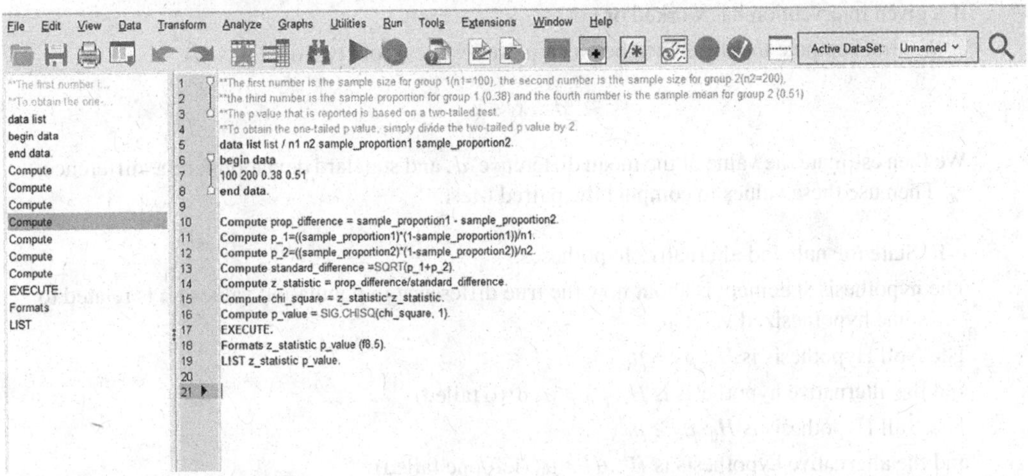

FIGURE 11.16 SPSS Syntax for computing Z-test Value for Two Population Proportions

  **Select the Syntax and click Run Selection
   The SPSS output will have the Z-test value for step 3 and p-value for Z-test for Step 4.

```
z_statistic  p_value

  -2.16501    .03039

Number of cases read:  1    Number of cases listed:  1
```

FIGURE 11.17 Z-test Value for the difference between Two Population Proportions

   We ignore the sign of the Z-test value, $Z - test = 2.165$
  Step 4: Determine the p-value (for t-test):
   p-Value: from the Z table estimate the extreme area under the curve using Z-test value = 2.165. We estimate the area from the center to Z value = 2.165. We then subtract the estimated area from 0.5 to have the p-value. If it's a two tailed test we multiply p-value by 2 and if it's one tailed we use the p-value as it is.
   From the SPSS output, the p-value = 0.03039 (two sided). Since its one tailed test. The p-value is divided by 2. Hence, p-value = 0.015195

Step 5: Decision on H_0 using p-value: We compare p-value = 0.015195 with $\alpha = 0.01$
To make decision: If p-value is less than α value, reject H_0; otherwise, do not reject H_0.
Since p-value = 0.015195 is greater than (>) α value, we fail to reject H_0.
Step 6: Make conclusions at a given level of confidence based on the decision in Step 5.
At 99% level of confidence, the company's claim that more women default on bank loan than men is true.

c. *Hypothesis Test for Two Population Means for Dependent/Paired/Matched Sample*

The **paired/matched-pairs t-test**, is appropriate when the test is conducted on paired data (data sets are *not* independent) and the sampling distribution is approximately normal.

The paired/matched t-test is used to determine if there is a statistically significantly difference between two quantitative/numerical variable for a given individual. Paired data are when we have two numerical/quantitative data from one individual. For example, to determine if there is statistically significant different between opening exam and final exam – to determine if a given intervention has worked or not.

We first compute the difference between paired values from two data sets (d).

$$d = x_1 - x_2$$

We then estimate the value of the mean difference, $\bar{d}$, and standard deviation for the difference, s_d. Then use these values to compute the paired t-test.

Step 1: State the null and alternative hypotheses.
The hypothesis statement is about how the true difference in population values μ_d is related to some hypothesized value μ_0
The Null Hypothesis is H_0: $\mu_d = \mu_0$
and the alternative hypothesis is H_a: $\mu_d \neq \mu_0$ (two tailed)
The Null Hypothesis is H_0: $\mu_d \geq \mu_0$
and the alternative hypothesis is H_a: $\mu_d < \mu_0$ (left/one tailed)
The Null Hypothesis is H_0: $\mu_d \leq \mu_0$
and the alternative hypothesis is H_a: $\mu_d > \mu_0$ (right/one tailed)
Note: $\mu_0 = 0$ unless stated
Step 2: Step 2: Decide on/State the significance level, α. If not given, the default $\alpha = 0.05$.
Step 3: We key the data into SPSS and then compute the value of the test statistic – **Paired t-test**.

$$t = \frac{\bar{d} - \mu_0}{S.E.}$$

$$S.E. = \frac{s_d}{\sqrt{n}}$$

with

$$df = n - 1$$

To compute paired t-test In SPSS we use the path:
Analyze > Compare Means > Paired Samples T test >There will be a pop up SPSS window for Paired Samples T test > Select **the paired variable of interest** as **the Test Variable** (the 2 numerical paired variables) > Click the **Option Tab** put the required **Confidence Interval Percentage** that matches the **given α – level** > Select the **t-test value** (t), **degree of freedom** (df) and **p-value** [Sig. (2-tailed)] from the SPSS Output.
Step 4: Determine the p-value from the SPSS output

Step 5: Make decision on H_0 using p-value approach: compare p-value with α

If p-value is less than (<) α, reject H_0; otherwise, do not reject H_0.

Step 6: Make conclusions at a given level of confidence based on the decision in Step 5.

Example

Consider the following data for 22 employees who underwent special training for customer service. Do these results provide evidence that the special training helped or hurt employees' performance? Use 0.05 level of significance.

Pair	Training	No Training
1	95	90
2	89	85
3	76	73
4	92	90
5	91	90
6	53	53
7	67	68
8	88	90
9	75	78
10	85	89
11	90	95
12	85	83
13	87	83
14	85	83
15	85	82
16	68	65
17	81	79
18	84	83
19	71	60
20	46	47
21	75	77
22	80	83

SOLUTION

Step 1: State the null and alternative hypotheses.

Null hypothesis: H_0: $\mu_d = 0$

Alternative hypothesis: H_0: $\mu_d \neq 0$

This is a two-tailed test.

Step 2: State the significance level, $\alpha = 0.05$.

Step 3: Compute the tests statistics: Paired t-test

We first get the difference in the observations $(d = x_1 - x_2)$, then use the data for the differences (d) to estimate the mean of the difference $(\bar{d})$ and standard deviation of the difference (s_d).

Pair	Training	No Training	d = Training – No Training
1	95	90	5
2	89	85	4
3	76	73	3
4	92	90	2
5	91	90	1

Pair	Training	No Training	d = Training – No Training
6	53	53	0
7	67	68	-1
8	88	90	-2
9	75	78	-3
10	85	89	-4
11	90	95	-5
12	85	83	2
13	87	83	4
14	85	83	2
15	85	82	3
16	68	65	3
17	81	79	2
18	84	83	1
19	71	60	11
20	46	47	-1
21	75	77	-2
22	80	83	-3

We then create the variable for the difference and key the difference data (d) into SPSS to compute mean and standard deviation values using SPSS via: Analyze > Descriptive Statistics > Frequencies > there will be SPSS pop up window for Frequencies > Select difference data as the Variable(s) > Click Statistics and select Mean, Standard Deviation and S.E. mean values > click Continue then OK (Figure 11.18).

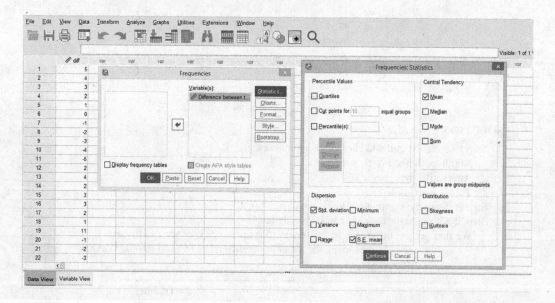

FIGURE 11.18 SPSS Frequency Menu for generating Mean, Standard Deviation and S.E. mean Values for Differences in Scores

The SPSS output will give Mean, Standard Deviation and S.E. mean values

Statistics		
Difference between training and no training		
N	Valid	22
	Missing	0
Mean		1.00
Std. Error of Mean		.764
Std. Deviation		3.586

FIGURE 11.19 Mean, Standard Deviation and S.E. mean Values for Differences in Scores

From the above SPSS output $\bar{d} = 1$, $s_d = 3.586$, $n = 22$, and $S.E. = \dfrac{s_d}{\sqrt{n}} = \dfrac{3.586}{\sqrt{22}} = 0.764$

$$t = \frac{\bar{d} - \mu_0}{S.E.}$$

$$S.E. = \frac{s_d}{\sqrt{n}}$$

with

$$df = n - 1$$

$$t = \frac{1 - 0}{0.764} = 1.3089$$

$$df = n - 1 = 22 - 1 = 21$$

To compute paired t-test in SPSS;
We key the data into SPSS and then compute the value of the test statistic – **Paired t-test**.
Variable View (Figure 11.20)

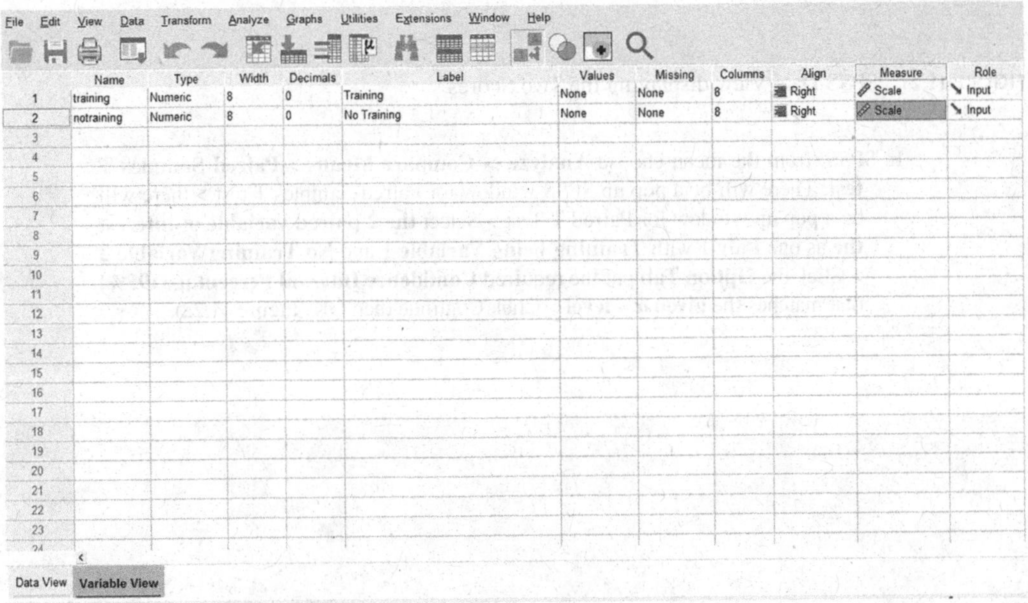

FIGURE 11.20 SPSS Variable View displaying the Two Scores

Data View (Figure 11.21)

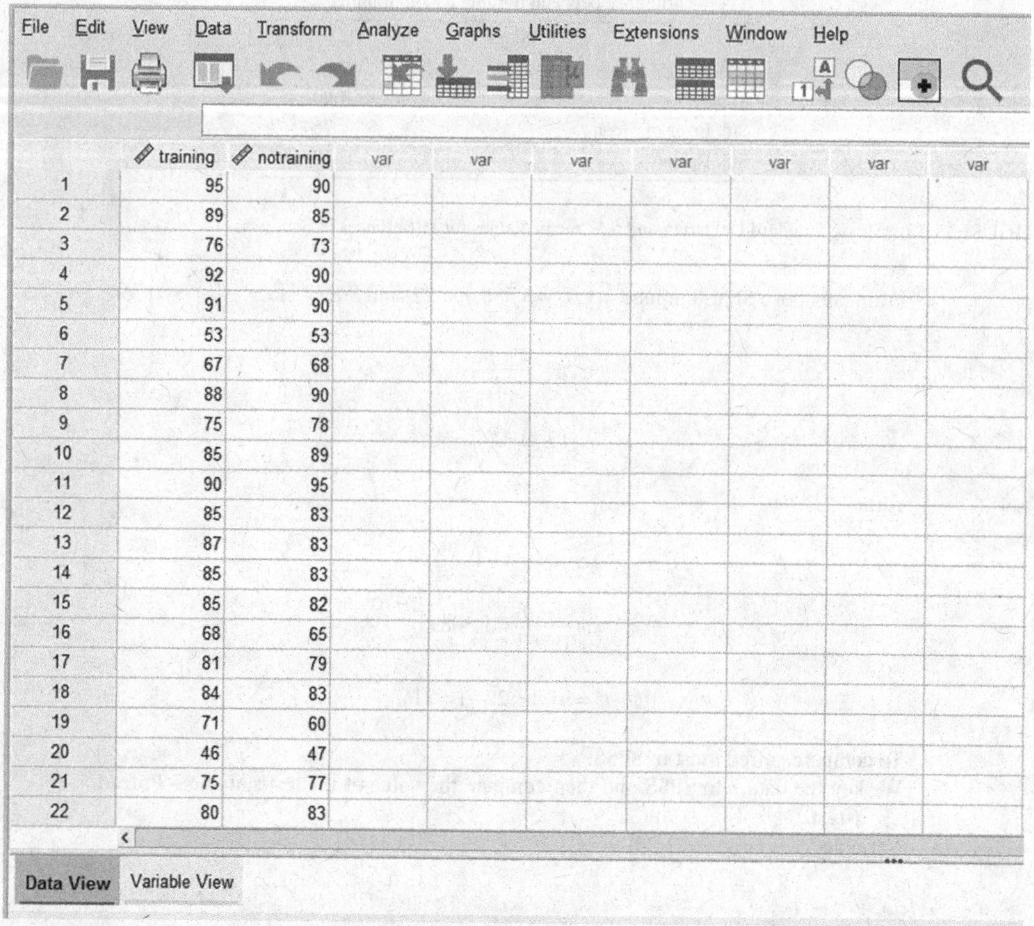

	training	notraining	var	var	var	var	var	var	var
1	95	90							
2	89	85							
3	76	73							
4	92	90							
5	91	90							
6	53	53							
7	67	68							
8	88	90							
9	75	78							
10	85	89							
11	90	95							
12	85	83							
13	87	83							
14	85	83							
15	85	82							
16	68	65							
17	81	79							
18	84	83							
19	71	60							
20	46	47							
21	75	77							
22	80	83							

Data View   Variable View

FIGURE 11.21   SPSS Data View displaying the Two Scores

In SPSS, from the menu choose **Analyze > Compare Means > Paired Samples T test**. There will be a pop up SPSS window for Paired-Samples T test > there will be a pop up window for Paired T Test > Select **the 2 paired variable of interest the** as one **Pair 1** with **Training being Variable 1** and **No Training Variable 2** > Click the **Option Tab** put the required **Confidence Interval Percentage (95%)** that matches the **given α – level** > Click Continue then OK (Figure 11.22)

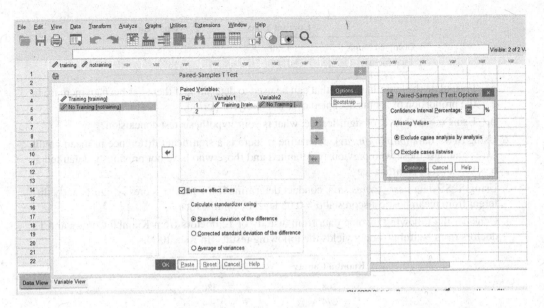

FIGURE 11.22 SPSS Paired-Samples t-Test Pop Up Window for generating t-Test Value for Paired Samples

SPSS Output

Paired Samples Test

				95% Confidence Interval of the Difference				
		Std. Deviation	Std. Error Mean			t	df	Sig. (2-tailed)
	Mean			Lower	Upper			
Pair 1 Training - No Training	1.000	3.586	.764	−.590	2.590	1.308	21	.205

FIGURE 11.23 t-Test Value for the Mean difference in Scores

Select the **t-test value** (t), **degree of freedom** (df) and **p-value** [Sig. (2-tailed)] from the SPSS Output.

Paired t-test value = 1.308, df = 21

Step 4: Determine the p-value (for t-test):

From the SPSS output, the p-value = 0.205

Step 5: Decision on H_0 using p-value: We compare p-value = 0.205 with $\alpha = 0.02$

To make decision: If p-value is less than α value, reject H_0; otherwise, do not reject H_0.

Since p-value = 0.205 is greater than (>) $\alpha = 0.05$ value, we fail reject H_0.

Step 6: Make conclusions at a given level of confidence based on the decision in Step 5.

At 95% level of confidence, there is no difference in the performance of the IQ tests and that special training doesn't helped or hurt student.

Practice Exercise

1. December holiday visits to Mombasa have been the trend in the past. Hotel occupancy is the measure of visitor volume and visitor activity. Hotel occupancy data for February in two consecutive years are as follows (Assume the two population proportions are different).

	Current Year	Previous Year
Occupied rooms	1580	1458
Total rooms	1860	1800

   a. Formulate the hypothesis test that can be used to determine if there has been an increase in the proportion of rooms occupied over the one-year period.
   b. Using $\alpha = 0.03$ level of significance, what is your hypothesis test conclusion?

2. Using SPSS data (*bankloan.sav*), determine if there is a significant difference in mean credit cards amongst those who previously defaulted and those who have not previously defaulted at $\alpha = 0.05$ level of significance

3. Using SPSS data (*test_scores.sav*), conduct determine if the pretest scores is significantly different from mean posttest scores at $\alpha = 0.05$ level of significance.

4. Consider the following income data from a study of 12 doctors from Kiambu County and 12 doctors from Nairobi County yields the following results (in Ksh. 1000):

Kiambu County	Nairobi County
268	380
274	364
282	300
291	364
234	322
235	403
261	384
272	238
330	342
371	300
245	244
301	271

At $\alpha = 0.02$, are the monthly income for doctors from Kiambu County are significantly different from the monthly income for doctors from Nairobi County?

12

Testing Hypothesis about More Than Two Populations

12.1 Introduction to Experimental Designs

An *experimental design* is *a plan and a structure to test hypotheses in which the researcher either controls or manipulates one or more variables*. It contains independent and dependent variables. In an experimental design, an *independent variable* may be either a treatment variable or a classification variable. A *treatment variable* is *a variable the experimenter controls or modifies in the experiment*. A *classification variable* is *some characteristic of the experimental subject that was present prior to the experiment and is not a result of the experimenter's manipulations or control*. Independent variables are sometimes also referred to as *factors*.

Experimental designs are analyzed statistically by a group of techniques referred to as *analysis of variance (ANOVA)*. There are several experimental designs including completely randomized design, randomized block design and matched pair design. In this case, we will consider completely randomized design (also known as one-way ANOVA).

12.2 The Completely Randomized Design (One-Way ANOVA)

The completely randomized design contains only one independent variable, with two or more treatment levels, or classifications. A completely randomized design is analyzed by a *one-way ANOVA*. The *one-way ANOVA*, provides methods for comparing the means of more than two populations. *ANOVA* uses F-test to compare means of more than two groups.

Assumptions (Conditions) for one-way ANOVA are as follows:

1. *Simple random samples*: The samples taken from the populations under consideration are simple random samples.
2. *Independent samples*: The samples taken from the populations under consideration are independent of one another.
3. *Normal populations*: For each population, the variable under consideration is normally distributed.
4. *Equal standard deviations*: The standard deviations of the variable under consideration are the same for all the populations.

12.3 Hypothesis Testing Procedure for One-Way ANOVA Using SPSS

The one-way ANOVA is used to determine if there is a statistically significantly difference in a given quantitative/numerical variable in more than two groups. For example, to determine if salary is statistically significant different between doctors, lecturers and lawyers.

Step 1: State the null and alternative hypotheses

$H_0: \mu_1 = \mu_2 = \ldots = \mu_k$ (All means for the groups are equal)

$H_a:$ *Not all the means are equal* (At least one group has a different mean)

DOI: 10.1201/9781003292654-15

Step 2: Decide/state the significance level, α. If not given the default $\alpha = 0.05$

Step 3: We key the data into SPSS and compute the value of the test statistic (F-test) and the df

$$F = \frac{MSTR}{MSE}$$

where the treatment mean square (MSTR) is computed as

$$MSTR = \frac{SSTR}{k-1}$$

where k denotes the number of groups and treatment sum of squares (SSTR) is computed as

$$SSTR = n_1 \left(\bar{x}_1 - \bar{x} \right)^2 + n_2 \left(\bar{x}_2 - \bar{x} \right)^2 + \cdots + n_k \left(\bar{x}_k - \bar{x} \right)^2$$

where $\bar{x}_1$ is the mean for group 1, $\bar{x}_2$ is the mean for group 2, n_1 is the sample for group 1, n_2 is the sample for group 2.

The error mean square (MSE) is computed as

$$MSE = \frac{SSE}{n-k}$$

where n denotes the total number of observations and

$$SSE = (n_1 - 1) s_1^2 + (n_2 - 1) s_2^2 + \ldots + (n_k - 1) s_k^2$$

where s_1 is the standard deviation for group 1, s_2 is the standard deviation for group 2.

The F-test is presented in one-way ANOVA table.

Source	SS	Degrees of Freedom	Mean Square	F-statistic	p-value
Treatment	SSTR	$k-1$	$MSTR = \dfrac{SSTR}{k-1}$	$F = \dfrac{MSTR}{MSE}$	
Error	Sum of squares due to regression	$n-k$	$MSE = \dfrac{SSE}{n-k}$		
Total	Sum of squares of errors	$n-1$			

To compute one-way ANOVA in SPSS we go **Analyze > Compare Means > One-Way ANOVA**. There will be SPSS pop up window for One-Way ANOVA > Select **Numerical Variable of interest** as **Dependent List** and **Grouping/Categorical Variable** as **Factor** > Click **Post-Hoc** select **Tukey** for Multiple comparisons > Click **Options** select **the appropriate confidence level** for Multiple comparisons > Click Continue then OK.

Step 4: Determine the p-value from the SPSS output. Alternatively, one may also compute the F critical value in SPSS using IDF.F(prob, k-1, n-k).

Step 5: Make decision on H_0 using p-value approach: compare p-value with α value

If p-value is less than (<) α, reject H_0; otherwise, do not reject H_0.

We can also use the critical value to make decision on H_0 i.e. if F-test > F critical value, reject H_0; otherwise fail to reject H_0.

Step 6: Make conclusions at a given level of confidence based on the decision in Step 5.

Note: If H_0 is rejected, conduct a Post-Hoc Test (Multiple Comparison Test) to determine which groups are significantly different from each other. Groups with p-values less than 0.05 are significantly different from each other. Show which groups are different and which groups are not different.

12.4 Multiple Comparison Tests

If the Null Hypothesis is rejected, then we conduct a multiple comparison test (post-hoc analyses) to determine which groups are significantly different from each other.

Tukey's HSD test is the simplest multiple comparison test that can be used to determine the mean of which groups are significantly different from each other.

Example

The data below shows residential energy consumption and expenditures from a recent energy consumption survey. Suppose that we want to decide whether a difference exists in mean annual energy consumption by households among the four counties. Use $\alpha = 0.05$.

Nairobi	Kisumu	Mombasa	Nakuru
15	17	11	10
10	12	7	12
13	18	9	8
14	13	13	7
13	15	9	12

Step 1: State the null and alternative hypothesis

$H_0: \mu_1 = \mu_2 = \mu_3 = \mu_4$ (mean energy consumptions across all the four counties are all equal)

Ha: Not all the means are equal (at least one county has a different mean energy consumption)

Step 2: State the significance level, $\alpha = 0.05$

Step 3: Compute the value of the test statistic (F-test) and the df

We key the data into SPSS (Figures 12.1 and 12.2)

Variable View

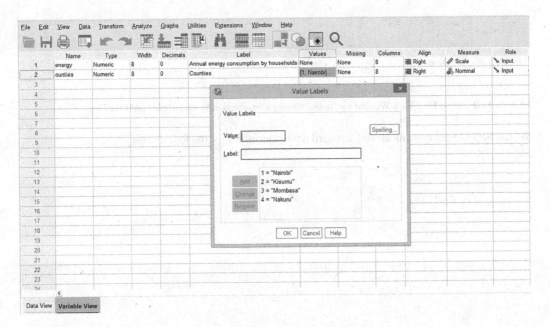

FIGURE 12.1 SPSS Variable View Window for Annual Energy Consumption by Households and Counties

Data View

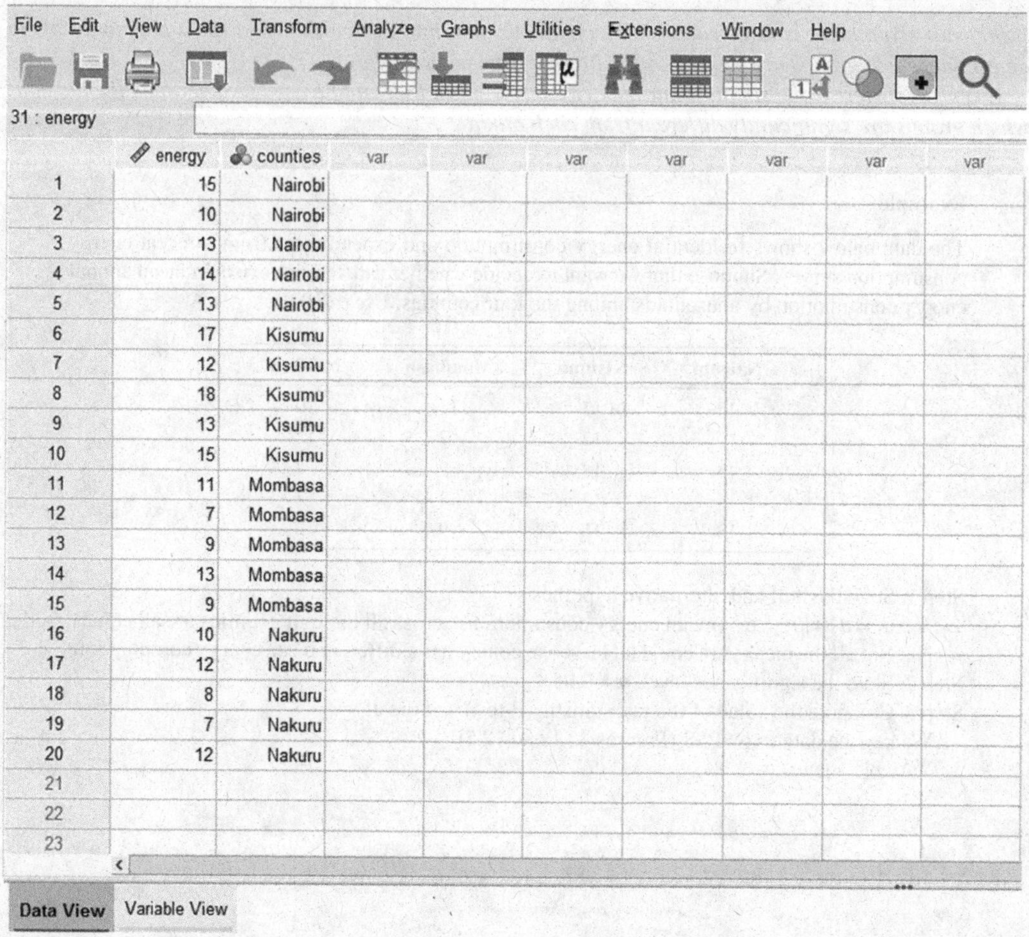

	energy	counties	var	var	var	var	var	var	var
1	15	Nairobi							
2	10	Nairobi							
3	13	Nairobi							
4	14	Nairobi							
5	13	Nairobi							
6	17	Kisumu							
7	12	Kisumu							
8	18	Kisumu							
9	13	Kisumu							
10	15	Kisumu							
11	11	Mombasa							
12	7	Mombasa							
13	9	Mombasa							
14	13	Mombasa							
15	9	Mombasa							
16	10	Nakuru							
17	12	Nakuru							
18	8	Nakuru							
19	7	Nakuru							
20	12	Nakuru							
21									
22									
23									

FIGURE 12.2 SPSS Data View Window for Annual Energy Consumption by Households and Counties

We use SPSS to compute mean and standard deviation values for the four groups.

Then we compute the mean and standard deviation values for the two groups using Analyze > Compare Means > Means > there will be a pop up window for Means > select Annual energy consumption by household as Dependent List and Counties as Independent List > Click Options and select Mean, Number of Cases and Standard Deviation as Cell Statistics then click Continue then OK (Figure 12.3).

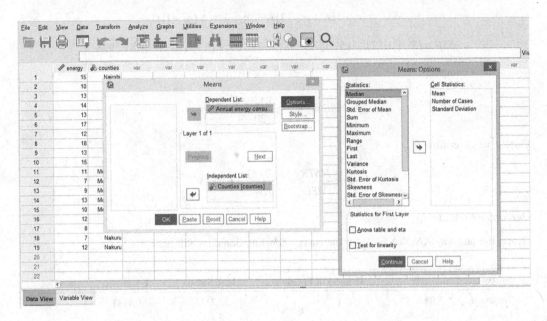

FIGURE 12.3 SPSS Frequency Menu for generating Mean and Standard Deviation Values for Annual Energy Consumption by Households per Counties

The SPSS output will give the sample, mean and standard deviation for each airline (Figure 12.4).

Report			
Annual Energy Consumption by Households			
Counties	**Mean**	**N**	**Std. Deviation**
Nairobi	13.00	5	1.871
Kisumu	15.00	5	2.550
Mombasa	9.80	5	2.280
Nakuru	9.80	5	2.280
Total	11.90	20	3.076

FIGURE 12.4 Mean and Standard Deviation and S.E. mean Values for Annual Energy Consumption by Households per Counties

We have $n_1 = 5$, $\bar{x}_1 = 13$, $s_1 = 1.871$, $n_2 = 5$, $\bar{x}_2 = 15$, $s_2 = 2.55$, $n_3 = 5$, $\bar{x}_3 = 9.8$, $s_3 = 2.28$, $k = 4$, $n_4 = 5$, $\bar{x}_4 = 9.8$, $s_4 = 2.28$, $n = 20$, $\bar{x} = 11.9$

$$SSTR = n_1(\bar{x}_1 - \bar{x})^2 + n_2(\bar{x}_2 - \bar{x})^2 + n_3(\bar{x}_3 - \bar{x})^2 + n_4(\bar{x}_4 - \bar{x})^2$$
$$SSTR = 5(13.0 - 11.9)^2 + 5(15 - 11.9)^2 + 5(9.8 - 11.9)^2 + 5(9.8 - 11.9)^2 = 98.2$$

So that we can compute

$$MSTR = \frac{SSTR}{k-1} = \frac{98.2}{4-1} = 32.733$$

$$SSE = (n_1 - 1)s_1^2 + (n_2 - 1)s_2^2 + (n_3 - 1)s_3^2 + (n_4 - 1)s_4^2$$

$$SSE = (5-1)1.871^2 + (5-1)2.55^2 + (5-1)2.28^2 + (5-1)2.2.28^2 = 81.5998$$

We can now compute

$$MSE = \frac{SSE}{n-k} = \frac{81.5998}{20-4} = 5.1$$

Finally, we determine F

$$F = \frac{MSTR}{MSE} = \frac{32.7333}{5.1} = 6.4183$$

One-way ANOVA table

Hence, the one-way ANOVA table for the energy consumption data will be:

Source	SS	Degrees of Freedom	Mean Square	F-statistic	p-value
Treatment	98.2	3	32.7333	6.4183	
Error	81.5998	16	5.1		
Total	79.7998	19			

To compute one-way ANOVA in SPSS we go **Analyze > Compare Means > One-Way ANOVA**. There will be SPSS pop up window for One-Way ANOVA > Select **Numerical Variable of interest (Annual energy consumption)** as **Dependent List** and **Grouping/Categorical Variable (Counties)** as **Factor** > Click **Post-Hoc** select **Tukey** for Multiple comparisons (Figure 12.5).

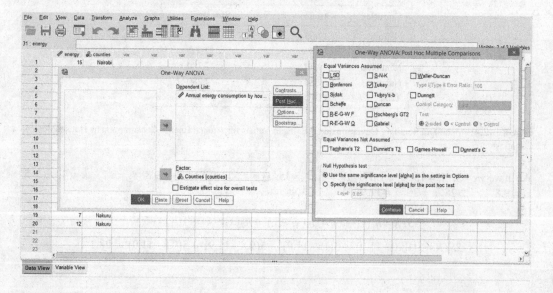

FIGURE 12.5   SPSS One-Way ANOVA Pop Up Window displaying Post-Hoc Multiple Comparisons Tests

Click **Options** select **the appropriate confidence level (0.95)** for Multiple comparisons > Click Continue then OK (Figure 12.6).

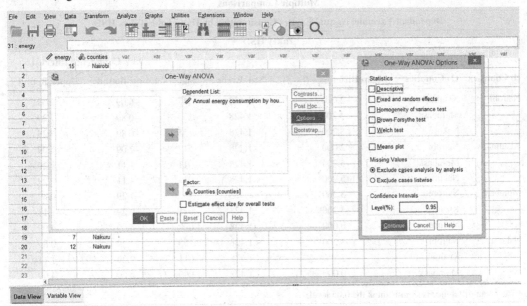

FIGURE 12.6 SPSS One-Way ANOVA Pop Up Window displaying One-Way ANOVA Options

The SPSS Output (Figure 12.7)

ANOVA					
Annual Energy Consumption by Households					
	Sum of Squares	df	Mean Square	F	Sig.
Between Groups	98.200	3	32.733	6.418	.005
Within Groups	81.600	16	5.100		
Total	179.800	19			

FIGURE 12.7 F-test for the One-way ANOVA

From the SPSS output select F-test and df
F-test = 6.418, df = 3, 16

Step 4: Determine the p-value (for F-test):

From the SPSS output, the p-value = 0.005. Using IDF.F(0.95, 3, 16) in SPSS we obtain F critical value = 3.2389.

Step 5: Decision on H_0 using p-value: We compare p-value = 0.005 with $\alpha = 0.05$

To make decision: If p-value is less than α value, reject H_0; otherwise, do not reject H_0.

Since p-value = 0.005 is less than (<) $\alpha = 0.05$ value, we reject H_0.

Decision on H_0 using critical value: We the F critical value = 3.2389 with F-test value = 6.418.

To make decision on H_0: if F-test > F critical value, reject H_0; otherwise fail to reject H_0.

Since F-test > F critical value, we reject H_0.

Step 6: Make conclusions at a given level of confidence based on the decision in Step 5.

At 95% level of confidence, at least one region has a different mean annual consumption. Hence, we conduct a post hoc to determine which regions have different annual energy consumption.

SPSS Post Hoc Test Output (Figure 12.8)

Post Hoc Tests

Multiple Comparisons

Dependent Variable: Annual Energy Consumption by Households Tukey HSD

Tukey HSD

(I) Counties	(J) Counties	Mean Difference (I – J)	Std. Error	Sig.	95% Confidence Interval	
					Lower Bound	Upper Bound
Nairobi	Kisumu	−2.000	1.428	.517	−6.09	2.09
	Mombasa	3.200	1.428	.155	−0.89	7.29
	Nakuru	3.200	1.428	.155	−0.89	7.29
Kisumu	Nairobi	2.000	1.428	.517	−2.09	6.09
	Mombasa	5.200ᵃ	1.428	.011	1.11	9.29
	Nakuru	5.200ᵃ	1.428	.011	1.11	9.29
Mombasa	Nairobi	−3.200	1.428	.155	−7.29	0.89
	Kisumu	−5.200ᵃ	1.428	.011	−9.29	−1.11
	Nakuru	0.000	1.428	1.000	−4.09	4.09
Nakuru	Nairobi	−3.200	1.428	.155	−7.29	0.89
	Kisumu	−5.200ᵃ	1.428	.011	−9.29	−1.11
	Mombasa	0.000	1.428	1.000	−4.09	4.09

ᵃ The mean difference is significant at the 0.05 level.

FIGURE 12.8 Post Hoc Test for the One-way ANOVA

From the SPSS post hoc test output, identify the mean differences with asterisk and those with no asterisk. Those with asterisk are statistically significant different while those without are not statistically significantly different.

From the post hoc table, the annual energy consumption for Midwest region is significantly different from the consumption of South region.

Annual energy consumption for Midwest is also significantly different from the consumption of West region. The rest of the regions have no difference in their energy consumptions.

Practice Exercise

1. Using bankloan.sav data, determine if there is a statistically significant difference in household income among the different levels of education. Use $\alpha = 0.01$ level of significance.

2. For the data set *car_insurance_claims.sav*, use determine if there is a statistical significant difference in average cost of claims among the various groups of vehicles. Use $\alpha = 0.05$.

3. The following data on paints drying time were obtained from the advertised five different paints.

Paint 1	Paint 2	Paint 3	Paint 4	Paint 5
119	145	144	151	144
117	133	144	152	126
126	143	137	155	113
123	137	136	150	135
121	131	141	153	129
130	142	139	147	138
134	124	142	160	144
122		145	165	115
119		149	154	129

At $\alpha = 0.01$ level of significance, determine if the type of paint has a statistical significant effect of drying time across the five different type of paint.

13

Testing Relationships about Categorical Data

13.1 The Chi-Square Distribution

A variable that has a chi-square distribution if its distribution has the shape of a special type of right-skewed curve is called a chi-square (χ^2) curve.

13.1.1 The Chi-Square Test

The chi-square test statistics is computed as follows:

$$\chi^2 = \sum_{i=1}^{c} \frac{(O_i - E_i)^2}{E_i}$$

where O_i are the given/observed values or frequencies, E_i are the expected or computed frequencies.

There are two types of chi-square tests:

1. Chi-square goodness-of-fit test
2. Chi-square test for independence

13.2 Chi-Square Goodness-of-Fit Test

The chi-square (χ^2) goodness of fit test statistic is a nonparametric test hat is used to make statistical inferences about the equality of population proportions for three or more populations. It can also be used to check if the data are consistent with the specified distribution. We will consider Chi-Square goodness of fit test to test if data are consistent with a specified distribution.

Assumptions of the chi-square goodness-of-fit test are as follows:

1. The variable must be a categorical variable (i.e. dichotomous, nominal or ordinal)
2. Observations must be independent, i.e. no relationships between any of the cases/participants
3. The groups of the categorical variable must be mutually exclusive
4. There must be at least five expected frequencies in each group of your categorical variable

13.2.1 Testing Hypothesis Steps for Chi-Square Goodness-of-Fit Test Using SPSS

Step 1: The null and alternative hypotheses are, respectively,

H_0: The variable has the specified distribution (variable/data are consistent with the specified distribution)

DOI: 10.1201/9781003292654-16

H_a: The variable does not have the specified distribution (variable/data are not consistent with the specified distribution)

Step 2: Decide on/State the significance level, α

Step 3: Compute the value of the test statistic

We create columns for observed values (O_i), expected frequencies $E_i = np_i$, $O - E$, $(O - E)^2$ and $(O - E)^2/E$

where p_i is the relative frequency or the proportion of each group.

x	O	$E_i = np_i$	$O - E$	$(O - E)^2$	$(O - E)^2/E$

Compute the chi-square test value as:

$$\chi^2 = \sum_{i=1}^{k} \frac{(O_i - E_i)^2}{E_i}$$

$df = k - 1$, where k is the number of possible values for the variable.

In Statistical Package for Social Science (SPSS), in the variable view create a variable for the categorical variables and code them then key the data (i.e. individual numbers for the different categories). Click Analyze > Nonparametric Tests > Legacy Dialogs > Chi-square. There will be a pop up window for Chi-square > Put the categorical variable into Test Variable List > Select All categorical equals under Expected Values **OR** To check for consistency with a given distribution, add the different proportions (in the order of the coded variables) under Expected Values > Click OK

Step 4: Determine the p-value or Chi-square critical value

From the SPSS Output select the p-value (Asymp. Sig). Alternatively, one may also compute the Ch-square critical value in SPSS using IDF.Chisq(prob, k-1).

Step 5: Decision on H_0 using p-value or critical value:

To obtain p-value: using the χ^2-table, we check along the *df* where the $\chi^2 - test$ value will fall. It will fall between some two values. Then check up the χ^2-table for the corresponding probability values.

SPSS will also give the p-value

Compare p-value with α

To make decision: If p-value $< \alpha$, reject H_0; otherwise, do not reject H_0.

We can also use the critical value to make decision on H_0 i.e. if Ch-square test > Chi-square critical value, reject H_0; otherwise fail to reject H_0.

Step 6: Make conclusions at a given level of confidence based on the decision in Step 5

Example

A random sample of 500 resignation reports from last year yielded the frequency distribution shown in the table below. At the 5% significance level, do the data provide sufficient evidence to conclude that last year's resignation distribution is different from the 2010 distribution?

Distribution of Reasons for Resignation from Work in the Nairobi Area in the Year 2010.

Reason for Resignation	Relative Frequency
Studies	0.011
New job	0.063
Personal	0.286
Retirement	0.640

Sample Results for 500 Randomly Selected Reasons for Resignation from Work From Last Year.

Reason for Resignation	Observed Frequency
Studies	3
New job	37
Personal	154
Retirement	306

Step 1: State the null and alternative hypotheses.

The null and alternative hypotheses are, respectively,

H_0: Last year's resignation distribution is the same as the 2000 distribution (is consistent).

H_a: Last year's resignation distribution is different from the 2000 distribution (is not consistent).

Step 2: State the significance level, α.

We are to perform the test at the 5% significance level, so $\alpha = 0.05$.

Step 3: Compute the value of the test statistic.

$$\chi^2 = \sum_{i=1}^{k} \frac{(O_i - E_i)^2}{E_i}$$

where O and E represent observed and expected frequencies, respectively.

Reason for Resignation x	Observed Frequency O	Expected Frequency $E_i = np_i$	Difference $O - E$	Square of Difference $(O-E)^2$	Chi-square sub-total $(O-E)^2 / E$
Studies	3	$500 \times 0.011 = 5.5$	−2.5	6.25	1.136
New job	37	$500 \times 0.063 = 31.5$	5.5	30.25	0.960
Personal	154	$500 \times 0.286 = 143$	11.0	121.00	0.846
Retirement	306	$500 \times 0.64 = 320$	−14.0	196.00	0.613
Total	$n = 500$	500			3.555

$$\chi^2 = \sum_{i=1}^{k} \frac{(O_i - E_i)^2}{E_i} = 3.555$$

$$df = k - 1 = 4 - 1 = 3$$

In SPSS, you first key the data into SPSS. There are two ways:

a. In SPSS, in the variable view create a variable for the categorical variables and code them, then key the data (i.e. individual numbers for the different categories) (Figures 13.1 and 13.2)

Variable View

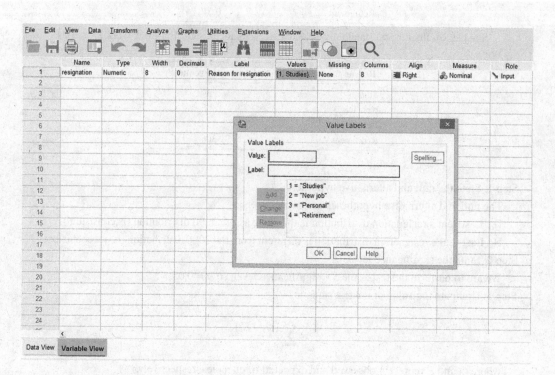

FIGURE 13.1   SPSS Variable View for Reasons for Resignation

Data View

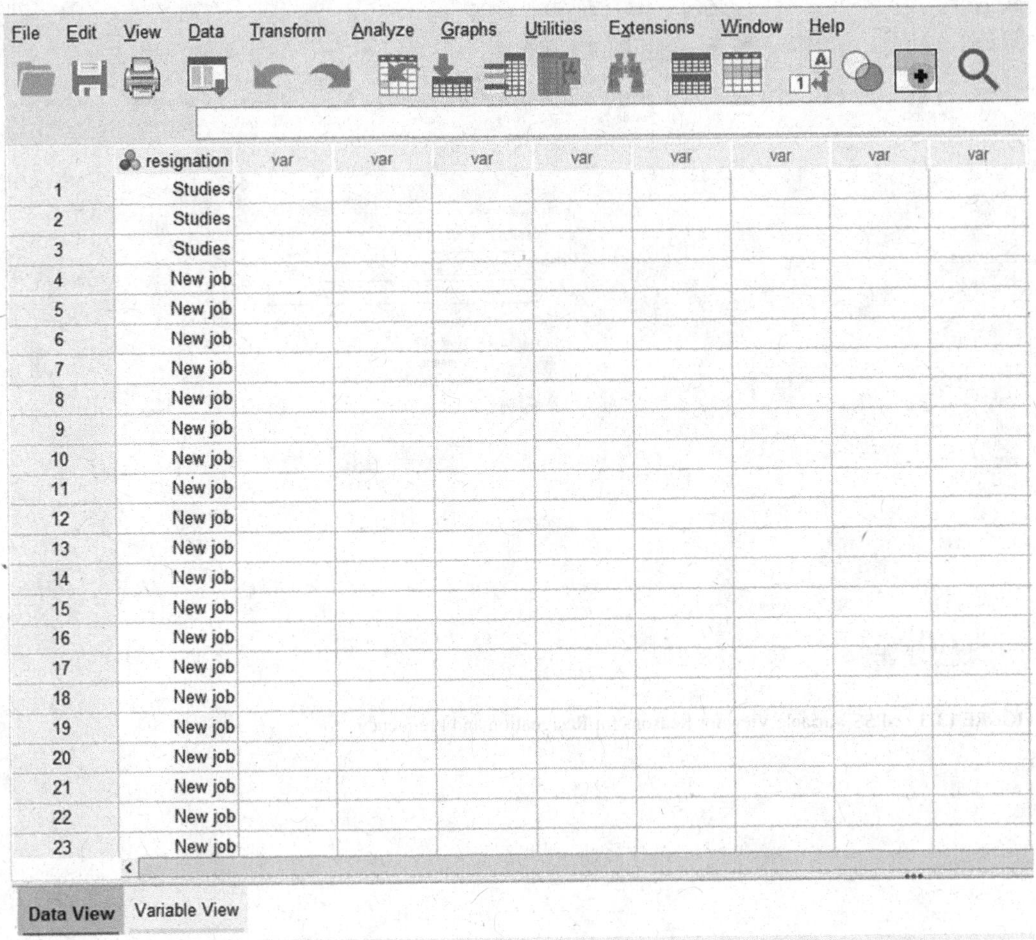

	resignation	var	var	var	var	var	var	var	var
	File **Edit** **View** **Data** **Transform** **Analyze** **Graphs** **Utilities** **Extensions** **Window** **Help**								
1	Studies								
2	Studies								
3	Studies								
4	New job								
5	New job								
6	New job								
7	New job								
8	New job								
9	New job								
10	New job								
11	New job								
12	New job								
13	New job								
14	New job								
15	New job								
16	New job								
17	New job								
18	New job								
19	New job								
20	New job								
21	New job								
22	New job								
23	New job								

Data View   Variable View

FIGURE 13.2 SPSS Data View for Reasons for Resignation

Alternatively, you can key data into SPSS as shown in the above frequency table.

b. Key in the data as shown in the given table. We create variable for reasons for resignation (coded 1 = studies, 2 = new job, 3 = personal, 4 = retirement), and variable for the given frequencies then key the data (Figures 13.3 and 13.4)

Variable View window

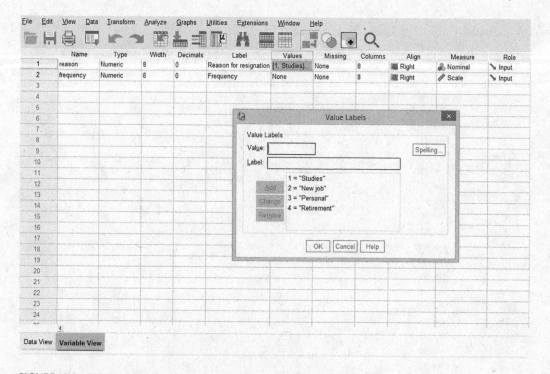

FIGURE 13.3   SPSS Variable View for Reasons for Resignation and Frequency

Data View window

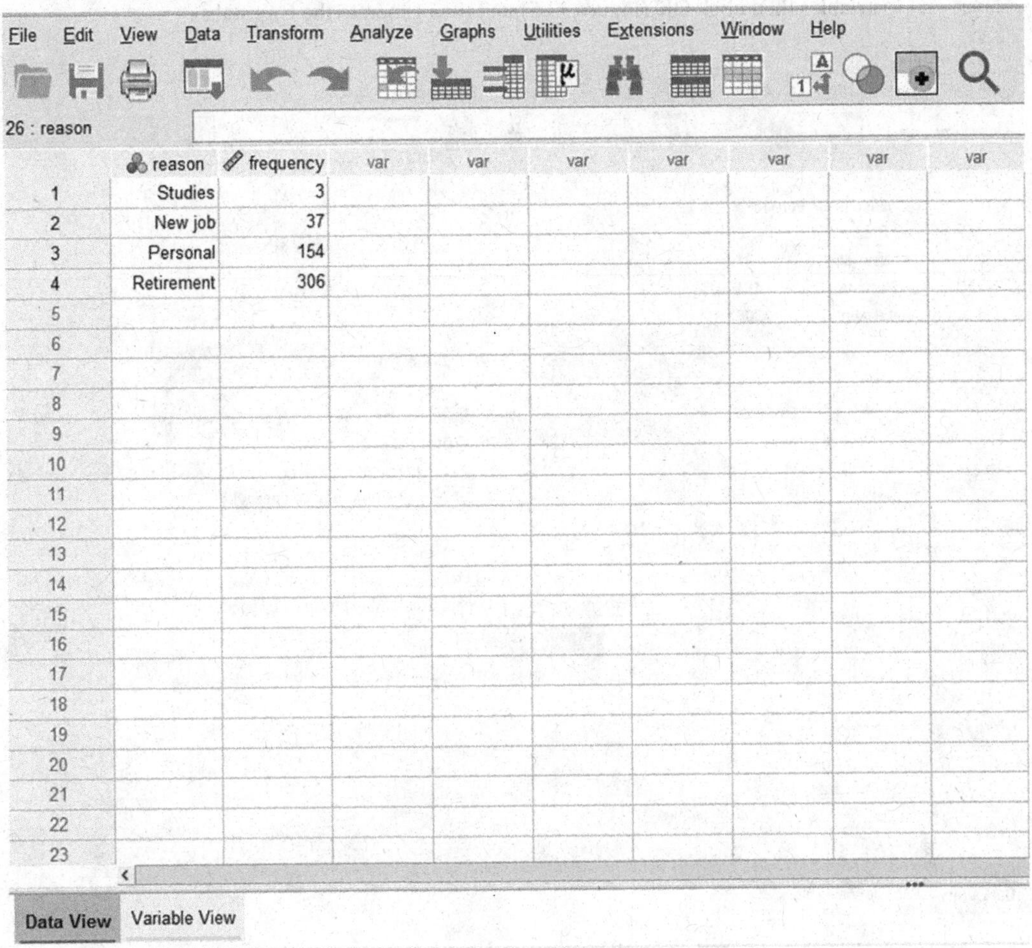
File Edit View Data Transform Analyze Graphs Utilities Extensions Window Help
26 : reason
reason frequency var var var var var var var
1 Studies 3
2 New job 37
3 Personal 154
4 Retirement 306
5
6
7
8
9
10
11
12
13
14
15
16
17
18
19
20
21
22
23
Data View Variable View

FIGURE 13.4 SPSS Data View for Reasons for Resignation and Frequency

Then weight the cases by going to Data > Weight cases > There will be a pop-up window for Weight Cases > Select Weight cases by > Transfer Frequencies to Frequency variables then click OK (Figure 13.5) and then generate the Crosstabs

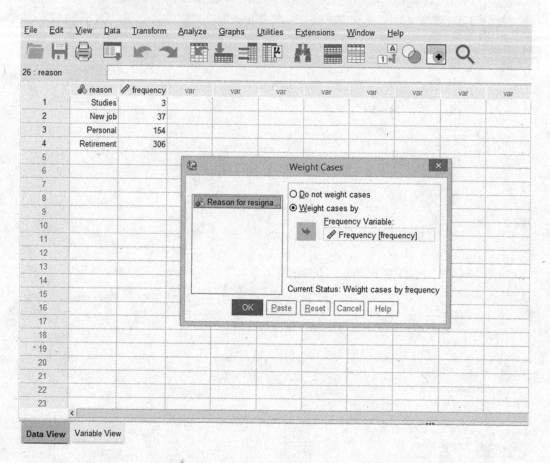

FIGURE 13.5   SPSS Pop Up Window for Weighting Cases by Frequencies

Using any of the keyed data sets (by method (a) and method (b)).

Click Analyze > Nonparametric Tests > Legacy Dialogs > Chi-square. There will be a pop up window for Chi-square > Put the categorical variable of interest (resignation) into Test Variable List > Select All categorical equals (for equal proportion) under Expected Values > Click OK (Figure 13.6).

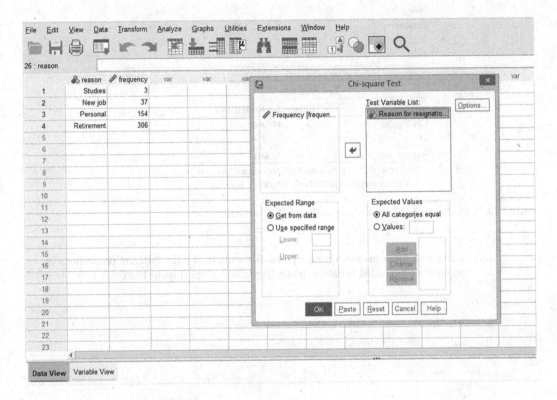

FIGURE 13.6 SPSS Chi-Square Test Pop Up Window for Equal Categories

The SPSS output will give chi-square test for equal proportions (Figure 13.7).

Chi-Square Test			
Frequencies			
Reason for resignation			
	Observed N	Expected N	Residual
Studies	3	125.0	−122.0
New job	37	125.0	−88.0
Personal	154	125.0	29.0
Retirement	306	125.0	181.0
Total	500		

Test Statistics	
	Reason for Resignation
Chi-Square	449.840[a]
df	3
Asymo. Sig.	<.001

[a] Zero cells (0.0%) have expected frequencies less than 5. The minimum expected cell frequency is 125.0.

FIGURE 13.7 Chi-Square Test Statistics Value for Reasons for Resignation

To check for consistency with a given distribution, add the different proportions (in the order of the coded variables) under Expected Values (Figure 13.8).

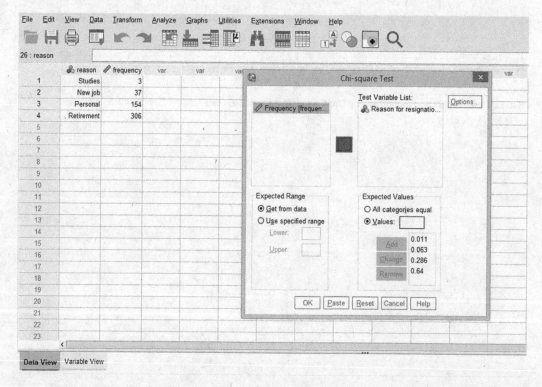

FIGURE 13.8 SPSS Chi-Square Test Pop Up Window for Different Categories

SPSS Output for equal proportions.

SPSS Output for consistency with a given distribution/different proportions (Figure 13.9).

Chi-Square Test			
Frequencies			
Reason for resignation			
	Observed N	Expected N	Residual
Studies	3	5.5	−2.5
New job	37	31.5	5.5
Personal	154	143.0	11.0
Retirement	306	320.0	−14.0
Total	500		

Test Statistics	
	Reason for Resignation
Chi-Square	3.555[a]
Df	3
Asymo. Sig.	.314

[a] Zero cells (0.0%) have expected frequencies less than 5. The minimum expected cell frequency is 5.5.

FIGURE 13.9   Chi-Square Test Statistics Value for Reasons for Resignation

Select the Chi-square test value, df and p-value

$$\chi^2 = 3.555, \text{df} = 3$$

Step 4: Determine the p-value or critical value.

From the SPSS output, the p-value (Asymp. Sig) = 0.314. Using IDF.Chisq(0.95, 3) in SPSS we obtain Ch-square critical value = 7.8147.

Step 5: Decision on H_0.

Decision using p-value: We compare p-value = 0.314 with α = 0.05.

To make decision: If the p-value is less than (<) α value, reject H_0; otherwise, do not reject H_0.

Since p-value is greater than (>) α value, we fail to reject H_0.

Decision on H_0 using critical value: We the Chi-square critical value = 7.8147 with Chi-square test value = 3.555.

To make decision on H_0: if Chi-square test > Chi-square critical value, reject H_0; otherwise fail to reject H_0.

Since Chi-square test < Chi-square critical value, we fail to reject H_0.

Step 6: Make conclusions at a given level of confidence based on the decision in Step 5.

At 95% level of confidence, last year's violent-crime distribution is same as the 2000 distribution.

13.3 Chi-Square Test for Independence

The chi-square test for independence is used to determine whether an association exists between two categorical variables, e.g. to determine if there is a statistical and significant relationship between gender (male or female) and employment status (i.e. employed or not employed). It is also known as chi-square test for association. Here we use joint table or contingency table.

Assumptions for chi-square test for independence are as follows:

- The two variables should be measured at an ordinal or nominal level (i.e., categorical data).
- The two variable should consist of two or more categorical, independent groups.

13.3.1 Hypothesis Testing Steps for Chi-Square Test for Independence

Step 1: The null and alternative hypotheses are, respectively,

H_0: Variable A is independent of Variable B.

H_a: Variable A is not independent of Variable B.

Step 2: Decide on/state the significance level, α.

Step 3: Compute the value of the test statistic.

We come up with a table of row totals, column totals and overall total.

Compute expected value of each observed value

$$E = \frac{(Row\ Total\ of\ a\ given\ Obs\ Value) \times (Column\ Totals\ of\ a\ given\ Obs\ Value)}{Overall\ Total}$$

$$\chi^2 = \sum_{i=1}^{k} \frac{(O_i - E_i)^2}{E_i}$$

$$df = (r - 1) \times (c - 1)$$

where $r = number\ rows$ and $c = number\ of\ columns$

Alternatively, we use SPSS if we have the dataset and not the contingency/joint table. If you only have the contingency table, then you first key in the individual observation for the two categorical variables.

Click Analyze – Descriptive Statistics – Crosstabs – Choose one categorical variable for row and the other categorical variable for columns, Under Statistics choose Chi-Square, Under Cells choose Observed and Expected.

From the SPSS output, select the chi-square test and df

Step 4: Determine the p-value.

From the SPSS output select the p-value (Asymp. Sig)

Step 5: Decision on H_0 using p-value:

For p-value: we compare p-value with α

To make decision: If p-value $< \alpha$, reject H_0; otherwise, do not reject H_0.

Step 6: Make conclusions at a given level of confidence based on the decision in Step 5.

Example

The table below shows the public opinion poll of 1000 surveyed voters. Respondents were classified by *gender (male or female)* and by *voting preference for candidates (X, Y or Z)*.

	Voting Preferences			
Gender	X	Y	Z	Total
Male	200	150	50	400
Female	250	300	50	600
Total	450	450	100	1000

Is there a gender gap? Do the men's voting preferences differ significantly from the women's preferences? Use a 0.05 level of significance.

SOLUTION:

Step 1: State the null and alternative hypotheses.

H_0: Gender is independent of voting preferences.

H_a: Gender is not independent of voting preferences.

Step 2: State the significance level, $\alpha = 0.05$.

Step 3: Compute the value of the test statistic.

We come up with a table of row totals, column totals and overall total.

Compute expected value of each observed value

$\dfrac{(400 \times 450)}{1000} = 180$	$\dfrac{(400 \times 450)}{1000} = 180$	$\dfrac{(400 \times 100)}{1000} = 40$
$\dfrac{(600 \times 450)}{1000} = 270$	$\dfrac{(600 \times 450)}{1000} = 270$	$\dfrac{(600 \times 100)}{1000} = 60$

$$\chi^2 = \sum_{i=1}^{k} \frac{(O_i - E_i)^2}{E_i}$$

$$\chi^2 = \frac{(200-180)^2}{180} + \frac{(150-180)^2}{180} + \frac{(50-40)^2}{40} + \frac{(250-270)^2}{270} + \frac{(300-270)^2}{270} + \frac{(50-60)^2}{60} = 16.2$$

$$df = (r-1)*(c-1) = (2-1)*(3-1) = 2$$

In SPSS, you first key the data into SPSS. There are two ways:

a. Key in the individual observation for the two categorical variables. We create variable for gender (coded 1 = male and 2 = female) and variable for preferred candidate (coded 1 = X, 2 = Y and 3 = Z) and then key the data (Figures 13.10 and 13.11)

Variable View

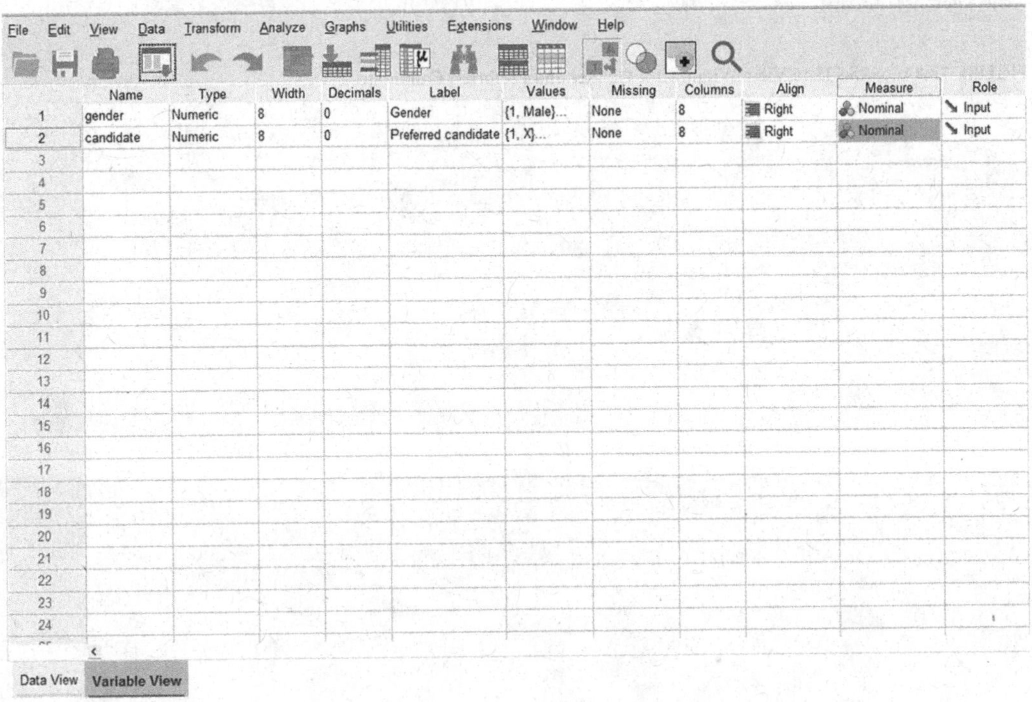

FIGURE 13.10 SPSS Variable View Window for Gender and Preferred Candidate

Data View

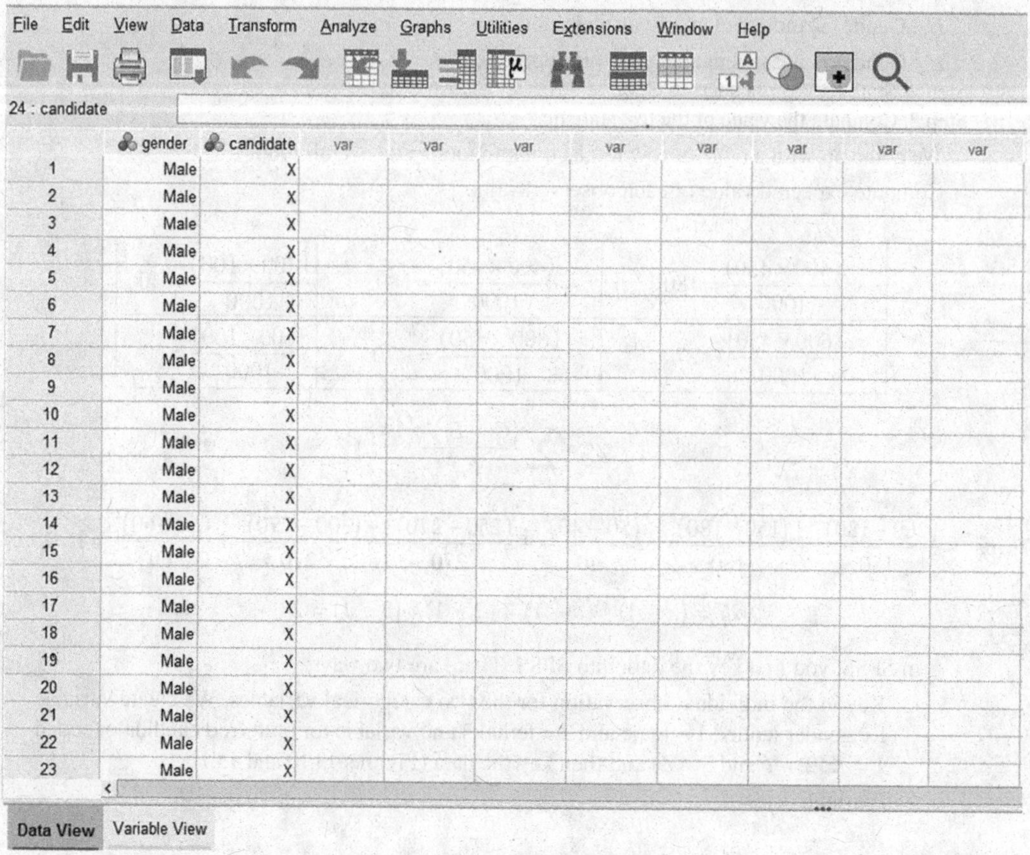

	gender	candidate	var	var	var	var	var	var	var	var
	File Edit View Data Transform Analyze Graphs Utilities Extensions Window Help										
24 : candidate											
1	Male	X								
2	Male	X								
3	Male	X								
4	Male	X								
5	Male	X								
6	Male	X								
7	Male	X								
8	Male	X								
9	Male	X								
10	Male	X								
11	Male	X								
12	Male	X								
13	Male	X								
14	Male	X								
15	Male	X								
16	Male	X								
17	Male	X								
18	Male	X								
19	Male	X								
20	Male	X								
21	Male	X								
22	Male	X								
23	Male	X								

Data View   Variable View

FIGURE 13.11   SPSS Data View Window for Gender and Preferred Candidate

Alternatively:

b. Key in the data as shown in the contingency table. We create variable for gender (coded 1 = male and 2 = female), variable for preferred candidate (coded 1 = X, 2 = Y and 3 = Z) and Scores/Totals for each of the two cases and then key the data (Figures 13.12 and 13.13)

Variable View

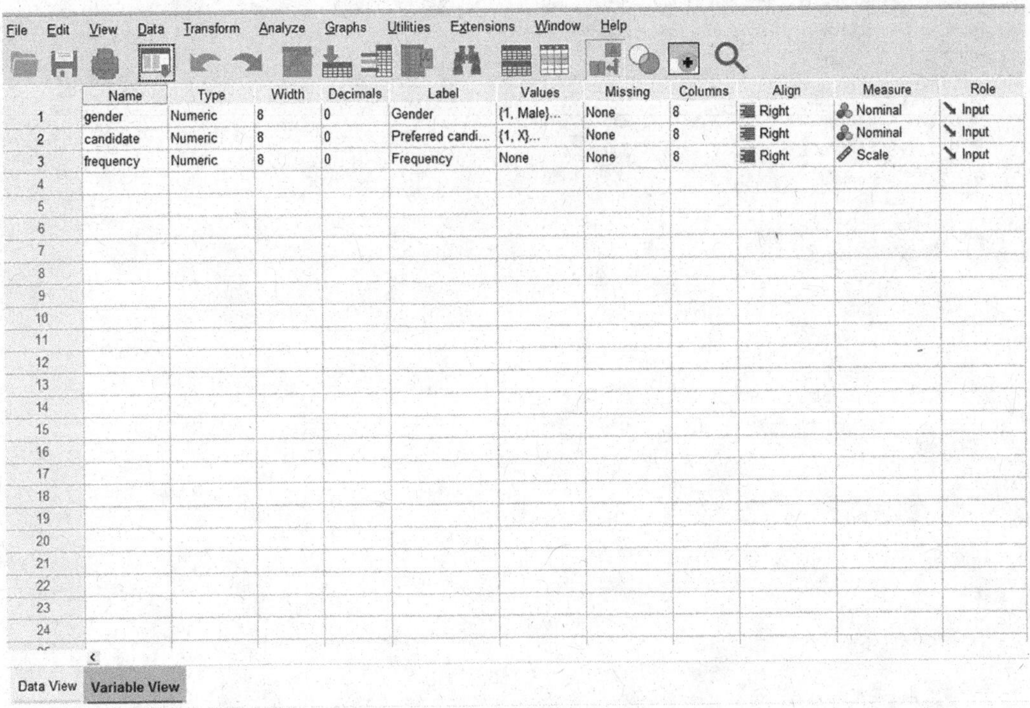

	Name	Type	Width	Decimals	Label	Values	Missing	Columns	Align	Measure	Role
1	gender	Numeric	8	0	Gender	{1, Male}...	None	8	Right	Nominal	Input
2	candidate	Numeric	8	0	Preferred candi...	{1, X}...	None	8	Right	Nominal	Input
3	frequency	Numeric	8	0	Frequency	None	None	8	Right	Scale	Input
4											
5											
6											
7											
8											
9											
10											
11											
12											
13											
14											
15											
16											
17											
18											
19											
20											
21											
22											
23											
24											

Data View **Variable View**

FIGURE 13.12 SPSS Variable View Window for Gender, Preferred Candidate and Frequency

Data View

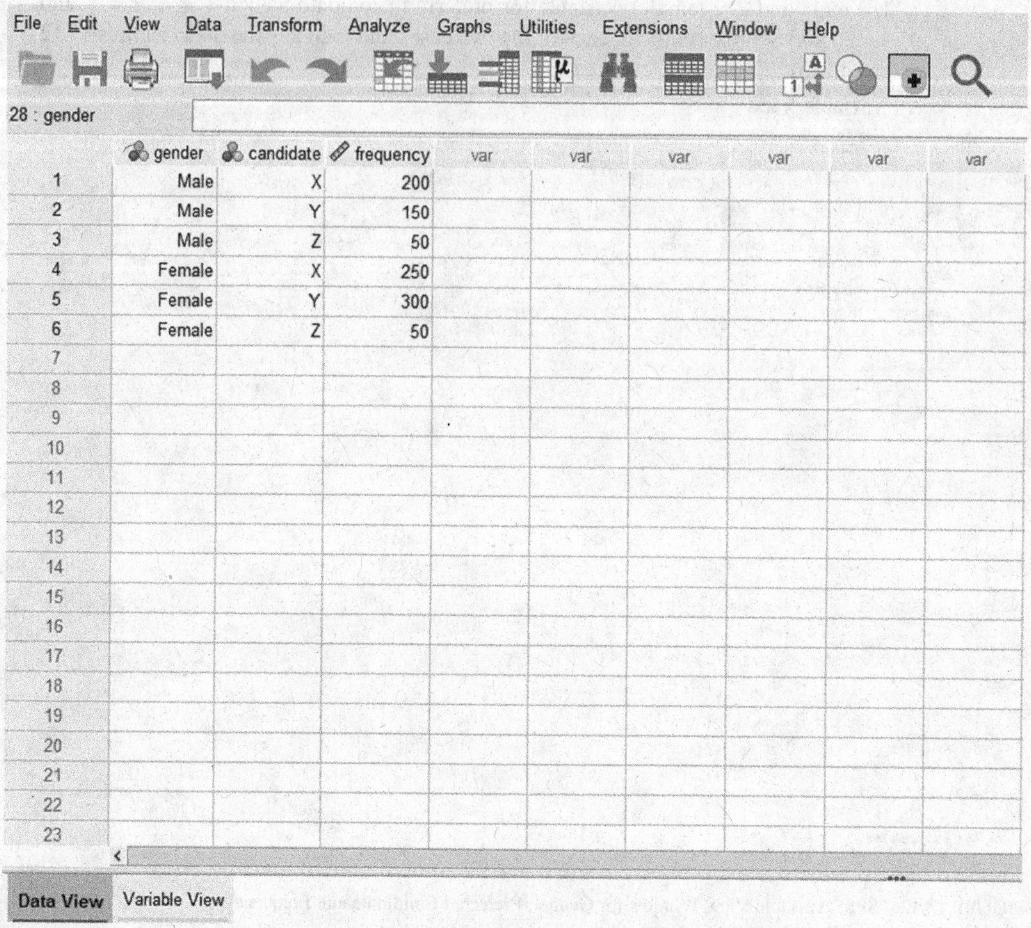

| File | Edit | View | Data | Transform | Analyze | Graphs | Utilities | Extensions | Window | Help |

28 : gender

	gender	candidate	frequency	var	var	var	var	var	var
1	Male	X	200						
2	Male	Y	150						
3	Male	Z	50						
4	Female	X	250						
5	Female	Y	300						
6	Female	Z	50						
7									
8									
9									
10									
11									
12									
13									
14									
15									
16									
17									
18									
19									
20									
21									
22									
23									

Data View Variable View

FIGURE 13.13   SPSS Data View Window for Gender, Preferred Candidate and Frequency

Then weight the cases by going to Data > Weight cases > There will be a pop-up window for Weight Cases > Select Weight cases by > Transfer Frequencies to Frequency variables then click OK and then generate the Crosstabs (Figure 13.14)

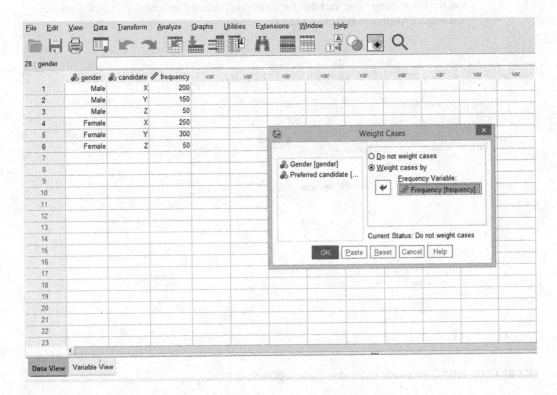

FIGURE 13.14 SPSS Pop Up Window for Weighting Cases by Frequencies

Using any of the keyed data sets (by method (a) and method (b)).

Click Analyze > Descriptive Statistics > Crosstabs > There will be a pop up SPSS window for Crosstabs: Cell Display > Choose one categorical variable for row (Gender) and the other categorical variable for columns (Preferred candidate) > Click Statistics and select Chi-Square (Figure 13.15)

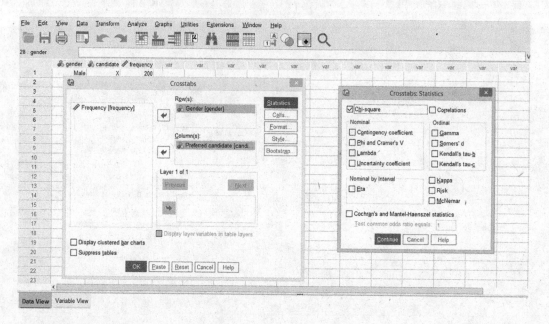

FIGURE 13.15   SPSS Crosstabs Pop Up Window displaying Chi-Square Test

Click Cells and select Observed and Expected > Click Continue then OK (Figure 13.16).

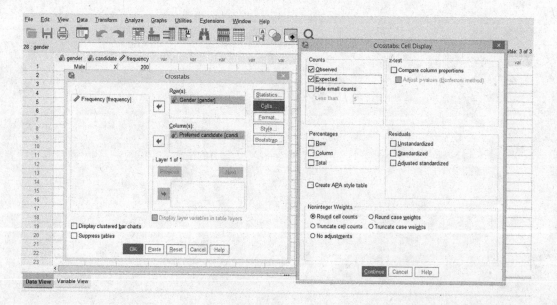

FIGURE 13.16   SPSS Crosstabs Pop Up Window displaying Cell Options

SPSS Output (Figure 13.17)

Gender ×* Preferred Candidate Crosstabulation

			Preferred candidate			
			X	Y	Z	Total
Gender	Male	Count	200	150	50	400
		Expected Count	180.0	180.0	40.0	400.0
	Female	Count	250	300	50	600
		Expected Count	270.0	270.0	60.0	600.0
Total		Count	450	450	100	1000
		Expected Count	450.0	450.0	100.0	1000.0

Chi-Square Tests

	Value	df	Asymptotic Significance (2-sided)
Pearson chi-square	16.204[a]	2	<.001
Likelihood ratio	16.266	2	<.001
Linear-by-linear association	0.974	1	.324
N of valid cases	1000		

[a] Zero cells (0.0%) have expected count less than 5. The minimum expected count is 40.00.

FIGURE 13.17   Chi-Square Test Statistics Value

From the SPSS output select the chi-square test, df and the p-value

$$\chi^2 = 16.204, df = 2$$

Step 4: Determine the p-value or critical value.

From the SPSS output, the p-value (Asymp. Sig) = <0.001. Using IDF.Chisq (0.95, 2) in SPSS we obtain Ch-square critical value = 5.9915.

Step 5: Decision on H_0 using p-value or critical value.

We compare p-value < 0.005 with $\alpha = 0.05$.

To make decision: If the p-value is less than (<) α value, reject H_0; otherwise, do not reject H_0.

Since p-value is less than (<) α value, we reject H_0.

Decision on H_0 using critical value: We the Chi-square critical value = 5.9915 with Chi-square test value = 16.204.

To make decision on H_0: if Chi-square test > Chi-square critical value, reject H_0; otherwise fail to reject H_0.

Since Chi-square test > Chi-square critical value, we reject H_0.

Step 6: Make conclusions at a given level of confidence based on the decision in Step 5.

At 95% level of confidence, gender is not independent of voting preferences, i.e. there is a relationship between gender and voting preference.

Example

Using *cereal.sav* data, test if *preference for breakfast* is independent of *gender* at $\alpha = 0.05$ level of significance.

SOLUTION:

Step 1: State the null and alternative hypotheses.

H_0: Gender is independent of preference for breakfast.

H_a: Gender is not independent of preference for breakfast.

Step 2: State the significance level, $\alpha = 0.05$.

Step 3: Compute the value of the test statistic.

First, we locate the cereal.sav data using the path:

For Windows

> Open **SPSS data** > Click **Open data document** folder > drive **(C):** > **Program Files (x86)** > **IBM** > **SPS** > **Statistics** > **27** > **Samples** > **English** > cereal.sav

For MAC

> Open **SPSS data** > Click **Open data document** folder > Applications > **IBM SPS Statistics 27** > **Samples** > **English** > cereal.sav

> Click Analyze > Descriptive Statistics > Crosstabs > There will be a pop up SPSS window for Crosstabs: Cell Display > Choose one categorical variable for Row (Gender) and the other categorical variable for Columns (Preferred breakfast) > Click Statistics and select Chi-Square (Figure 13.18)

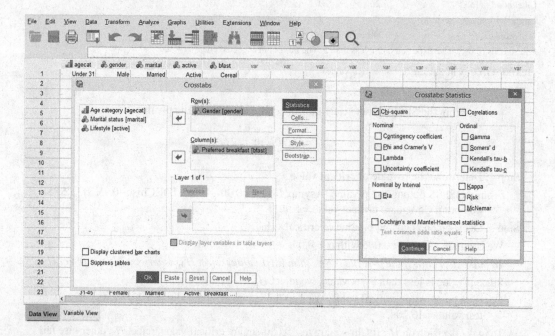

FIGURE 13.18   SPSS Crosstabs Pop Up Window displaying Chi-Square Test

Click Cells and select Observed and Expected > Click Continue then OK (Figure 13.19)

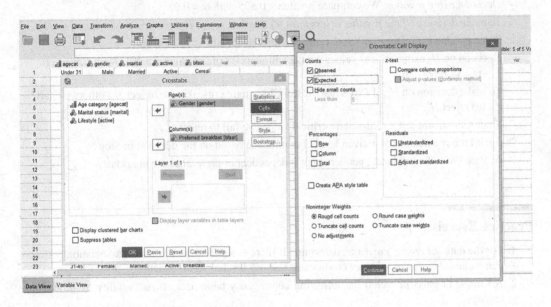

FIGURE 13.19 SPSS Crosstabs Pop Up Window displaying Cell Options

SPSS Output (Figure 13.20)

			Breakfast Bar	Oatmeal	Cereal	Total
Gender × Preferred Breakfast Crosstabulation						
			Preferred Breakfast			
Gender	Male	Count	104	155	165	424
		Expected Count	111.3	149.4	163.3	424.0
	Female	Count	127	155	174	456
		Expected Count	119.7	160.6	175.7	456.0
Total		Count	231	310	339	880
		Expected Count	231.0	310.0	339.0	880.0

Chi-Square Tests			
	Value	df	Asymptotic Significance (2-sided)
Pearson chi-square	1.367[a]	2	.505
Likelihood ratio	1.369	2	.504
Linear-by-linear association	0.577	1	.447
N of valid cases	800		

[a] Zero cells (0.0%) have expected count less than 5. The minimum expected count is 111.30.

FIGURE 13.20 Chi-Square Test Statistics Value

From the SPSS output select the chi-square test, df and p-value

$$\chi^2 - test = 1.367, \, df = 2$$

Step 4: Determine the p-value or critical value.

The p-value is the Asymp. Sig. (2-sided): p-value = 0.505. Using IDF.Chisq(0.95, 2) in SPSS we obtain Ch-square critical value = 5.9915.

Step 5: Decision on H_0 using p-value or critical value.

   Decision using p-value: We compare p-value $= 0.505$ with $\alpha = 0.05$.

   To make decision: If the p-value is less than (<) α value, reject H_0; otherwise, do not reject H_0.

   Since p-value is greater than (>) α value, we fail reject H_0.

   Decision on H_0 using critical value: We the Chi-square critical value $= 5.9915$ with Chi-square test value $= 1.367$.

   To make decision on H_0: if Chi-square test > Chi-square critical value, reject H_0; otherwise fail to reject H_0.

   Since Chi-square test < Chi-square critical value, we fail to reject H_0.

Step 6: Make conclusions at a given level of confidence based on the decision in Step 5.

   At 95% level of confidence, gender is not independent of preference for breakfast.

Practice Exercise

1. For the data set *credit_card.sav*, determine if there is a statistical significant association between gender and the type of primary credit card. Use 5% level of significance.

2. A sample of parts provided the following contingency table data on part quality by production batches.

Batch	Quality	
	Number Good	Number Defective
1	458	84
2	283	68
3	167	82

   Using $\alpha = 0.025$ level of significance is there any association between quality and production batch? What is your conclusion? What is the p-value? What is the critical value?

3. The population per county by the National Bureau together with their relative-frequency distribution per county in 2020 was as follows:

County	Kisumu	Mombasa	Nairobi	Nakuru
Relative frequency	0.17	0.231	0.372	0.227

   A simple random sample of this year's county population gave the following frequency distribution.

County	Kisumu	Mombasa	Nairobi	Nakuru
Frequency	56	41	97	72

   At 5% significance level, do the data provide sufficient evidence to conclude that this year's county population distribution has changed from the 2020 distribution?

14

Testing Relationships about Numerical Data

14.1 Inferences in Correlation Analysis Using SPSS

Correlation analysis is used to determine the strength and direction of relationship/association between two numerical/quantitative variables. For example, to determine if there is a statistical and significant relationship between salary and years of experience. There are two types: Pearson product moment and Spearman rank. In Chapter 4, we learnt about correlation as a measure of association, whereby we discussed different approaches for studying bivariate data including scatter plot and computing coefficient of correlation measure.

The Pearson product moment correlation coefficient (when using ratio data) is computed as:

$$r_{xy} = \frac{\frac{1}{n}\Sigma xy - \overline{xy}}{\sqrt{\left(\frac{1}{n}\Sigma x^2 - \overline{x}^2\right)\left(\frac{1}{n}\Sigma y^2 - \overline{y}^2\right)}}$$

Spearman rank coefficient of correlation (when data are ranked) is computed as:

$$\rho = 1 - \frac{6\Sigma d_i^2}{n(n^2-1)}$$

p-value is calculated using t distribution with n-2 degree of freedom.

In this chapter, we will be discussing inferences in correlation using Statistical Package for Social Sciences (SPSS). To make inferences about correlation analysis we use either Pearson coefficient of correlation or spearman rank coefficient of correlation.

In SPSS, we can generate Karl Pearson and Spearman Rank Coefficient of Correlation values using the path:

Analyze > Correlate > Bivariate > There will be a pop up SPSS window for Bivariate Correlations > Select the two numerical variables of interest and drag them to Variables Window > Click Pearson Correlation Coefficient and Spearman Correlation Coefficients then click OK.

Hypothesis Testing Steps in Correlation Analysis

    Step 1: The null and alternative hypotheses,

        H_0: No significant association (relationship) between Variable and Variable B

        H_a: There is a significant association (relationship) between Variable and Variable B

    Step 2: Decide on/state the significance level, α.

    Step 3: Compute the value of the test statistic – Pearson or Spearman rank coefficient of correlation.

DOI: 10.1201/9781003292654-17

The Pearson product moment correlation coefficient (when using ratio data) is computed as:

$$r_{xy} = \frac{\frac{1}{n}\sum xy - \overline{xy}}{\sqrt{\left(\frac{1}{n}\sum x^2 - \overline{x}^2\right)\left(\frac{1}{n}\sum y^2 - \overline{y}^2\right)}}$$

Spearman rank coefficient of correlation (when data are ranked) is computed as:

$$\rho = 1 - \frac{6\sum d_i^2}{n(n^2 - 1)}$$

In SPSS, we first enter the data into SPSS or locate the SPSS inbuilt data. Once you have the data

Using SPSS we go to Analyze > Correlate > Bivariate > There will be a pop up SPSS window for Bivariate Correlations > Select the two numerical variables of interest and drag them to Variables Window > Click Pearson Correlation Coefficient and Spearman Correlation Coefficients then click OK.

Step 4: Determine the p-value.

From the SPSS output, select the p-value

Step 5: Decision on H_0 using p-value:

For p-value: we compare p-value with α

To make decision: If p-value < α, reject H_0; otherwise, do not reject H_0.

Step 6: Make conclusions at a given level of confidence based on the decision in Step 5.

Note: Interpretation is in terms of strength and direction of relationship

Example: Determine if there is a significant association between *income* and *credit card debt* at $\alpha = 0.05$ level of significance (use Pearson coefficient of correlation)

Individual	Income	Credit Card Debt
1	1182	1116.5
2	1353	1005
3	1004	933.3
4	373	541.5
5	409	476.2
6	327	359.3

Step 1: State the null and alternative hypotheses,
  H_0: No significant association (relationship) between income and credit card debt
  H_a: There is a significant association (relationship) between income and credit card debt

Step 2: State the significance level, α. $\alpha = 0.05$

Step 3: Compute the value of the test statistic – Pearson or Spearman coefficient of correlation

We assign one variable x and the other variable y. In this case let income be x and credit be y. Then we generate a table for computing the values to be substituted into the formula

Individual	x	y	xy	x^2	y^2
1	1182	1116.5	1319703	1397124	1246572
2	1353	1005	1359765	1830609	1010025
3	1004	933.3	937033.2	1008016	871048.9
4	373	541.5	201979.5	139129	293222.3
5	409	476.2	194765.8	167281	226766.4
6	327	359.3	117491.1	106929	129096.5
Total	4648	4431.8	4130738	4649088	3776731

We can see that $\sum x = 4648$, $\sum y = 4431.8$, $\sum xy = 4130738$, $\sum x^2 = 4648$ and $\sum y^2 = 3776731$. We can compute means as $\bar{x} = \frac{\sum x}{n} = \frac{4648}{6} = 774.67$ and $\bar{y} = \frac{\sum y}{n} = \frac{4431.8}{6} = 738.63$. Substituting the values in to the formula

$$r_{xy} = \frac{\frac{1}{n}\sum xy - \overline{xy}}{\sqrt{\left(\frac{1}{n}\sum x^2 - \bar{x}^2\right)\left(\frac{1}{n}\sum y^2 - \bar{y}^2\right)}} = \frac{\frac{1}{6}\times 4130738 - (774.67 \times 735.63)}{\sqrt{\left(\frac{1}{6}\times 4648 - 774.67^2\right)\left(\frac{1}{6}\times 3776731 - 738.63^2\right)}} = 0.96033$$

Computing Spearman rank coefficient of correlation

Individual	x	y	x_i	y_i	$d_i = x_i - y_i$	d_i^2
1	1182	1116.5	2	1	1	1
2	1353	1005	1	2	−1	1
3	1004	933.3	3	3	0	0
4	373	541.5	5	4	1	1
5	409	476.2	4	5	−1	1
6	327	359.3	6	6	0	0
Total						4

We can see that $\sum d_i^2 = 4$
Substituting the values into the formula

$$\rho(x,y) = 1 - \left(\frac{6\sum d_i^2}{n(n^2-1)}\right) = 1 - \left(\frac{6\times 4}{6(6^2-1)}\right) = 0.88571$$

In SPSS, we first key the data into SPSS (Figure 14.1)
Variable View

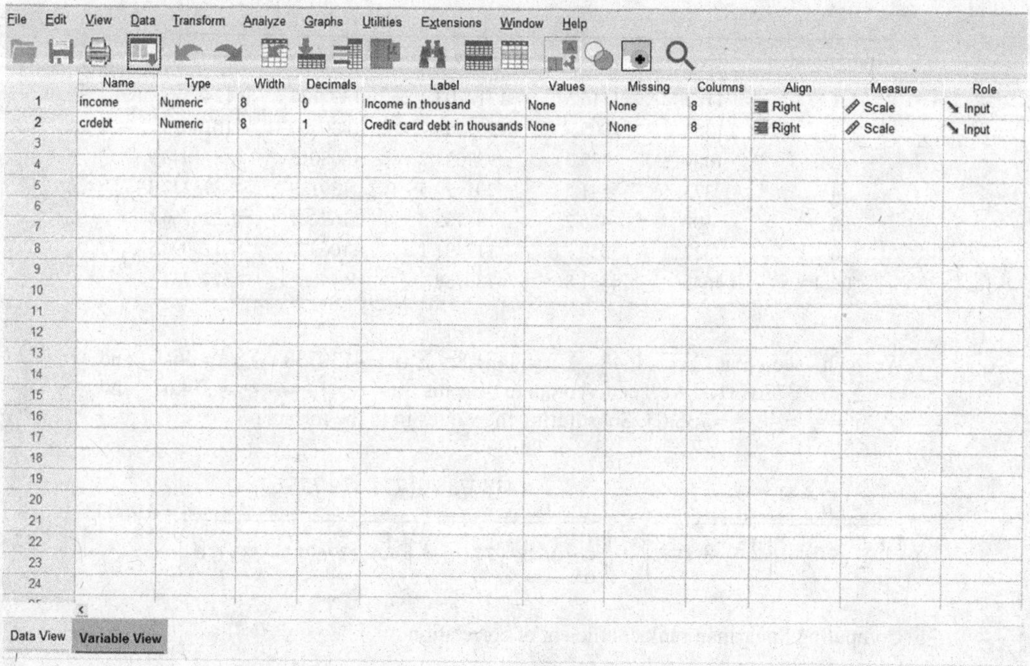
File Edit View Data Transform Analyze Graphs Utilities Extensions Window Help
Name Type Width Decimals Label Values Missing Columns Align Measure Role
1 income Numeric 8 0 Income in thousand None None 8 Right Scale Input
2 crdebt Numeric 8 1 Credit card debt in thousands None None 8 Right Scale Input
Data View
Variable View

FIGURE 14.1 SPSS Variable View Window for Income in Thousands and Credit Card Debt in Thousands

Data View (Figure 14.2)

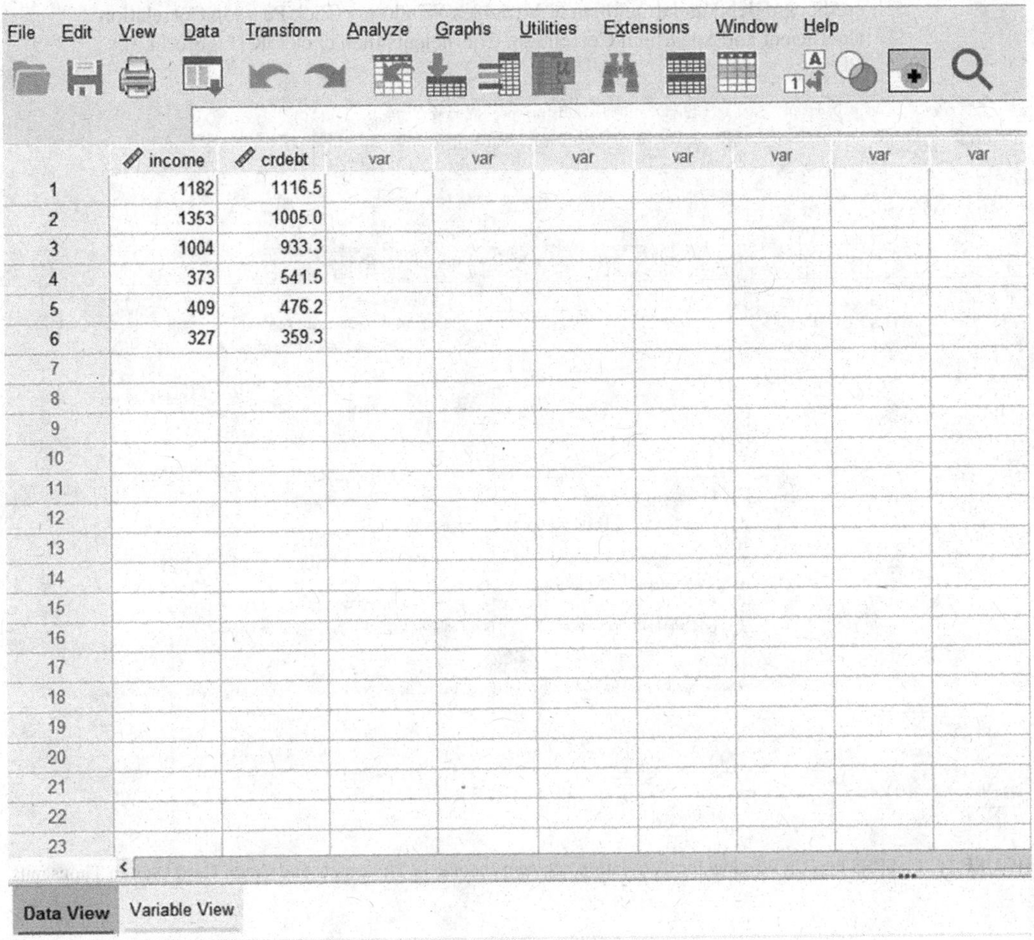

	income	crdebt	var	var	var	var	var	var	var
1	1182	1116.5							
2	1353	1005.0							
3	1004	933.3							
4	373	541.5							
5	409	476.2							
6	327	359.3							
7									
8									
9									
10									
11									
12									
13									
14									
15									
16									
17									
18									
19									
20									
21									
22									
23									

Data View   Variable View

FIGURE 14.2 SPSS Data View Window for Income in Thousands and Credit Card Debt in Thousands

We go to Analyze > Correlate > Bivariate > There will be a pop up SPSS window for
Bivariate Correlations > Select the two numerical variables of interest (income and
credit card debt) and drag them to Variables Window > Click Pearson Correlation
Coefficient and Spearman Correlation Coefficients then click OK (Figure 14.3).

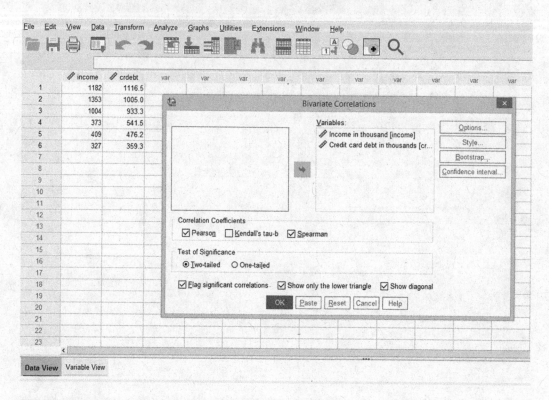

FIGURE 14.3   SPSS Pop Up Window for Correlation between Income in Thousands and Credit Card Debt in Thousands

SPSS Output (Figure 14.4)

Correlations		Income in Thousand	Credit Card Debt in Thousands
Income in thousand	Pearson correlation	–	
	N	6	
Credit card debt in thousands	Pearson correlation	.960**	–
	Sig. (2-tailed)	.002	
	N	6	6

** Correlation is significant at the 0.01 level (2-tailed).

Nonparametric Correlations			
Correlations			
		Income in Thousand	Credit Card Debt in Thousands
Spearman's rho	Income in thousand	Correlation coefficient	–	
		Sig. (2-tailed)		
		N	6	
	Credit card debt in thousands	Correlation coefficient	.886*	–
		Sig. (2-tailed)	.019	
		N	6	6

* Correlation is significant at the 0.05 level (2-tailed).

FIGURE 14.4 Correlation Values between Income in Thousands and Credit Card Debt in Thousands

From the SPSS Output select Pearson correlation value or Spearman rank correlation value together with their p-values
From the SPSS Output select Pearson correlation value

$$r(6) = 0.960$$

Step 4: Determine the p-value.
From the SPSS Output, select the p-value: This is the value of Sig.
p-value = 0.002
Step 5: Decision on H_0 using p-value:
For p-value: we compare p-value = 0.002 with $\alpha = 0.05$
<To make decision: If p-value < α, reject H_0; otherwise, do not reject H_0.
Since p-value < α, reject H_0
Step 6: Make conclusions at a given level of confidence based on the decision in Step 5.
At 95% level of confidence, there is a significant strong and positive association between income and credit card debt.

Example

Using bankloan.sav data, determine if there is a significant association between *household income* and *credit card debt* at $\alpha = 0.05$ level of significance (use Pearson coefficient of correlation).

SOLUTION:

Step 1: State the null and alternative hypotheses,

H_0: No significant association (relationship) between household income and credit card debt

H_a: There is a significant association (relationship) between household income and credit card debt

Step 2: State the significance level, α. $\alpha = 0.05$

Step 3: Compute the value of the test statistic – Pearson coefficient of correlation

First, we locate the bankloan.sav data using the path:

For Windows

Open **SPSS data** > Click **Open data document** folder > drive **(C):** > **Program Files (x86)** > **IBM** > **SPS** > **Statistics** > **27** > **Samples** > **English** > bankloan.sav

For MAC

Open **SPSS data** > Click **Open data document** folder > Applications > **IBM SPS Statistics 27** > **Samples** > **English** > bankloan.sav

Go to Analyze > Correlate > Bivariate > There will be a pop up SPSS window for Bivariate Correlations > Select the two numerical variables of interest (household income and credit card debt) and drag them to Variables Window > Click Pearson Correlation Coefficient and Spearman Correlation Coefficients then click OK (Figure 14.5).

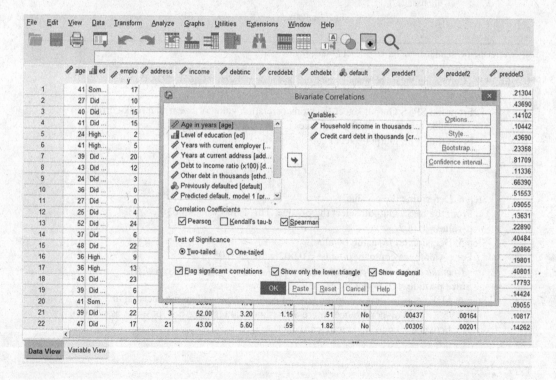

FIGURE 14.5 SPSS Pop Up Window for Correlation between Income in Thousands and Credit Card Debt in Thousands

SPSS Output (Figure 14.6)

		Correlations	
		Household Income in Thousand	**Credit Card Debt in Thousands**
Household income in thousand	Pearson correlation	–	
	N	850	
Credit card debt in thousands	Pearson correlation	.552**	–
	Sig. (2-tailed)	<.001	
	N	850	850

** Correlation is significant at the 0.01 level (2-tailed).

Nonparametric Correlations			
		Correlations	
		Household Income in Thousand	**Credit Card Debt in Thousands**
Spearman's rho	Household income in thousand	Correlation coefficient	–	
		Sig. (2-tailed)	.	
		N	850	
	Credit card debt in thousands	Correlation coefficient	.503**	–
		Sig. (2-tailed)	<.001	
		N	850	850

** Correlation is significant at the 0.01 level (2-tailed).

FIGURE 14.6 Correlation Values between Income in Thousands and Credit Card Debt in Thousands

From the SPSS Output, select Pearson or Spearman correlation value and p-value. Using Pearson coefficient of correlation

$$r(850) = 0.552$$

Step 4: Determine the p-value.
   From the SPSS Output select the p-value: This is the value of Sig.
   p-value < 0.001.
Step 5: Decision on H_0 using p-value:
   For p-value: we compare p-value < 0.001 with $\alpha = 0.05$
   To make decision: If p-value < α, reject H_0; otherwise, do not reject H_0.
   Since p-value < α, reject H_0
Step 6: Make conclusions at a given level of confidence based on the decision in Step 5.
   At 95% level of confidence, there is a significant strong and positive association between household income and credit card debt.

14.2 Regression Analysis Using SPSS

Regression analysis is a mathematical measure of the average relationship between dependent and independent(s) numerical variables. For example, salary and years of experience. In linear regression, there are two types of variable: independent and dependent. *Dependent variable* is the outcome variable and is usually denoted by y, e.g. salary. *Independent variable* is the variable that is resulting into the outcome and is usually denoted by x, e.g. years of experience. There are two types of linear regression analysis: *simple linear* and *multiple linear regression analysis*. Simple linear regression: there is one dependent and one independent variable. The mathematical expression for simple linear

$$y = \beta_0 + \beta_1 X + \varepsilon$$

Multiple linear regression: there is one dependent and several independent variables. The mathematical expression for simple linear

$$y = \beta_0 + \beta_1 X_1 + \beta_2 X_2 + \cdots + \beta_n X_n + \varepsilon$$

We will first consider simple linear regression and then multiple linear regression.

Assumptions of the Regression Model

The following are the assumptions of simple regression analysis.

1. The two variables should be measured at the continuous level (i.e. they are either interval or ratio variables).
2. There needs to be a linear relationship between the two variables. In this case, we conduct test for linearity (deviation from linearity test) or use a scatter plot to visually check for linearity.
3. There should be no significant outliers. This can be done by visually inspecting the box plot.
4. The observations are independent. This can easily be assess using the Durbin-Watson statistic.
5. The data must not show multicollinearity. Multicollinearity is tested by inspecting the Tolerance/Variance Inflation Factor (VIF) values.
6. The data needs to show homoscedasticity. This is where the variances along the line of best fit remain similar as you move along the line, i.e. constant variances. This can be easily be done using heteroscedasticity scatter plot and statistical test for equality of variance such Levene's test for equality of variance.
7. The residuals (errors) are approximately normally distributed. This can be assessed using a histogram (with a superimposed normal curve) or a Normal P-P plot. We can also use Shapiro wiki test for normality, skewness and kurtosis values.

Testing for the assumptions of linear regression

1. There needs to be a linear relationship between the two variables. In this case we conduct test for linearity (deviation from linearity test) or use a scatter plot to visually check for linearity

Testing for linearity using deviation from linearity test

In SPSS Data View Window: Go to Analyze > Compare Means > Means > There will be SPSS pop up window for Means > put the *dependent numerical variable* under Dependent List box and the *independent numerical variable* under Independent List box > Click Options then under Statistics for First Layer select Test for linearity > under Cell Statistics select Mean, Number of Cases and Standard Deviation then click Continue then OK. Go to SPSS output and select the ANOVA Table and interpret the relationships between the two numerical variables using F-test and p-value of the F-test.

Testing for linearity using scatter plot

Go to Graphs > Legacy Dialogs > Scatter/ Dot > select Simple then click Define > select *the first numerical variable of interest* and put it in Y axis, then select *the second numerical variable of interest* and put on the X axis then click OK. Go to SPSS output and select the scatter plot then interpret the relationship between the two variables.

2. There should be no significant outliers. This can be done by visually inspecting the boxplot

Testing for outliers using boxplot

Analyze > Descriptive Statistics > Explore > Select *the numerical variable of interest* as Dependent List > Click Plots and put uncheck Histogram and Stem-and-Leaf Plot. Under Boxplot select Factor levels together and click Continue > Under Display select Plots and then click OK.

Alternatively

Graphs > Legacy Dialogs > Boxplot > select Simple and select Summaries for separate variables then click Define > select *the numerical variable of interest* and put it under Boxplot Represent then click OK.

3. The observations are independent. This can easily be assess using the Durbin-Watson statistic.

We go to Analyze > Regression > Linear > there will be SPSS pop up window for Linear Regression> select the *dependent numerical variable of interest* and put it under Dependent and the independent numerical variable of interest under Independent(s)> click Statistics and select Durbin-Watson under residuals. Under Regression Coefficients select Estimates and Model fit > Click Continue then OK. Go to SPSS output window and select the Durbin-Watson test statistics value from the Model Summary table.

4. The data must not show multicollinearity. Use SPSS to detect for multicollinearity by inspecting the Tolerance/VIF values.

We go to Analyze > Regression > Linear > there will be SPSS pop up window for Linear Regression> select the *dependent numerical variable of interest* and put it under Dependent and the independent numerical variable of interest under Independent(s)> click Statistics and select Estimates, Model fit and Collinearity diagnostics > Click Continue then OK. Go to SPSS output window and select the VIF value from the coefficients table.

5. The data needs to show homoscedasticity. This is where the variances along the line of best fit remain similar as you move along the line i.e. constant variances. This can be easily be done using heteroscedasticity scatter plot and statistical test for equality of variance such levene's test for equality of variance.

To generate heteroscedasticity scatter plot we go to: Analyze > Regression > Linear > there will be SPSS pop up window for Linear Regression> select the *dependent numerical variable of interest* and put it under Dependent and the independent numerical variable of interest under Independent(s) > click Statistics and select Estimates and Model fit > Click Continue then put ZRESID (residual values) in the Y box and ZPRED (predicted values) in the X box then click continue then OK. Go to SPSS output window and select the Scatterplot.

6. The residuals (errors) are approximately normally distributed. This can be assessed using a histogram (with a superimposed normal curve) or a Normal P-P Plot. We can also use shapiro wiki test for normality, skewness and kurtosis values

To generate Normal PP Plot to test for the normality assumption go to: Analyze > Regression > Linear > there will be SPSS pop up window for Linear Regression> select the *dependent numerical variable of interest (outcome variable)* and put it under Dependent and the *independent numerical variable of interest (predictor variable)* under Independent(s) > click Statistics and select Estimates and Model fit > Click Continue then click Plots and put ZRESID (residual values) in the Y box and ZPRED (predicted values) in the X box. Check Normal probability plot then click continue then OK. Go to SPSS output window and select the Normal PP plot.

Alternatively

Go to Analyze > Descriptive Statistics > Explore > there will be a pop up window for Explore > put both the *dependent (outcome) numerical variable of interest* and the *dependent (outcome) numerical variable of interest* under Dependent List > click Plots then select None under Boxplot and Histogram under Descriptive. Check Normality plots with tests then click Continue then under Display select Both then OK. Go to SPSS output and select Descriptives table, Test for Normality table, Histogram and Normal QQ plot.

Example

Using advert.sav data, tests for the assumptions of linear regression with Detrended sales as the dependent variable and Advertising spending as the independent variable.

SOLUTION:

First we locate the advert.sales data

For Windows
Open **SPSS data** > Click **Open data document** folder > drive **(C):** > **Program Files (x86)** > **IBM** > **SPS** > **Statistics** > **27** > **Samples** > **English** > advert.sav

For MAC
Open **SPSS data** > Click **Open data document** folder > Applications > **IBM SPS Statistics 27** > **Samples** > **English** > advert.sav

While at the Data View Window, go to Analyze > Regression > Linear > there will be SPSS pop up window for Linear Regression> put the *dependent numerical variable of interest* (Detrended sales) in the Dependent box and the *independent numerical variable of interest* (Advertising spending) in the Independent(s) > click Statistics button and under Regression Coefficients select Estimates, Model fit and Collinearity diagnostics. Under Residuals select Durbin-Watson and click Continue (Figure 14.7)

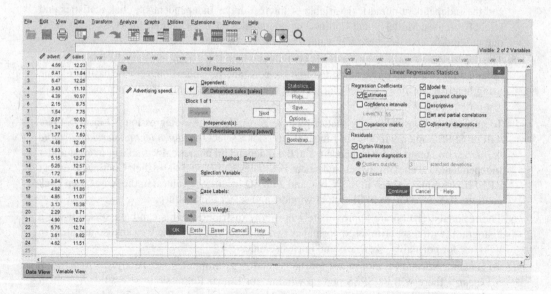

FIGURE 14.7 SPSS Pop Up Window Displaying Options for Statistics Button for Linear Regression Analysis

Click Plots button and put ZRESID (residual values) in the Y box and ZPRED (predicted values) in the X box. Check Normal probability plot then click Continue then OK (Figure 14.8).

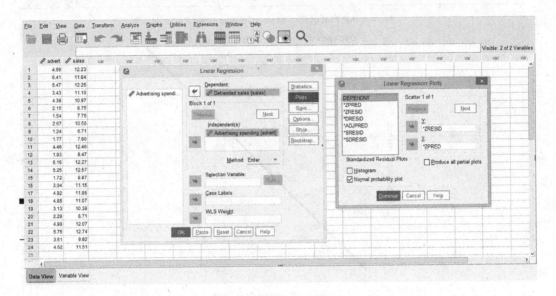

FIGURE 14.8 SPSS Pop Up Window Displaying Options for Plots Button for Linear Regression Analysis

Go to SPSS output window and select Model Summary table (Figure 14.9), Coefficients table (Figure 14.10), Normal PP plot (Figure 14.11) and Scatterplot (Figure 14.12).

Model Summary[b]

Model	R	R Square	Adjusted R Square	Std. Error of the Estimate	Durbin-Watson
1	.916[a]	.839	.832	.73875	2.389

a. Predictors: (Constant), Advertising spending
b. Dependent Variable: Detrended sales

FIGURE 14.9 Model Summary Table for Linear Regression between Detrended Sales and Advertising Spending

The Durbin-Watson statistics is 2.389 indicating that the observations are independent

Coefficients[a]

Model	Unstandardized Coefficients B	Unstandardized Coefficients Std. Error	Standardized Coefficients Beta	t	Sig.	Collinearity Statistics Tolerance	Collinearity Statistics VIF
1 (Constant)	6.584	.402		16.391	<.001		
Advertising spending	1.071	.100	.916	10.703	<.001	1.000	1.000

a. Dependent Variable: Detrended sales

FIGURE 14.10 Coefficients Table for Linear Regression between Detrended Sales and Advertising Spending

The VIF value is below 10 indicating no multicollinearity.

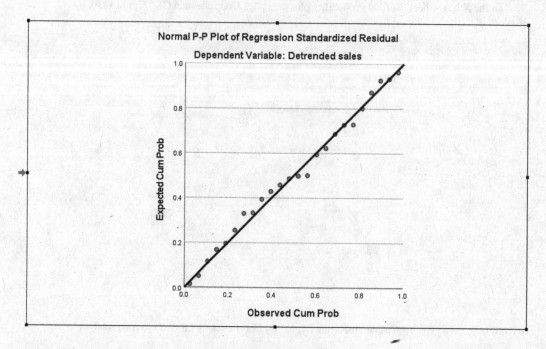

FIGURE 14.11   Normal PP Plot of Regression Standardized Residuals

The little circles are following the normality line indicating that the normality assumption has been met.

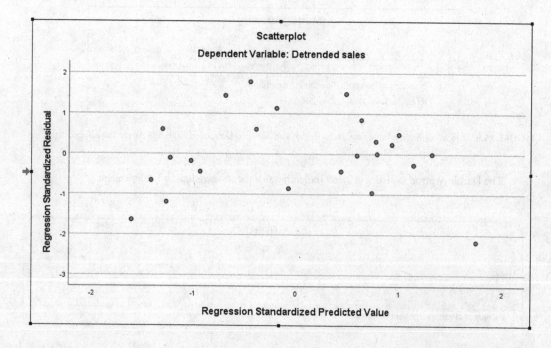

FIGURE 14.12   Heteroscedasticity Scatterplot for Linear Regression between Detrended Sales and Advertising Spending

There is no clear pattern in the scatterplot indicating that the assumption of homoscedasticity has been met.

Testing for outliers using boxplot

Graphs > Legacy Dialogs > Boxplot > select Simple and select Summaries for separate variables then click Define > select *the numerical variable of interest* (Detrended sales and Advertising spending) and put it under Boxplot Represent then click OK (Figure 14.13).

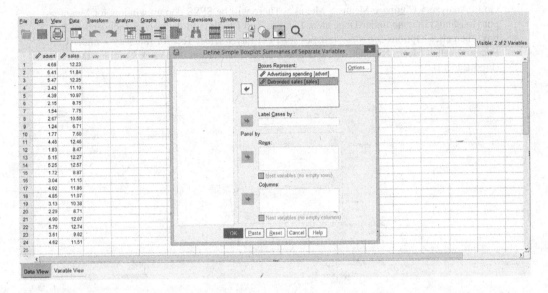

FIGURE 14.13   SPSS Pop UP Window Displaying Options for Boxplot

Go to SPSS Output and select the boxplot (Figure 14.14)

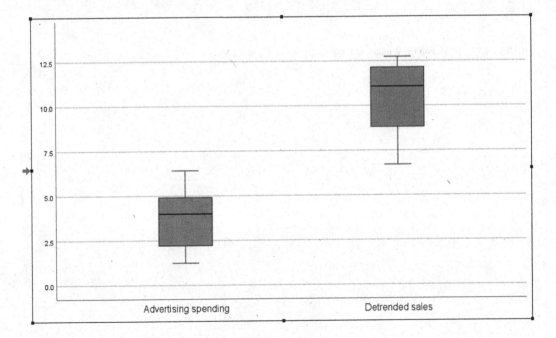

FIGURE 14.14   Boxplot for Advertising Spending and Detrended Sales

The boxplot shows that there are no outliers for Advertising Spending variable and Detrended Sales variable

To Test for linearity

Example: Using bankloan.sav data test for linearity between household income (dependent) and credit card debt

Using deviation from linearity test
Go to Analyze > Compare Means > Means > There will be SPSS pop up window for Means > put Household income under Dependent List box and Credit card debt under Independent List box then click Options > under Statistics for First Layer select Test for linearity > under Cell Statistics select Mean, Number of Cases and Standard Deviation then click Continue then OK (Figure 14.15).

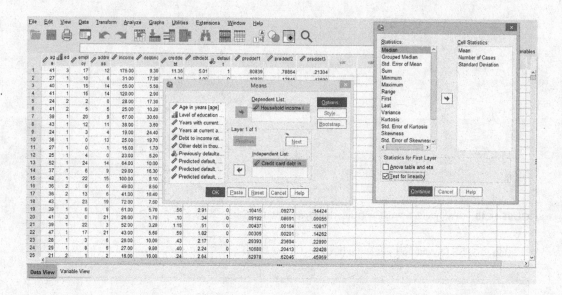

FIGURE 14.15   SPSS Pop Up Window Displaying Options for Test for Means

Go to SPSS output and select the ANOVA Table (Figure 14.16) and interpret the relationships between the two numerical variables using F-test and p-value of the F-test.

ANOVA Table

			Sum of Squares	df	Mean Square	F	Sig.
Household income in thousands * Credit card debt in thousands	Between Groups	(Combined)	1260253.881	845	1491.425	6.011	.045
		Linearity	383632.278	1	383632.278	1546.125	<.001
		Deviation from Linearity	876621.604	844	1038.651	4.186	.084
	Within Groups		992.500	4	248.125		
	Total		1261246.381	849			

FIGURE 14.16 ANOVA Table for Linearity Test

The p-value for Deviation from Linearity Test is 0.084 which is greater than 0.05. This shows that there is a linear relationship between the Household income and Credit card debt.

Testing for linearity using scatter plot

Go to Graphs > Legacy Dialogs > Scatter/ Dot > select Simple then click Define > select *Household income* and put it in Y axis, then select *Credit card debt* and put on the X axis then click OK (Figure 14.17).

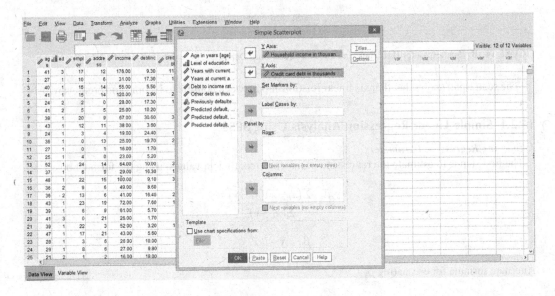

FIGURE 14.17 SPSS Pop Up Window Displaying Options for Scatterplot

Go to SPSS output (Figure 14.18) and select the scatter plot then interpret the relationship between the two variables.

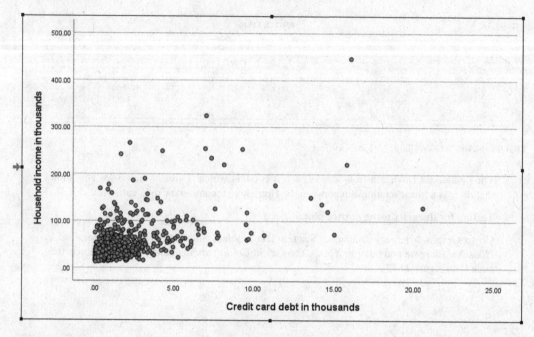

FIGURE 14.18   Scatterplot for Household Income and Credit Card Debt

The scatterplot shows that there is linear relationship between household income and credit card debt.

14.2.1 Simple Linear Regression Analysis Using SPSS

Estimating Regression Parameters

The first step is to estimate the regression parameters (coefficient values) using the formula below:

$$\beta_1 = \frac{\sum(x_i - \overline{x})(y_i - \overline{y})}{\sum(x_i - \overline{x})^2}$$

$$\beta_0 = \overline{y} - b_1\overline{x}$$

Alternate formula for estimating β_1

$$\beta_1 = \frac{\sum xy - \dfrac{(\sum x \sum y)}{n}}{\sum x^2 - \dfrac{(\sum x)^2}{n}}$$

SPSS will compute the regression coefficient values once you have entered your data in to SPSS.

In SPSS we go to Analyze > Regression > Linear > Choose the numerical variables of interest for Dependent variable and another numerical variable of interest for Independent variable > Click Statistics and select Estimates > Continue then OK > In the SPSS output use the coefficient table to select the values for β_0 and β_1.

Once we have the values for β_0 and β_1 we substitute the values into the equation $\hat{y} = \beta_0 + \beta_1 X$ and have a linear regression model. The estimated regression equation can be used to predict future values.

Interpretation:

- β_0 is the constant: without any value of x, y will be at β_0 value.
- β_1 is the slope or the rate of change in y with respect to x:- for every unit increase/decrease in x, y will be increasing or decreasing by β_1 value. Decrease is when we have a negative β_1 value and increase is when we have a positive β_1 value.

Example:

Develop an estimated regression equation that can be used to predict credit card debt given the income. Interpret the values of β_0 and β_1.

Individual	Income	Credit Card Debt
1	1182	1116.5
2	1353	1005
3	1004	933.3
4	373	541.5
5	409	476.2
6	327	359.3

Credit card debt is the outcome/dependent variable while income is the independent/explanatory variable. We assign income be x and credit be y. Then we generate a table for computing the values to be substituted into the formula.

Individual	x	y	xy	x^2	y^2
1	1182	1116.5	1319703	1397124	1246572
2	1353	1005	1359765	1830609	1010025
3	1004	933.3	937033.2	1008016	871048.9
4	373	541.5	201979.5	139129	293222.3
5	409	476.2	194765.8	167281	226766.4
6	327	359.3	117491.1	106929	129096.5
Total	4648	4431.8	4130738	4649088	3776731

We can see that $\sum x = 4648$, $\sum y = 4431.8$, $\sum xy = 4130738$, and $\sum x^2 = 4649088$. We can compute means as $\bar{x} = \frac{\sum x}{n} = \frac{4648}{6} = 774.67$ and $\bar{y} = \frac{\sum y}{n} = \frac{4431.8}{6} = 738.63$. Substituting the values in to the formula.

We first compute β_1

$$\beta_1 = \frac{\sum xy - \frac{(\sum x \sum y)}{n}}{\sum x^2 - \frac{(\sum x)^2}{n}} = \frac{4130738 - \frac{(4648 \times 4431.8)}{6}}{4649088 - \frac{4648^2}{6}} = 0.6653$$

$$\beta_0 = \bar{y} - b_1 \bar{x}$$

$$\beta_0 = 738.63 - (0.6653 \times 774.67) = 223.242$$

In SPSS, we first key the data into SPSS
Variable View

	Name	Type	Width	Decimals	Label	Values	Missing	Columns	Align	Measure	Role
1	income	Numeric	8	0	Income in thousand	None	None	8	Right	Scale	Input
2	crdebt	Numeric	8	1	Credit card debt in thousands	None	None	8	Right	Scale	Input
3											
4											
5											
6											
7											
8											
9											
10											
11											
12											
13											
14											
15											
16											
17											
18											
19											
20											
21											
22											
23											
24											

Data View   Variable View

SPSS Variable View Window for Income in Thousands and Credit Card Debt in Thousands

Data View

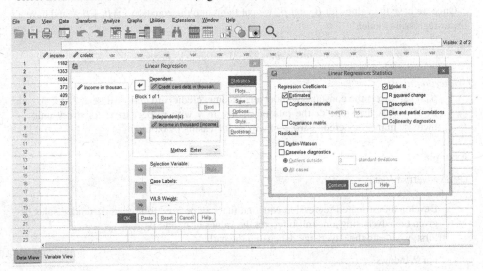

SPSS Data View Window for Income in Thousands and Credit Card Debt in Thousands

We go to Analyze > Regression > Linear > There will be a pop up SPSS window for regression > Choose Credit card debt for Dependent and Income for Independent(s) > Click Statistics and select Estimates > Continue then OK (Figure 14.19)

FIGURE 14.19 SPSS Pop Up Window for Linear Regression between Income in Thousands and Credit Card Debt in Thousands

SPSS Output (Figure 14.20)

Model Summary

Model	R	R Square	Adjusted R Square	Std. Error of the Estimate
1	.960[a]	.922	.903	98.9107

a. Predictors: (Constant), Income in thousand

ANOVA[a]

Model		Sum of Squares	df	Mean Square	F	Sig.
1	Regression	464122.846	1	464122.846	47.440	.002[b]
	Residual	39133.267	4	9783.317		
	Total	503256.113	5			

a. Dependent Variable: Credit card debt in thousands
b. Predictors: (Constant), Income in thousand

Coefficients[a]

Model		Unstandardized Coefficients		Standardized Coefficients	t	Sig.
		B	Std. Error	Beta		
1	(Constant)	223.215	85.032		2.625	.058
	Income in thousand	.665	.097	.960	6.888	.002

a. Dependent Variable: Credit card debt in thousands

FIGURE 14.20   Linear Regression Values for Income in Thousands and Credit Card Debt in Thousands

In the SPSS output (Figure 14.21), use the coefficient table to select the values for β_0 and β_1.

$$\beta_0 = 223.215 \text{ and } \beta_1 = 0.665$$

Once we have the values for β_0 and β_1, we put them into the equation $\hat{y} = \beta_0 + \beta_1 X$ and have a linear regression model.

$$Credit\ card\ debt = 223.215 + 0.6653 \times Income$$

Interpretation:
- $\beta_0 = 223.215$ is the constant: without any income, credit card debt will be at 223.215.
- $\beta_1 = 0.665$ is the slope or the rate of change in y with respect to x: for every unit increase in income, credit card debt will be increasing by 0.665.

Example

Using advert.sav data:

   a. Develop an estimated regression equation that can be used to predict detrended sales given the advertising amount. Interpret the values of β_0 and β_1
      First, we locate the advert.sav data using the path:

      For Windows
      Open **SPSS data** > Click **Open data document** folder > drive **(C)**: > **Program Files (x86)** > **IBM** > **SPS** > **Statistics** > **27** > **Samples** > **English** > advert.sav

For MAC

Open **SPSS data** > Click **Open data document** folder > Applications > **IBM SPS Statistics 27** > **Samples** > **English** > advert.sav

We go to Analyze > Regression > Linear > There will be a pop up SPSS window for regression > Choose the numerical variables of interest (Detrended sales) for Dependent and another numerical variable of interest (Advertising spending) for Independent(s) > Click Statistics and select Estimates > Continue then OK (Figure 14.21)

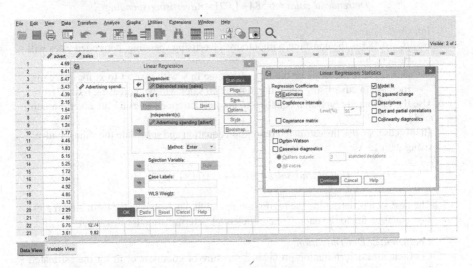

FIGURE 14.21 SPSS Pop Up Window for Linear Regression between Detrended Sales and Advertising Spending

SPSS Output (Figure 14.22)

Model Summary

Model	R	R Square	Adjusted R Square	Std. Error of the Estimate
1	.916[a]	.839	.832	.73875

a. Predictors: (Constant), Advertising spending

ANOVA[a]

Model		Sum of Squares	df	Mean Square	F	Sig.
1	Regression	62.514	1	62.514	114.548	<.001[b]
	Residual	12.006	22	.546		
	Total	74.520	23			

a. Dependent Variable: Detrended sales

b. Predictors: (Constant), Advertising spending

Coefficients[a]

Model		Unstandardized Coefficients B	Std. Error	Standardized Coefficients Beta	t	Sig.
1	(Constant)	6.584	.402		16.391	<.001
	Advertising spending	1.071	.100	.916	10.703	<.001

a. Dependent Variable: Detrended sales

FIGURE 14.22 Linear Regression Values for Detrended Sales and Advertising Spending

In the SPSS output (Figure 14.22), use the coefficient table to select the values for β_0 and β_1.

$$\beta_0 = 6.584 \text{ and } \beta_1 = 1.071$$

Once we have the values for β_0 and β_1, we put them into the equation $\hat{y} = \beta_0 + \beta_1 X$ and have a linear regression model.

$$Detrendend \ sales = 6.584 + 1.071 \times Advertising \ spending$$

Interpretation:
- $\beta_0 = 6.584$ is the constant: without any advertising spending, detrended sales will be at 6.584.
- $\beta_1 = 1.071$ is the slope or the rate of change in y with respect to x: for every unit increase in advertising spending, detrended sales will be increasing by 1.071.

b. Use the estimated regression equation to predict detrended sales for an advert amount of 13.

In this case, we use the estimated regression equation and substitute the value of advertising with 13

$$Detrendend \ sales = 6.584 + (1.071 \times 13) = 20.507$$

The next step is determine the regression model fit, i.e. how bets the estimated regression equation fits to the data. Use coefficient of determination (or R square)

Coefficient of Determination
Coefficient of determination provides a measure of goodness of fit for the estimated regression line/equation. It is denoted by r^2. r^2 is computed as:

$$r^2 = \frac{SSR}{SST}$$

where the total sum of squares (SST) is given by

$$SST = \sum (y_i - \overline{y})^2$$

sum of squares due to regression

$$SSR = \sum (\widehat{y_i} - \overline{y})^2$$

SSR is a measure of how much the y values on the estimated regression line deviates from y.
Sum of squares due to error (SSE) is given by

$$SSE = \sum (y_i - \widehat{y_i})^2$$

SSE is a measure of the error in using the estimated regression equation to estimate the values of the dependent variable.

Interpretation of r^2: The coefficient of determination, or r^2, is the proportion of the total variation in Y "explained" by the regression of Y on X. Usually, R square value greater than 0.5 implies a better model fit.

Example: Consider the following data for income and credit card debt. Estimate and interpret the coefficient of determination value for the regression equation with credit card debt as the outcome variable.

Individual	Income	Credit Card Debt
1	1182	1116.5
2	1353	1005
3	1004	933.3
4	373	541.5
5	409	476.2
6	327	359.3

Credit card debt is the outcome/dependent variable, while income is the independent/explanatory variable. We assign income be x and credit be y. Then we generate a table for computing the values to be substituted into the formula.

Individual	x	y	y^2
1	1182	1116.5	1246572
2	1353	1005	1010025
3	1004	933.3	871048.9
4	373	541.5	293222.3
5	409	476.2	226766.4
6	327	359.3	129096.5
Total	4648	4431.8	3776731

We have already estimated the values for β_0 and β_1 and the regression equation is given by

$$Credit\ card\ debt = 223.215 + 0.6653 \times Income$$

We can now use the estimated regression equation to estimate the SSR, SSE and SST values

$$SST = \Sigma(y_i - \bar{y})^2 = \Sigma y^2 - \frac{(\Sigma y)^2}{n} = 3776731 - \frac{4431.8^2}{6} = 503255.7933$$

SSR is given by

$$SSR = \Sigma(\hat{y}_i - \bar{y})^2$$

We have

$$\bar{y} = \frac{\Sigma y}{n} = \frac{4431.8}{6} = 738.63$$

We create a table to generate SSR values

Individual	x	y	$\hat{y} = 223.215 + 0.6653x$	$\hat{y}_i - \bar{y}$	$(\hat{y}_i - \bar{y})^2$
1	1182	1116.5	1009.627	270.9966	73439.15721
2	1353	1005	1123.393	384.7629	148042.4892
3	1004	933.3	891.2032	152.5732	23278.58136
4	373	541.5	471.3989	−267.2311	71412.46081
5	409	476.2	495.3497	−243.2803	59185.30437
6	327	359.3	440.7951	−297.8349	88705.62766
Total	4648	4431.8			464063.6206

$$SSR = \Sigma \left(\hat{y}_i - \bar{y} \right)^2 = 464063.6206$$

We can now compute r^2 value as

$$r^2 = \frac{SSR}{SST} = \frac{464063.6206}{503255.7933} = 0.9221$$

The values for SSR, SSE and SST can be found in the regression Analysis of Variance (ANOVA) table from the SPSS output. While the r^2 value can be obtained from the model summary table.

Model Summary

Model	R	R Square	Adjusted R Square	Std. Error of the Estimate
1	.960[a]	.922	.903	98.9107

a. Predictors: (Constant), Income in thousand

ANOVA[a]

Model		Sum of Squares	df	Mean Square	F	Sig.
1	Regression	464122.846	1	464122.846	47.440	.002[b]
	Residual	39133.267	4	9783.317		
	Total	503256.113	5			

a. Dependent Variable: Credit card debt in thousands
b. Predictors: (Constant), Income in thousand

Coefficients[a]

Model		Unstandardized Coefficients		Standardized Coefficients		
		B	Std. Error	Beta	t	Sig.
1	(Constant)	223.215	85.032		2.625	.058
	Income in thousand	.665	.097	.960	6.888	.002

a. Dependent Variable: Credit card debt in thousands

Linear Regression Values for Income in Thousands and Credit Card Debt in Thousands

Model Summary

Model	R	R Square	Adjusted R Square	Std. Error of the Estimate
1	.916[a]	.839	.832	.73875

a. Predictors: (Constant), Advertising spending

ANOVA[a]

Model		Sum of Squares	df	Mean Square	F	Sig.
1	Regression	62.514	1	62.514	114.548	<.001[b]
	Residual	12.006	22	.546		
	Total	74.520	23			

a. Dependent Variable: Detrended sales
b. Predictors: (Constant), Advertising spending

Coefficients[a]

Model		Unstandardized Coefficients		Standardized Coefficients	t	Sig.
		B	Std. Error	Beta		
1	(Constant)	6.584	.402		16.391	<.001
	Advertising spending	1.071	.100	.916	10.703	<.001

a. Dependent Variable: Detrended sales

Linear Regression Values for Detrended Sales and Advertising Spending

Using the above SPSS output (Figure 14.22)
From the above SPSS output, R square = 0.839. This implies that the 83.9% of the total variation in detrended sales is explained by the regression of detrended sales and advertising spending. Since R square > 0.5, it implies that the estimated regression equation fits very well with the data set.
We can now conduct hypothesis tests about the regression model

Testing Hypothesis for Regression Model
In linear regression analysis, there are usually two tests: t-test and F-test. The F-test in the regression ANOVA is used to determine if there is a statistical significant linear relationship between dependent/outcome variable (quantitative) and independent/predictor variable (quantitative). For example, if there is a statistical and significant linear relationship between salary and years of experience. The t-test in the regression coefficient table is used to determine if there is a statistical significant effect/influence of independent/predictor variable (quantitative) on the dependent/outcome variable (quantitative). For example, if years of experience statistical and significantly affect/influence/predict salary. The values from the regression coefficient table are used to develop the regression model equation that shows the relationship between dependent and independent variables and how the independent variable affect/predicts/influences the dependent variable.

Hypothesis Tests for the Slope/Regression Coefficient
Hypothesis test for the slope/regression coefficient is also known as t-test for the statistical significance of the regression coefficient. The simple linear regression model is $y = \beta_0 + \beta_1 x + \varepsilon$, if x and y are linearly related, we must have that $\beta_1 \neq 0$. t-test is used to test the hypothesis about β_1. It is used to test for the significance of the regression coefficient, i.e. *if the independent variable*

has a significant effect/influence on the dependent variable. The purpose is to perform a hypothesis test to decide whether a predictor variable is useful for making predictions.

Step 1: The null and alternative hypotheses are, respectively
   H_0: $\beta_1 = 0$ (Independent variable has no significant effect/influence on the dependent variable.)
   H_a: $\beta_1 \neq 0$ (Independent variable has a significant effect/influence on the dependent variable.)
Step 2: Decide on/state the significance level, α.
Step 3: Compute the value of the test statistic. t-test is computed as

$$t = \frac{\beta_1}{s_{b_1}}$$

where

$$s_{b_1} = \frac{s}{\sqrt{\sum(x_i - \bar{x})^2}}$$

$$\sum(x_i - \bar{x})^2 = \sum x^2 - \frac{(\sum x)^2}{n}$$

$$s = \sqrt{MSE} = \sqrt{\frac{SSE}{n-2}}$$

$$df = n - 2$$

SPSS will compute the tests statistics
   In SPSS go to: Analyze > Regression > Linear > There will be a pop up SPSS window for regression > Choose the numerical variables of interest for Dependent and another numerical variable of interest for Independent(s) > Click Statistics and select Estimates > Continue then OK > go to SPSS Output > In the SPSS output select the t value for the independent variable (β_1) from the Coefficient Table
Step 4: Determine the p-value,
   In the SPSS output select the p-value (the value of Sig.)
Step 5: Decision on H_0 using p-value:
   For p-value: we compare p-value with α
   To make decision: If p-value < α, reject H_0; otherwise, do not reject H_0.
Step 6: Make conclusions at a given level of confidence based on the decision in Step 5.
   The confidence interval for β_1 is given by

$$\beta_1 \pm t_{\alpha/2} \times s_{b_1}$$

Example

Using the above income and credit card debt data, determine if the income has a significant effect on credit card debt at $\alpha = 0.05$ level of significance.

Step 1: State the null and alternative hypotheses, respectively,
   H_0: $\beta_1 = 0$ (income has no significant effect/influence on credit card debt)
   H_a: $\beta_1 \neq 0$ (income has a significant effect/influence on credit card debt).
Step 2: State the significance level, α. $\alpha = 0.05$
Step 3: Compute the value of the test statistic

$$\sum(x_i - \bar{x})^2 = \sum x^2 - \frac{(\sum x)^2}{n} = 4649088 - \frac{4648^2}{6} = 1048437.33$$

$$SSE = SST - SSR = 503255.7933 - 464063.6206 = 39192.1727$$

$$s = \sqrt{MSE} = \sqrt{\frac{SSE}{n-2}} = \sqrt{\frac{39192.1727}{6-2}} = 98.985065$$

$$s_{b_1} = \frac{s}{\sqrt{\sum(x_i - \bar{x})^2}} = \frac{98.985065}{\sqrt{1048437.33}} = 0.09667$$

$$t = \frac{\beta_1}{s_{b_1}} = \frac{0.6653}{0.09667} = 6.8821$$

$$df = n - 2 = 6 - 2 = 4$$

In SPSS we go to Analyze > Regression > Linear > There will be a pop up SPSS window for regression > Choose Credit card debt for Dependent and Income for Independent(s) > Click Statistics and select Estimates > Continue then OK.

Model Summary

Model	R	R Square	Adjusted R Square	Std. Error of the Estimate
1	.960[a]	.922	.903	98.9107

a. Predictors: (Constant), Income in thousand

ANOVA[a]

Model		Sum of Squares	df	Mean Square	F	Sig.
1	Regression	464122.846	1	464122.846	47.440	.002[b]
	Residual	39133.267	4	9783.317		
	Total	503256.113	5			

a. Dependent Variable: Credit card debt in thousands
b. Predictors: (Constant), Income in thousand

Coefficients[a]

Model		Unstandardized Coefficients		Standardized Coefficients	t	Sig.
		B	Std. Error	Beta		
1	(Constant)	223.215	85.032		2.625	.058
	Income in thousand	.665	.097	.960	6.888	.002

a. Dependent Variable: Credit card debt in thousands

Linear Regression Values for Income in Thousands and Credit Card Debt in Thousands

SPSS Output
In the SPSS output, use the coefficient table
Then from the SPSS output use the coefficient table to locate the t-test and p-value for the independent variable (income)

$$t - test = 6.888, \ df = 4$$

Step 4: Determine the p-value,
From the SPSS output select the p-value for (β_1)/income: p-value for advertising = 0.002
Step 5: Decision on H_0 using p-value:
For p-value: we compare p-value =0.002 with $\alpha = 0.05$
To make decision: *If p-value < α, reject H_0; otherwise, do not reject H_0.*
Since p-value < α, reject H_0
Step 6: Make conclusions at a given level of confidence based on the decision in Step 5.
At 95% level of significance, income has a significant effect/influence on the credit card debt.

Example

Using advert.sav data, determine if the advertising amount has a significant effect on detrended sales at $\alpha = 0.05$ level of significance.

Step 1: State the null and alternative hypotheses, respectively,
H_0: $\beta_1 = 0$ (advertising spending has no significant effect/influence on the detrended sales)
H_a: $\beta_1 \neq 0$ (advertising spending has a significant effect/influence on the detrended sales).
Step 2: State the significance level, α. $\alpha = 0.05$
Step 3: Compute the value of the test statistic
First, we locate the advert.sav data using the path:

For Windows
Open **SPSS data** > Click **Open data document** folder > drive **(C)**: > **Program Files (x86)** > **IBM** > **SPS** > **Statistics** > **27** > **Samples** > **English** > advert.sav

For MAC
Open **SPSS data** > Click **Open data document** folder > Applications > **IBM SPS Statistics 27** > **Samples** > **English** > advert.sav
We go to Analyze > Regression > Linear > There will be a pop up SPSS window for regression > Choose the numerical variables of interest (Detrended sales) for Dependent and another numerical variable of interest (Advertising spending) for Independent(s) > Click Statistics and select Estimates > Continue then OK

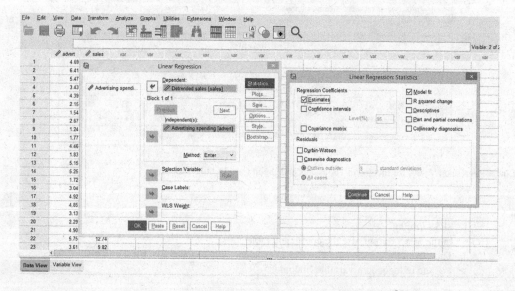

SPSS Pop Up Window for Linear Regression between Detrended Sales and Advertising Spending

SPSS output, select the coefficient table (Figure 14.23)

Coefficients[a]

Model		Unstandardized Coefficients		Standardized Coefficients	t	Sig.
		B	Std. Error	Beta		
1	(Constant)	6.584	.402		16.391	<.001
	Advertising spending	1.071	.100	.916	10.703	<.001

a. Dependent Variable: Detrended sales

FIGURE 14.23 Regression Coefficient Values for Detrended Sales and Advertising Spending

In the SPSS output (Figure 14.23) select the t value for the independent variable (β_1) from the Coefficients table. This is the t-test value for advertising spending

$$t-test = 16.391, \ df = 22$$

Step 4: Determine the p-value,
  From the SPSS output (Figure 14.23) select the p-value for (β_1)/advertising spending:
    p-value for advertising <0.001
Step 5: Decision on H_0 using p-value:
  For p-value: we compare p-value <0.001 with $\alpha = 0.05$
  To make decision: *If p-value < α, reject H_0; otherwise, do not reject H_0.*
  Since p-value < α, reject H_0
Step 6: Make conclusions at a given level of confidence based on the decision in Step 5.
  At 95% level of significance, advertising spending has a significant effect/influence on the detrended sales.

Hypothesis Tests for the Regression Model
Hypothesis tests for the regression model is also known as **F-test for the statistical significance of the regression model.** An F-test, based on the F probability distribution, is used to test for significance in regression, i.e. if there is a significant linear relationship between independent variable and dependent variable. But with more than one independent variable, only the F test can be used to test for an overall significant relationship.

Step 1: The null and alternative hypotheses are, respectively,
  H_0: $\beta_1 = 0$ (there is no significant linear relationship between the independent variable and the dependent variable)
  H_a: $\beta_1 \neq 0$ (there is a significant linear relationship between the independent variable and the dependent variable).
Step 2: Decide on/state the significance level, α.

Step 3: Compute the value of the test statistic: F-test

$$F = \frac{MSR}{MSE}$$

where

$$MSE = \frac{SSE}{n-2}$$

$$MSR = \frac{SSR}{Number\ of\ independent\ variable = 1}$$

$$df = \left\{ \begin{array}{l} Numerator = 1 \\ Denominator = n-2 \end{array} \right\}$$

SPSS will compute the tests statistics

In SPSS go to: Analyze > Regression > Linear > There will be a pop up SPSS window for regression > Choose the numerical variables of interest for Dependent and another numerical variable of interest for Independent(s) > Click Statistics and select Estimates > Continue then OK > go to SPSS Output > In the SPSS output select the F value from the Regression ANOVA table

Step 4: Determine the p-value,

From the SPSS output, select the p-value from the ANOVA table

Step 5: Decision on H_0 using p-value:

For p-value: we compare p-value with α

To make decision: *If p-value $< \alpha$, reject H_0; otherwise, do not reject H_0.*

Step 6: Make conclusions at a given level of confidence based on the decision in Step 5.

Example

Using the above income and credit card debt data, determine if there is a significant linear relationship between income and credit card debt at $\alpha = 0.05$ level of significance.

Step 1: State the null and alternative hypotheses, respectively,

H_0: $\beta_1 = 0$ (there is no significant linear relationship between income and credit card debt)

H_a: $\beta_1 \neq 0$ (there is a significant linear relationship between income and credit card debt).

Step 2: State the significance level, α. $\alpha = 0.05$

Step 3: Compute the value of the test statistic: F-test

$$MSE = \frac{SSE}{n-2} = \frac{39192.1727}{6-2} = 9798.0432$$

$$MSR = \frac{SSR}{Number\ of\ independent\ variable = 1} = \frac{464063.6206}{1} = 464063.6206$$

$$F = \frac{MSR}{MSE} = \frac{464063.6206}{9798.0432} = 47.36289$$

$$df = \left\{ \begin{array}{l} Numerator = 1 \\ Denominator = n-2 = 6-2 = 4 \end{array} \right.$$

In SPSS we go to Analyze > Regression > Linear > There will be a pop up SPSS window for regression > Choose Credit card debt for Dependent and Income for Independent(s) > Click Statistics and select Estimates > Continue then OK. In the SPSS output select the F value from the Regression ANOVA table

SPSS Output (Figure 14.24)

ANOVA[a]

Model		Sum of Squares	df	Mean Square	F	Sig.
1	Regression	464122.846	1	464122.846	47.440	.002[b]
	Residual	39133.267	4	9783.317		
	Total	503256.113	5			

a. Dependent Variable: Credit card debt in thousands

b. Predictors: (Constant), Income in thousand

FIGURE 14.24   Regression ANOVA Values for Credit Card Debt in Thousands and Income in Thousands

In the SPSS output use the Regression ANOVA Table to locate the F-test and p-value for the F-test

$$F-test = 47.440, \quad df = 1,4$$

Step 4: Determine the p-value,
   From the SPSS output select the p-value for the F-test: p-value = 0.002
Step 5: Decision on H_0 using p-value:
   For p-value: we compare p-value = 0.002 with $\alpha = 0.05$
   To make decision: *If p-value < α, reject H_0; otherwise, do not reject H_0.*
   Since p-value < α, reject H_0
Step 6: Make conclusions at a given level of confidence based on the decision in Step 5.
   At 95% level of significance, there is a significant linear relationship between income and credit card debt.

Example

Using advert.sav data, determine if there is a significant linear relationship between advertising amount and detrended sales at $\alpha = 0.05$ level of significance.
   Step 1: State the null and alternative hypotheses, respectively,
      H_0: $\beta_1 = 0$ (there is no significant linear relationship between the advertising spending and detrended sales)
      H_a: $\beta_1 \neq 0$ (there is a significant linear relationship between advertising spending and detrended sales).
   Step 2: State the significance level, α. $\alpha = 0.05$
   Step 3: Compute the value of the test statistic: F-test
      SPSS will compute the tests statistics
      In SPSS go to: Analyze > Regression > Linear > There will be a pop up SPSS window for regression > Choose the numerical variables of interest for Dependent and another numerical variable of interest for Independent(s) > Click Statistics and select Estimates > Continue then OK > go to SPSS Output > In the SPSS output select the F value from the Regression ANOVA table

SPSS Output (Figure 14.25)

		ANOVA[a]				
Model		Sum of Squares	df	Mean Square	F	Sig.
1	Regression	62.514	1	62.514	114.548	<.001[b]
	Residual	12.006	22	.546		
	Total	74.520	23			

a. Dependent Variable: Detrended sales

b. Predictors: (Constant), Advertising spending

FIGURE 14.25   Regression ANOVA Values for Detrended Sales and Advertising Spending

From the SPSS output (Figure 14.25) select the F value from the Regression ANOVA table

$$F-test = 114.548, \quad df = 1, 22$$

Step 4: Determine the p-value,
    From the SPSS output, select the p-value from the ANOVA table: p-value for F-test <0.001
Step 5: Decision on H_0 using p-value:
    For p-value: we compare p-value <0.00 with $\alpha = 0.05$
    To make decision: If p-value < α, reject H_0; otherwise, do not reject H_0.
    Since p-value < α, reject H_0
Step 6: Make conclusions at a given level of confidence based on the decision in Step 5.
    At 95% level of significance, there is a significant linear relationship between advertising spending and detrended sales.

14.2.2 Multiple Linear Regression Analysis Using SPSS

Multiple Linear regression involves having one dependent variable and several independent variables. The same procedure for simple linear regression is applied only that you will have more than one independent variable. SPSS is used to estimate the regression coefficients, multiple coefficient of determination and hypothesis testing.
    Assumptions

- The dependent variable should be measured on a continuous scale (i.e. it is either an interval or ratio variable).
- You have two or more independent variables, which can be either continuous (i.e. an interval or ratio variable) or categorical (i.e. an ordinal or nominal variable)
- You should have independence of observations (i.e. independence of residuals). This can easily check using the Durbin-Watson statistic using SPSS.
- There needs to be a linear relationship between (a) the dependent variable and each of your independent variables, and (b) the dependent variable and the independent variables collectively. This can be assessed using scatterplots and partial regression plots using SPSS.
- The data needs to show homoscedasticity, which is where the variances along the line of best fit remain similar as you move along the line.

- The data must not show multicollinearity, which occurs when you have two or more independent variables that are highly correlated with each other. Multicollinearity can be detected by inspecting the correlation coefficients and Tolerance/VIF values.
- There should be no significant outliers, high leverage points or highly influential points. You can use Cook's distance.
- The residuals (errors) are approximately normally distributed (we explain these terms in our enhanced multiple regression guide). One can check this assumption include using: (a) a histogram (with a superimposed normal curve) and a Normal P-P Plot; or (b) a Normal Q-Q Plot of the studentized residuals.

In SPSS we use the path:
We go to Analyze > Regression > Linear > There will be a pop up SPSS window for regression > Choose the one numerical variable of interest for Dependent and two or more other numerical variables of interest for Independent(s) > Click Statistics and select Estimates > Continue then OK > in the SPSS Output select the appropriate table

14.2.2.1 Estimating Regression Parameters

Once we have the values for β_0, β_1, β_2 ... β_n we substitute the values into the equation $\hat{y} = \beta_0 + \beta_1 X_1 + \beta_2 X_2 + \cdots + \beta_n X_n$ and have a multiple linear regression model. The estimated regression equation can be used to predict future values.

Interpretation:

- β_0 is the constant: without any value of x_1, x_2, ... upto x_n, y will be at β_0 value.
- To interpret each slope we hold other slopes constant. For example, to interpret β_1 value: Holding β_2 ... β_n constant, for every unit increase/decrease in x, y will be increasing or decreasing by β_1 value. Decrease is when we have a negative β_1 value and increase is when we have a positive β_1 value. The same is repeated for other slopes (β_2 ... β_n).

14.2.2.2 Multiple Coefficient of Determination

Multiple coefficient of determination/adjusted R square (R^2_{Adj}), provides a measure of goodness of fit for the estimated multiple regression line/equation. It is denoted by *Adjusted* r^2. *Adjusted* r^2 is computed as:

$$R^2_{Adj} = 1 - \left[\frac{(1 - R^2)(n - 1)}{n - k - 1} \right]$$

where n is the number of data points in the data and k is the number of independent variables in the model.

Interpretation of R^2_{Adj}: The adjusted R square tells you the percentage of variation explained by only the independent variables that actually affect the dependent variable. The multiple coefficient of determination is the proportion of the total variation in Y "explained" by the regression of Y on several X variables. Usually Adjusted R Square value greater than 0.5 implies a better model fit.

14.2.2.3 Hypothesis Tests for the Slope/Regression Coefficient

Hypothesis tests for each of the slope/regression coefficient is also known as t-test for the statistical significance of the regression coefficient. This will depend on the number of independent variables in the multiple linear regression model. The six steps are the same as that of testing the slope of simple linear regression model.

14.2.2.4 Hypothesis Tests for the Overall Significant Relationship

Hypothesis tests for the regression model is also known as **F-test for the overall statistical significance of the regression model.** For more than one independent variable, only the F test can be used to test for an overall significant relationship. The six steps are the same as that of testing the regression model of simple linear regression.

Example

Using SPSS data set *bankloan.sav*, develop an estimated regression equation relating household income to credit card debt and other debt. Interpret the values of β_0, β_1 and β_2

First, we locate the bankloan.sav data using the path:

For Windows
Open **SPSS data** > Click **Open data document** folder > drive **(C):** > **Program Files (x86)** > **IBM** > **SPS** > **Statistics** > **27** > **Samples** > **English** > bankloan.sav

For MAC
Open **SPSS data** > Click **Open data document** folder > Applications > **IBM SPS Statistics 27** > **Samples** > **English** > bankloan.sav

We go to Analyze > Regression > Linear > There will be a pop up SPSS window for regression > Choose the numerical variables of interest (household income) for Dependent and the other two numerical variable of interest (credit card debt and other debt) for Independent(s) > Click Statistics and select Estimates > Continue then OK (Figure 14.26)

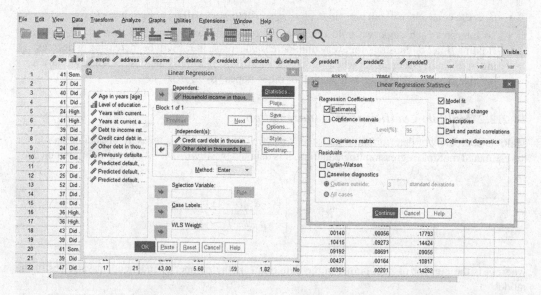

FIGURE 14.26 SPSS Pop Up Window for Linear Regression between Credit Card Debt in Thousands, Other Debts in Thousands and Household Income in Thousands

SPSS Output (Figure 14.27)

Model Summary

Model	R	R Square	Adjusted R Square	Std. Error of the Estimate
1	.640[a]	.409	.408	29.66098

a. Predictors: (Constant), Other debt in thousands, Credit card debt in thousands

ANOVA[a]

Model		Sum of Squares	df	Mean Square	F	Sig.
1	Regression	516078.176	2	258039.088	293.302	<.001[b]
	Residual	745168.205	847	879.774		
	Total	1261246.381	849			

a. Dependent Variable: Household income in thousands

b. Predictors: (Constant), Other debt in thousands, Credit card debt in thousands

Coefficients[a]

Model		Unstandardized Coefficients		Standardized Coefficients	t	Sig.
		B	Std. Error	Beta		
1	(Constant)	23.922	1.389		17.222	<.001
	Credit card debt in thousands	5.041	.627	.278	8.045	<.001
	Other debt in thousands	4.809	.392	.424	12.270	<.001

a. Dependent Variable: Household income in thousands

FIGURE 14.27 Linear Regression Values for Credit Card Debt in Thousands, Other Debts in Thousands and Household Income in Thousands

In the SPSS output (Figure 14.27) use the Coefficients table to select the values for β_0, β_1 and β_2.

$$\beta_0 = 23.922, \ \beta_1 = 5.041 \text{ and } \beta_2 = 4.809$$

Once we have the values for β_0, β_1 and β_2 we put them into the equation $\hat{y} = \beta_0 + \beta_1 X_1 + \beta_2 X_2$ and have a linear regression model.

$$Household \ Income = 23.922 + 5.041 \times Credit \ Card \ Debt + 4.809 \times Other \ Debt$$

Interpretation:
- $\beta_0 = 23.922$ is the constant: without any credit card debt and other debt, household income will be at 23.922.
- $\beta_1 = 5.041$ is the slope 1 or the rate of change in y with respect to x1: holding other debt constant, for every unit increase in credit card debt, household income will be increasing by 5.041.
- $\beta_2 = 4.809$ is the slope 2 or the rate of change in y with respect to x2: holding credit card debt constant, for every unit increase in other debt, household income will be increasing by 4.809.

Practice Exercise

1. Using *car_sales.sav* data, fuel capacity would predict sales in thousands
   a. Generate a scatter plot showing the relationship sales in thousands and the fuel capacity. Interpret the output
   b. Interpret the model prediction power
   c. Fit the estimated regression model showing the relationship between sales in thousands and the fuel capacity. Let fuel capacity be the independent variable. Interpret the regression coefficients
   d. Interpret the level of association between sales in thousands and the fuel capacity
   e. Test for the overall significance of the estimated regression model
   f. Test for the significance of regression coefficient
2. Using customer_subset.sav data,
   a. Fit the estimated linear regression equation showing the relationship between credit card debt in thousands and combined factors of age in years and years of education. Interpret the regression coefficients
   b. Interpret the multiple coefficient of determination
   c. Determine if years of education would significantly affect the credit card debt in thousands

Part IV

Special Topics in Statistical Analysis

15

Nonparametric Tests

15.1 Nonparametric Test

A nonparametric test is a statistical test, in which specific assumptions are not made about the population parameter. The researcher has no idea regarding the population parameters. It is not based on underlying assumptions, and it does not require a knowledge of a population's distribution. The test is mainly based on differences in medians. Nonparametric tests are also known as distribution-free tests. Nonparametric tests are used when your data are not normal.

The main nonparametric tests are: one-sample sign test/one-sample Wilcoxon signed rank test, Wilcoxon rank sum test for paired samples, Mann-Whitney U test and Kruskal-Wallis test.

15.2 One-Sample Sign Test Using SPSS

Sign test is used to test the null hypothesis that the median of a distribution is equal to some hypothesized value k. The test is based on the direction or the data are recorded as plus and minus signs rather than numerical magnitude, hence it is called Sign test.

One-sample sign test is used to estimate the median of a population and compare it to a reference value or target value. The one-sample sign test simply computes a significance test of a hypothesized median value for a single data set.

The one-sample sign test is to compare the total number of observations less than (–ve) or greater than (+) the hypothesized value. The one-sample sign test is similar to *the one-sample Wilcoxon signed rank test*.

The one-sample sign test is a nonparametric version of one-sample t-test. Similar to one-sample t-test, the sign test for a population median can be one-tailed (right or left tailed) or two-tailed distribution based on the hypothesis.

Assumptions

- Data are non-normally distributed
- A random sample of independent measurements for a population with unknown median
- The variable of interest is continuous
- One-sample test handles nonsymmetric data set that means skewed either to the right or the left

15.2.1 Hypothesis Tests for the One-Sample Wilcoxon Signed Rank Test

Step 1: The null and alternative hypotheses are, respectively,

H_0: *Median* $= m_o$ Median is not different from the hypothesized value and the alternative is
H_a: *Median* $\neq m_o$ Median is different from the hypothesized value (two tailed)

Or

The Null Hypothesis is H_0: *Median* $\geq m_0$ and the alternative hypothesis is H_a: *Median* $< m_0$ (left/one tailed)

Or

The Null Hypothesis is H_0: $Median \leq m_0$ and the alternative hypothesis is H_a: $Median > m_0$ (right/one tailed)

Step 2: Decide on/state the significance level, α

Step 3: Compute the value of the test statistic

Assign positive signs to observations with values $\doteq m_0$ and negative signs to the sample data with values $\neq m_0$, and determine the sample size (n) – n is the sum of positive and negative signs

Compute the test statistic:

- If $n \leq 25$ (approx), use x. Where x is the smaller number of positive and negative signs
- For larger sample size, if $n > 25$, use

$$z = \frac{(x+0.5) - \dfrac{n}{2}}{\dfrac{\sqrt{n}}{2}}$$

SPSS will compute the tests statistics

In Statistical Package for Social Science (SPSS) go to: Analyze > Nonparametric Test > One Sample > There will be a pop-up SPSS window for One Sample Nonparametric Test > Click Fields > Remove all the variables from Test Fields > Select the categorical variable of interest and move it to Test Fields > Click Settings > Select Customized tests > Select compare median to hypothesized (Wilcoxon signed rank test) and put the hypothesized median value > Click Run > there will be One-Sample Wilcoxon Signed Rank Test in the SPSS Output

Select the One-Sample Wilcoxon Signed Rank Test from the SPSS Output

Step 4: Determine the p-value,

From the SPSS output, select the p-value [Asymptotic Sig. (2-sided test)]

Step 5: Decision on H_0 using p-value:

For p-value: we compare p-value with α

To make decision: *If p-value < α, reject H_0; otherwise, do not reject H_0.*

Step 6: Make conclusions at a given level of confidence based on the decision in Step 5.

Example

Using bankloan.sav data, can we conclude that the median age in years is different from 33 at $\alpha = 0.05$.

Example

Consider the following data on household income in thousands.

1.7   1.5   2.6   2.2   2.4   2.3   2.6   3.0   1.4   2.3   1.9   2.1   1.6

At $\alpha = 0.05$ level of significance, can we conclude that the median income is different from 2?

Step 1: The null and alternative hypotheses are, respectively,
H_0: *Median* = 2 Median income is not different from 2 and the alternative is
H_a: *Median* ≠ 2 Median is different from 2 (two tailed)
Step 2: Decide on/state the significance level, $\alpha = 0.05$
Step 3: Compute the value of the test statistic
Key the data into SPSS (Figures 15.1 and 15.2)
Variable View

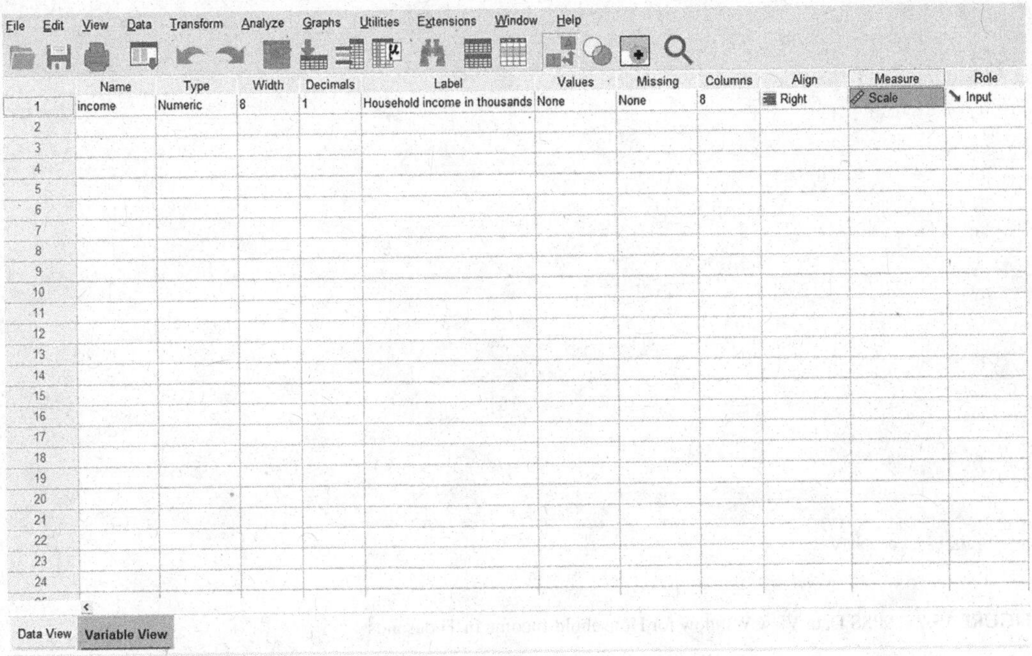

FIGURE 15.1 SPSS Variable View Window for Household Income in Thousands

Data View

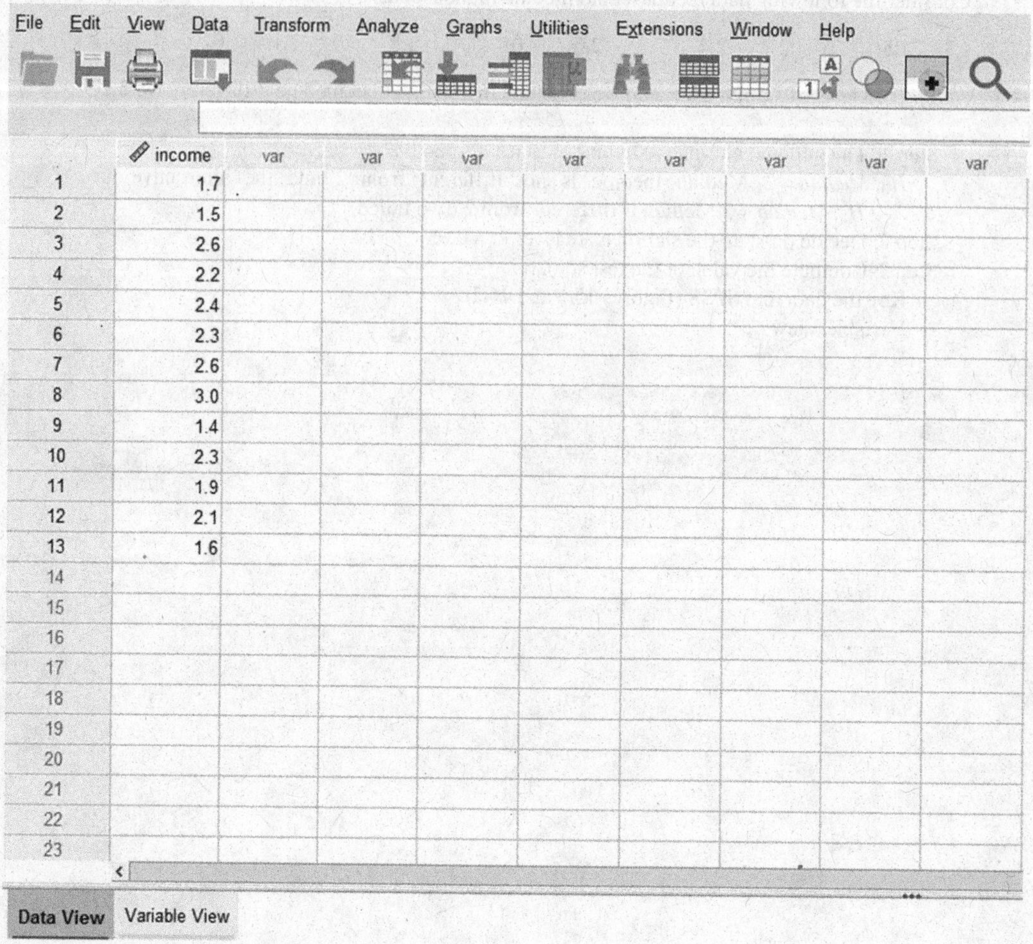

	income	var	var	var	var	var	var	var	var
1	1.7								
2	1.5								
3	2.6								
4	2.2								
5	2.4								
6	2.3								
7	2.6								
8	3.0								
9	1.4								
10	2.3								
11	1.9								
12	2.1								
13	1.6								
14									
15									
16									
17									
18									
19									
20									
21									
22									
23									

Data View   Variable View

FIGURE 15.2   SPSS Data View Window for Household Income in Thousands

In SPSS go to: Analyze > Nonparametric Test > One Sample > There will be a pop-up SPSS window for One Sample Nonparametric Test > Click Fields > Put the variable of interest (income) in Test Fields (Figure 15.3).

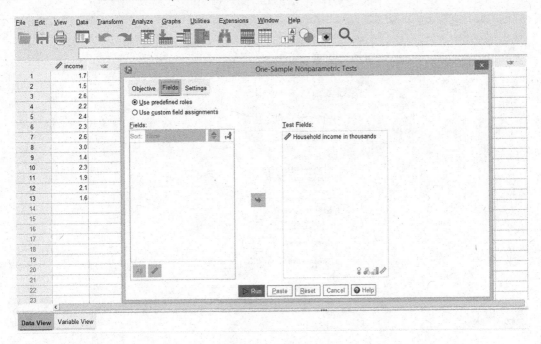

FIGURE 15.3 SPSS One-Sample Nonparametric Tests Pop Up Window

Click Settings > Under Choose Tests select Customized tests > Select compare median to hypothesized (Wilcoxon signed rank test) and put the hypothesized median value (2) (Figure 15.4).

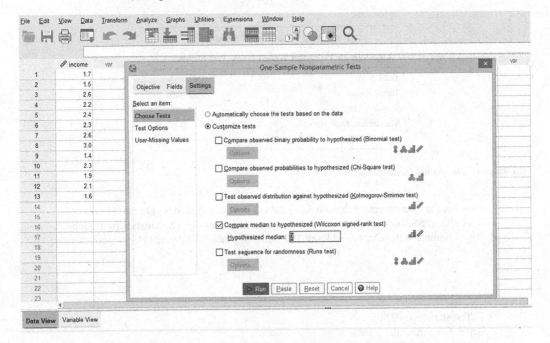

FIGURE 15.4 SPSS One-Sample Nonparametric Tests Pop Up Window for displaying Wilcoxon Signed-Rank Test

Click Settings > Under Tests Options put the confidence level and the significance level > Click Run (Figure 15.5)

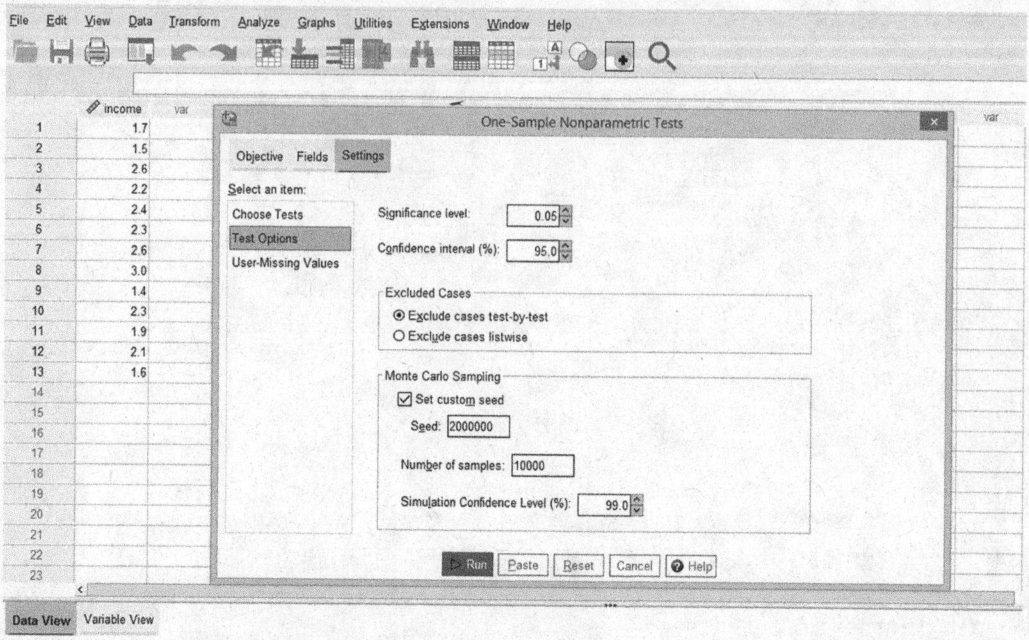

FIGURE 15.5 SPSS One-Sample Nonparametric Tests Pop Up Window for displaying Test Options

The SPSS Output will have the One-Sample Wilcoxon Signed Rank Test results (Figure 15.6)

Household Income in Thousands	
One-Sample Wilcoxon Signed Rank Test Summary	
Total N	13
Test statistic	57.000
Standard error	14.265
Standardized test statistic	.806
Asymptotic sig. (2-sided test)	.420

FIGURE 15.6 One-Sample Wilcoxon Signed Rant Test

Select the One-Sample Wilcoxon Signed Rank Test from the SPSS Output
In the SPSS, output selects the standardized test statistics (Z statistic) from One-Sample Wilcoxon Signed Rank Test Summary table and p-value [Asymptotic Sig. (2-tailed)]

$$z = 0.806$$

Test statistics = 0.806
Step 4: Determine the p-value,
From the SPSS output, select the p-value = 0.420 [Asymptotic Sig. (2-sided test)]

Step 5: Decision on H_0 using p-value:
  For p-value: we compare p-value = 0.420 with $\alpha = 0.05$
  To make decision: *If p-value < α, reject H_0; otherwise, do not reject H_0.*
  Since *p-value > α, we fail to reject H_0*
Step 6: Make conclusions at a given level of confidence based on the decision in Step 5.
  At 95% level of confidence, the median household income in thousands is not different
    from 2.

15.3 Wilcoxon Signed Rank Test for Matched/Paired Samples Using SPSS

Wilcoxon signed rank test is the nonparametric test equivalent to the dependent t-test. It is used to compare two sets of scores that come from the same participants.
  Assumptions

- The dependent variable should be measured at the ordinal or continuous level.
- The independent variable should consist of two categorical, "related groups" or "matched pairs." "Related groups" indicates that the same subjects are present in both groups.
- The distribution of the differences between the two-related groups needs to be symmetrical in shape.

15.3.1 Hypothesis Tests for the Paired or Matched Sample Signed Rank Test

Step 1: The null and alternative hypotheses are, respectively,
  H_0: No difference in median of the signed differences/the median difference is zero and the alternative is
  H_a: There is a difference in the m of the signed difference/the median difference is not zero (two tailed)
  Or
  The Null Hypothesis is H_0: The median difference is greater than or equals to zero and the alternative hypothesis is H_a: The median difference is less than zero (left/one tailed)
  Or
  The Null Hypothesis is H_0: The median difference is less than or equals to zero and the alternative hypothesis is H_a: The median difference is more than zero (right/one tailed)
Step 2: Decide on/state the significance level, α.
Step 3: Compute the value of the test statistic

15.3.1.1 For Sign Test

Compute the differences in scores for each observation
  Capture the signs of the difference in scores
  The test statistics is the number of negative signs (i.e. the test statistic for the Sign Test is the smaller of the number of positive or negative signs)

15.3.1.2 For Wilcoxon Singed Rank Test

We compute difference scores for each observation
  Rank the ordered absolute values of the difference scores
  Attach the signs ("+" or "–") of the observed differences to each rank
  The test statistic for the Wilcoxon Signed Rank Test is W, defined as the smaller of W+ and W– which are the sums of the positive and negative ranks, respectively.

In SPSS, we go to Analyze > Nonparametric Tests > Legacy Dialogs > 2 Related Samples > There will be a pop-up window for Two-Related-Samples Test > Transfer the paired variables interests into the Test Pairs: box where one of the variables will be in Variable 1 and the other variable in Variable 2 > Select Wilcoxon under Test Type > Click Continue then OK > In the SPSS output select the Wilcoxon signed-ranks test using the Z statistic from the Test Statistics table and p-value (Asymp. Sig. (2-tailed))

Step 4: Determine the p-value,

From the SPSS output, select the p-value [Asymptotic Sig. (2-sided test)]

Step 5: Decision on H_0 using p-value:

For p-value: we compare p-value with α

To make decision: *If p-value < α, reject H_0; otherwise, do not reject H_0.*

Step 6: Make conclusions at a given level of confidence based on the decision in Step 5.

Example

Scores in the first and final round for a sample of eight golfers who completed golf tournament as shown below:

Golfer	First Round	Final Round
1	73	78
2	68	75
3	70	66
4	67	71
5	71	73
6	70	68
7	68	65
8	73	72

At 2% level of significance, do these results suggest a significant difference between the scores?

Step 1: The null and alternative hypotheses are, respectively,
  H_0: No difference in median of the signed differences/the median difference is zero
    and the alternative is
  H_a: There is a difference in the median of the signed difference/the median difference
    is not zero (two tailed)
Step 2: State the significance level, $\alpha = 0.05$.
Step 3: Compute the value of the test statistic
  Key the data into SPSS (Figures 15.7 and 15.8)

Variable View

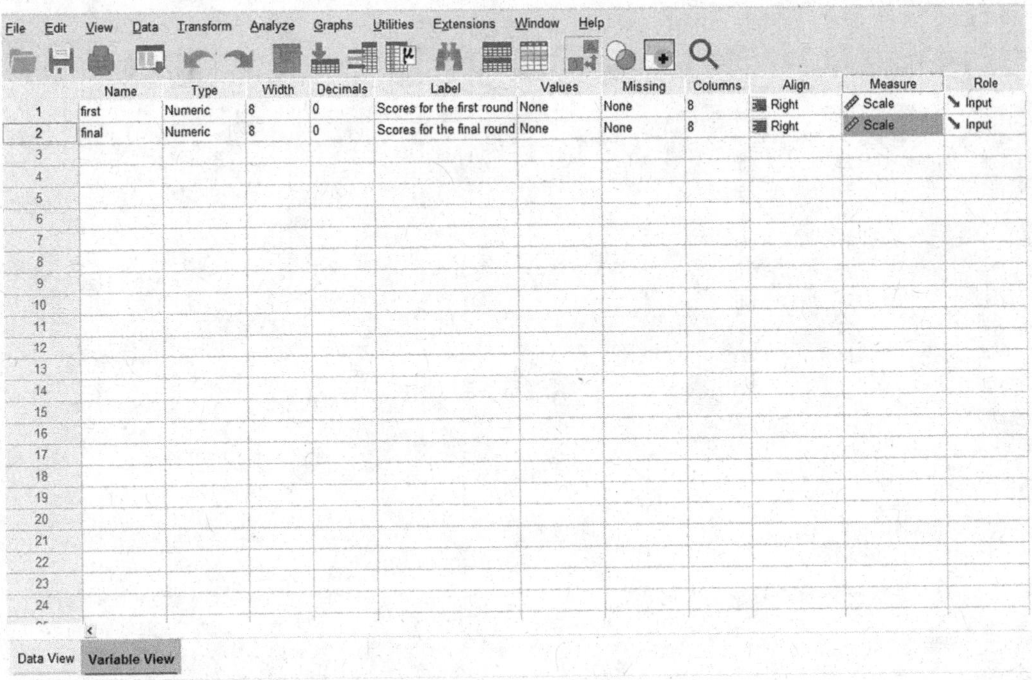

FIGURE 15.7 SPSS Variable View Window for First and Final Scores

Data View

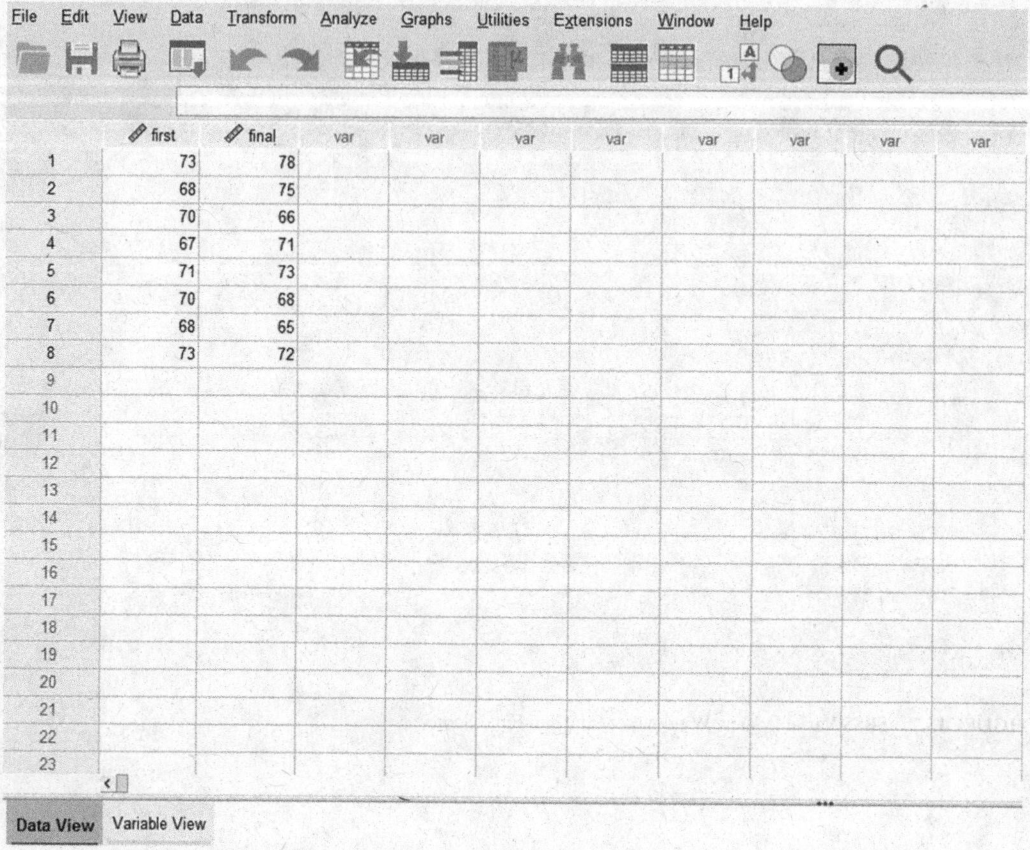
File Edit View Data Transform Analyze Graphs Utilities Extensions Window Help
first final var var var var var var var var
1 73 78
2 68 75
3 70 66
4 67 71
5 71 73
6 70 68
7 68 65
8 73 72
9
10
11
12
13
14
15
16
17
18
19
20
21
22
23
Data View Variable View

FIGURE 15.8 SPSS Data View Window for First and Final Scores

In SPSS, we go to Analyze > Nonparametric Tests > Legacy Dialogs > 2 Related Samples > There will be a pop-up window for Two-Related-Samples Test > Transfer the paired variables interests (Scores in the first round and scores in the final round) into the Test Pairs: box where one of the variables (Scores in the first round) will be in Variable 1 and the other variable (scores in the final round) in Variable 2 > Select Wilcoxon under Test Type > Click Continue then OK (Figure 15.9).

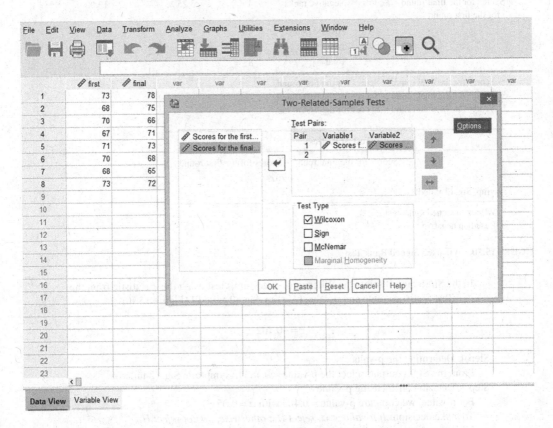

FIGURE 15.9 SPSS Two-Related-Samples Tests Window for Computing Wilcoxon Test

SPSS Output

Wilcoxon Signed Ranks Test

Ranks

		N	Mean Rank	Sum of Ranks
Scores for the final round – Scores for the first round	Negative ranks	4[a]	3.25	13.00
	Positive ranks	4[b]	5.75	23.00
	Ties	0[c]		
	Total	8		

[a] Scores for the final round < Scores for the first round.
[b] Scores for the final round > Scores for the first round.
[c] Scores for the final round = Scores for the first round.

Test Statistics[a]

	Scores for the final round – Scores for the first round
Z	$-.702$[b]
Asymp. Sig. (2-tailed)	.483

[a] Wilcoxon signed ranks test.
[b] Based on negative ranks.

FIGURE 15.10 Wilcoxon Signed Rank Test

In the SPSS output select the Wilcoxon signed-ranks test using the Z statistic from the Test Statistics table and p-value [Asymp. Sig. (2-tailed)] (Figure 15.10)

$$z = -0.702$$

Step 4: Determine the p-value,
   From the SPSS output, select the p-value = 0.483 [Asymptotic Sig. (2-tailed)]
Step 5: Decision on H_0 using p-value:
   For p-value: we compare p-value = 0.483 with $\alpha = 0.05$
   To make decision: *If p-value < α, reject H_0; otherwise, do not reject H_0.*
   Since *p-value > α, we fail to reject H_0*
Step 6: Make conclusions at a given level of confidence based on the decision in Step 5.
   At 95% level of confidence, there is no difference in the median of the signed difference/the median difference is zero.

15.4 Mann-Whitney U Test Using SPSS

Mann-Whitney U test is used to compare differences between two independent groups when dependent variables are either ordinal or continuous but not normally distributed. It is often considered the nonparametric alternative to the independent samples t-test. Mann-Whitney test will helps to compare the medians of the two populations. The Mann-Whitney test is usually performed as a two-sided test
   Assumptions

- The sample drawn from the population is random
- Independence within the samples and also mutual independence is assumed
- The two variables are **not normally distributed**
- Ordinal or continuous measurement scale is assumed

15.4.1 Hypothesis Tests for Mann-Whitney U Test

Step 1: The null and alternative hypotheses are, respectively,

H_0: No difference in medians for the two populations and the alternative is

H_a: There is a difference in medians for the two populations (two tailed)

Or

The Null Hypothesis is H_0: The median for one population is greater than or equals to median of the other population and the alternative hypothesis is H_a: The median for one population is less than or equals to median of the other population (left/one tailed)

Or

The Null Hypothesis is H_0: The median for one population is less than or equals to median of the other population and the alternative hypothesis is H_a: The median for one population is more than or equals to median of the other population (right/one tailed)

Step 2: Decide on/state the significance level, α.

Step 3: Compute the value of the test statistic

Assign ranks and to do so we order the data from smallest to largest. This is done on the combined or total sample (i.e. pooling the data from the two groups)

We sum the ranks in each group

The test statistic for the Mann Whitney U Test is denoted **U** and is the **smaller** of U_1 and U_2, defined below.

$$U_1 = n_1 n_2 + \frac{n_1(n_1+1)}{2} - R_1$$

$$U_2 = n_1 n_2 + \frac{n_2(n_2+1)}{2} - R_2$$

where R_1 is the sum of the ranks for group 1 and R_2 is the sum of the ranks for group 2.

For large samples ($n > 20$), the value of U approaches a normal distribution, and so the null hypothesis can be tested by a Z-test.

In SPSS, we go to Analyze > Nonparametric Tests > Legacy Dialogs > 2 Independent Samples > There will be a pop-up window for Tests for Two Independent Samples Test > Transfer the numerical/dependent variable of interest into Test Variable List > Put the categorical/independent variable of interest into Grouping Variable and click Define Ranges to put the 2 codes for each of the two groups of interest > Select/check Mann-Whitney U as the Test Type > Click Continue then OK > In the SPSS output select the Mann-Whitney U test using the U statistic from the Test Statistics table and p-value (Asymp. Sig. (2-tailed))

Step 4: Determine the p-value,

From the SPSS output, select the p-value (Asymptotic Sig. (2-sided test)

Step 5: Decision on H_0 using p-value:

For p-value: we compare p-value with α

To make decision: *If p-value < α, reject H_0; otherwise, do not reject H_0.*

Step 6: Make conclusions at a given level of confidence based on the decision in Step 5.

Example

Consider the following data on the ratings for the measure of job satisfaction for lawyers and doctors.

Lawyers	Physical Therapists
55	44
42	78
80	79
53	60
50	59
64	55
74	80
55	84
60	79
52	78

Do these results suggest that there is a significant difference in job satisfaction amongst the lawyers and doctors? Use $\alpha = 0.05$.

Step 1: The null and alternative hypotheses are, respectively,
     H_0: No difference in median job satisfaction for the lawyers and doctors; and the alternative is
     H_a: There is a difference in median job satisfaction for the lawyers and doctors (two tailed)
Step 2: State the significance level, $\alpha = 0.05$.
Step 3: Compute the value of the test statistic
     Key the data into SPSS (Figures 15.11 and 15.12)
     Variable View

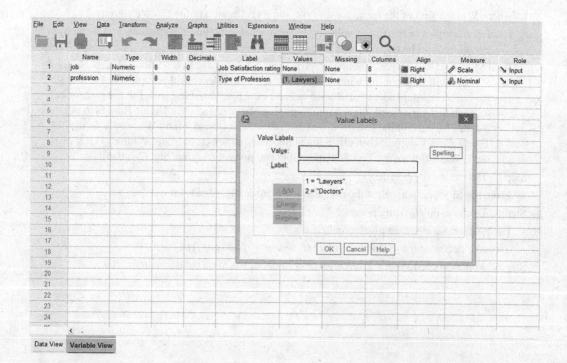

FIGURE 15.11 SPSS Variable View Window for Job Satisfaction Ratings and different Types of Professions

Data View

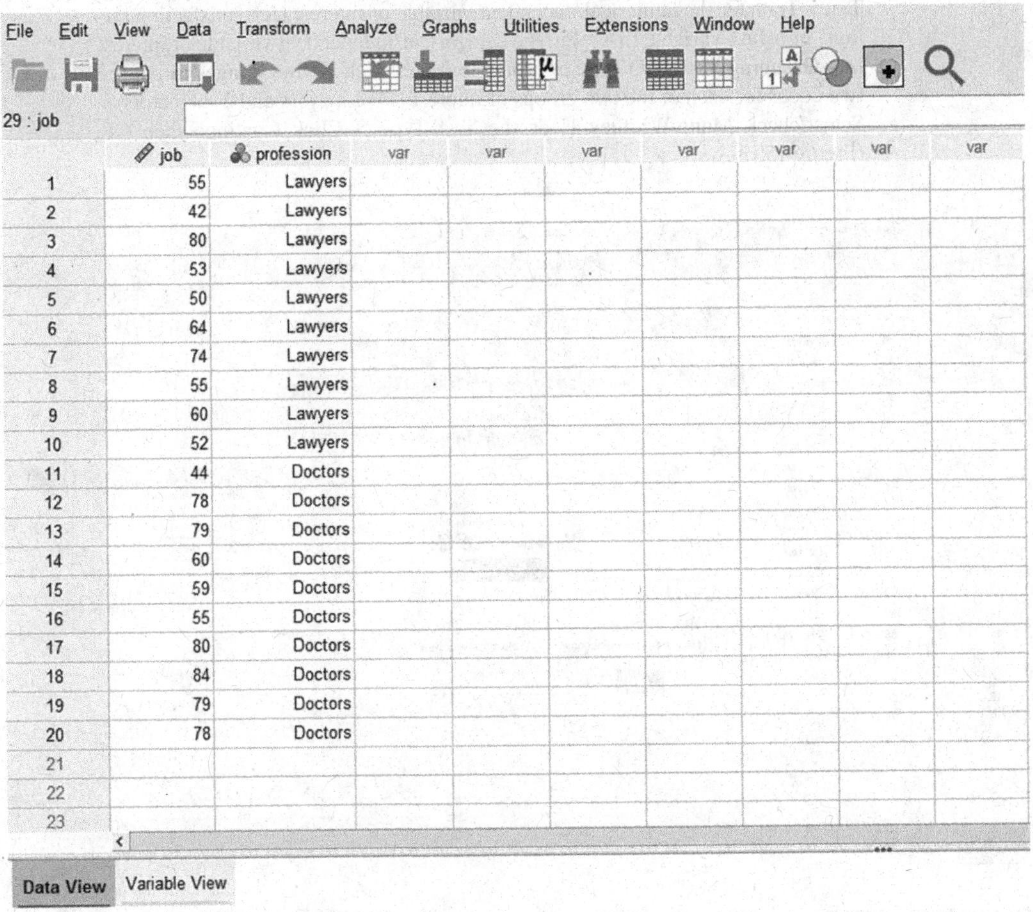

	job	profession	var	var	var	var	var	var	var
1	55	Lawyers							
2	42	Lawyers							
3	80	Lawyers							
4	53	Lawyers							
5	50	Lawyers							
6	64	Lawyers							
7	74	Lawyers							
8	55	Lawyers							
9	60	Lawyers							
10	52	Lawyers							
11	44	Doctors							
12	78	Doctors							
13	79	Doctors							
14	60	Doctors							
15	59	Doctors							
16	55	Doctors							
17	80	Doctors							
18	84	Doctors							
19	79	Doctors							
20	78	Doctors							
21									
22									
23									

FIGURE 15.12 SPSS Data View Window for Job Satisfaction Ratings and different Types of Professions

In SPSS, we go to Analyze > Nonparametric Tests > Legacy Dialogs > 2 Independent Samples > There will be a pop-up window for Tests for Two Independent Samples Test > Transfer the numerical/dependent variable of interest (Job satisfaction ratings) into Test Variable List > Put the categorical/independent variable of interest into Grouping Variable (Type of profession) and click Define Ranges to put the two codes for each of the two groups of interest (1 = Lawyers and 2 = Doctors) > Select/check Mann-Whitney U as the Test Type > Click Continue then OK (Figure 15.13).

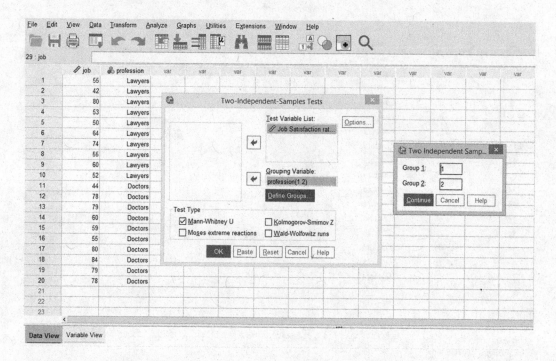

FIGURE 15.13 SPSS Two-Independent Samples Tests Pop Up Window for generating Mann-Whitney U Test

SPSS Output (Figure 15.14).

Mann-Whitney Test

Ranks

	Type of Profession	N	Mean Rank	Sum of Ranks
Job Satisfaction rating	Lawyers	10	8.10	81.00
	Doctors	10	12.90	129.00
	Total	20		

Test Statistics[a]

	Job Satisfaction Rating
Mann-Whitney U	26.000
Wilcoxon W	81.000
Z	−1.820
Asymp. sig. (2-tailed)	.069
Exact sig. [2*(1-tailes sig.)]	.075[b]

[a] Group variable: type of profession.
[b] Not corrected for ties.

FIGURE 15.14   Mann-Whitney Test

In the SPSS output, select the Mann-Whitney U test using the U statistic from the Test Statistics table and p-value (Asymp. Sig. (2-tailed))

$$U = 26$$

With samples this large, the value of U approaches a normal distribution, and so the null hypothesis can be tested by a Z-test

$$Z = -1.820$$

Step 4: Determine the p-value,
   From the SPSS output, select the p-value = 0.069 (Asymptotic Sig. (2-sided test))
Step 5: Decision on H_0 using p-value:
   For p-value: we compare p-value = 0.069 with $\alpha = 0.05$
   To make decision: If p-value < α, reject H_0; otherwise, do not reject H_0.
   Since p-value > α, we fail to reject H_0
Step 6: Make conclusions at a given level of confidence based on the decision in Step 5.
   At 95% level of confidence, the median job satisfaction scores for lawyers is not different from the median job satisfaction scores for doctors.

15.5 Kruskal-Wallis H Test Using SPSS

Kruskal-Wallis H test is used instead of a one-way analysis of variance (ANOVA) to find out if two or more medians are different. Ranks of the data points are used for the calculations, rather than the data points themselves. It is a nonparametric version of ANOVA and an extension of the Mann-Whitney U test to allow the comparison of more than two independent groups. It is generally used when the measurement variable does not meet the normality assumptions of one-way ANOVA.

Assumptions

- All samples are randomly drawn from their respective population.
- There should be independence in the observations, i.e. there is no relationship between the observations in each group or between the groups themselves.
- The measurement scale is at least ordinal or continuous level (i.e. interval or ratio).
- Mutual independence among the various samples.
- The independent variable should consist of two or more categorical, independent groups .

15.5.1 Hypothesis Tests for Kruskal-Wallis H Test

Step 1: The null and alternative hypotheses are, respectively,

H_0: Population medians are equal and the alternative is

H_a: Population medians are not equal

Step 2: Decide on/state the significance level, α.

Step 3: Compute the value of the test statistic

First pool all the data across the groups.

Rank the data from 1 for the smallest value of the dependent variable and next smallest variable rank 2 and so on ... to N for across the groups (if any value ties, in that case it is advised to use mid-point), N being the highest variable.

Find T_i, the total of the ranks for each group

Compute the test statistic (H) by substituting the values into the formula

The test statistic for the Kruskal Wallis test (denoted by H) is computed as

$$H = \frac{12}{N(N+1)} \sum \frac{T_i^2}{N_i} - 3(N+1)$$

where T_i is the rank sum for the ith sample, $i = 1,2,3,...k$ and N_i is the number of participants in each group.

In SPSS, we go to Analyze > Nonparametric Tests > Legacy Dialogs > k Independent Samples > There will be a pop-up window for Tests for k Independent Samples Test > Transfer the numerical/dependent variable of interest into Test Variable List > Put the categorical/ independent variable of interest into Grouping Variable and click Define Ranges to put the values represent the range of codes you gave the groups of the independent variable groups of interest where Minimum is the lowest code and Maximum the highest code> Select/ check Kruskal Wallis H as the Test Type > Click Continue then OK > In the SPSS output select the Kruskal-Wallis H test using the Chi-square statistic from the Test Statistics table and p-value (Asymp. Sig. (2-tailed))

Step 4: Determine the p-value,

From the SPSS output, select the p-value (Asymptotic Sig. (2-sided test))

Step 5: Decision on H_0 using p-value:

For p-value: we compare p-value with α

To make decision: *If p-value < α, reject H_0; otherwise, do not reject H_0.*

Step 6: Make conclusions at a given level of confidence based on the decision in Step 5.

Example

The following data show the peak levels of ozone pollution in four cities (Nairobi, Kisumu, Mombasa and Eldoret) for ten selected dates in 2020.

	City			
Data	Nairobi	Kisumu	Mombasa	Eldoret
Jan 9	18	20	18	14
Jan 17	23	31	22	30
Jan 18	19	25	22	21
Jan 31	29	36	28	35
Feb 1	27	31	28	24
Feb 6	26	31	31	25
Feb 14	31	24	19	25
Feb 17	31	31	28	28
Feb 20	33	35	35	34
Feb 29	20	42	42	21

At 95% level of confidence are there any statistical significant differences in the peak ozone levels among the four cities?

Step 1: The null and alternative hypotheses are, respectively,
H_0: Population medians across all the four cities are equal and the alternative is
H_a: Population medians across all the four cities are not equal
Step 2: State the significance level, $\alpha = 0.05$.
Step 3: Compute the value of the test statistic
Key the data into SPSS (Figures 15.15 and 15.16)
Variable View

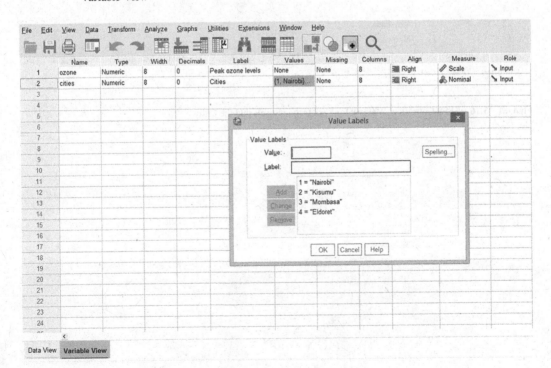

FIGURE 15.15 SPSS Variable View Window for Peak Ozone Levels and the Cities

Data View

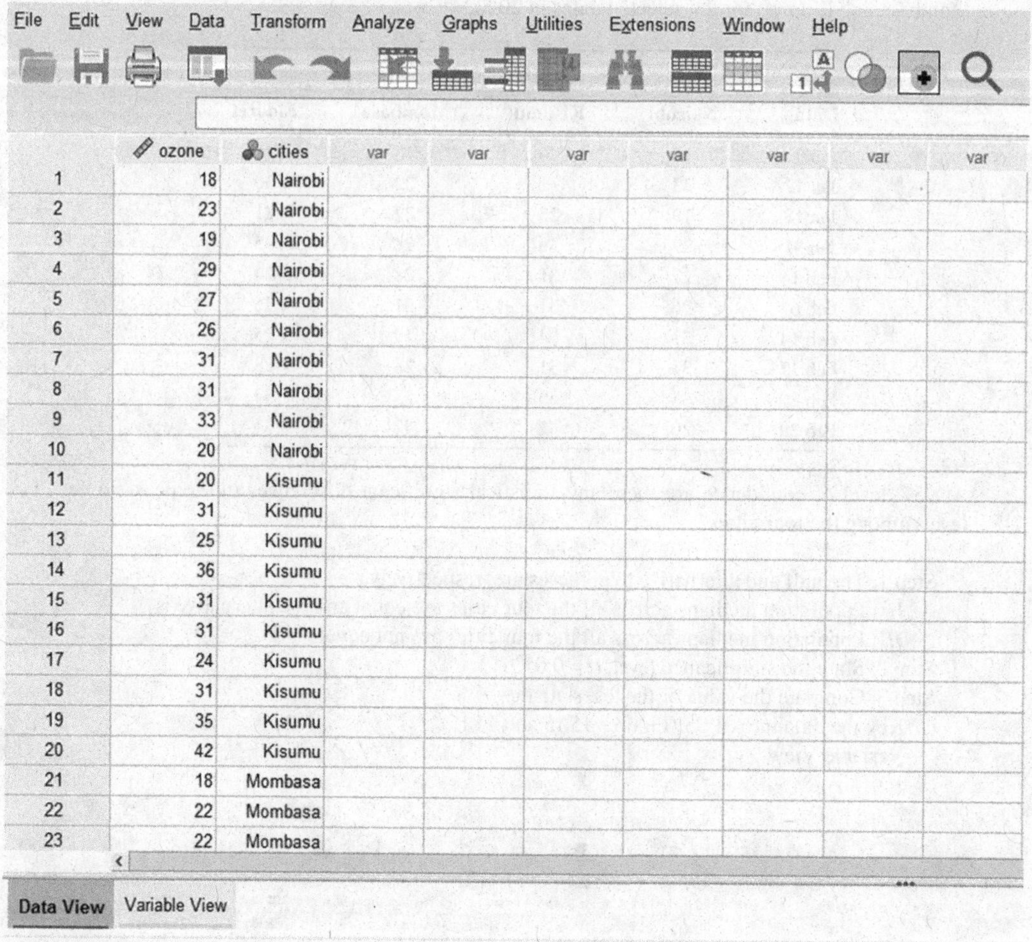

	🖊 ozone	🎚 cities	var	var	var	var	var	var	var
1	18	Nairobi							
2	23	Nairobi							
3	19	Nairobi							
4	29	Nairobi							
5	27	Nairobi							
6	26	Nairobi							
7	31	Nairobi							
8	31	Nairobi							
9	33	Nairobi							
10	20	Nairobi							
11	20	Kisumu							
12	31	Kisumu							
13	25	Kisumu							
14	36	Kisumu							
15	31	Kisumu							
16	31	Kisumu							
17	24	Kisumu							
18	31	Kisumu							
19	35	Kisumu							
20	42	Kisumu							
21	18	Mombasa							
22	22	Mombasa							
23	22	Mombasa							

Data View | Variable View

FIGURE 15.16   SPSS Data View Window for Peak Ozone Levels and the Cities

In SPSS, we go to Analyze > Nonparametric Tests > Legacy Dialogs > k Independent Samples > There will be a pop-up window for Tests for k Independent Samples Test > Transfer the numerical/dependent variable of interest (Peak ozone levels) into Test Variable List > Put the categorical/independent variable of interest (Cities) into Grouping Variable and click Define Ranges to put the values represent the range of codes you gave the groups of the independent variable groups of interest where Minimum (1) is the lowest code and Maximum (4) the highest code > Select/check Kruskal Wallis H as the Test Type > Click Continue then OK (Figure 15.7)

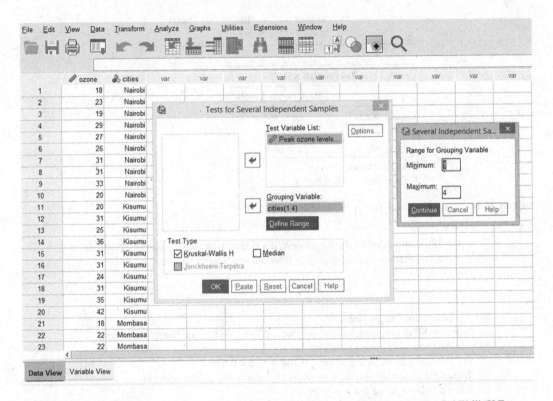

FIGURE 15.17 SPSS Tests for Several Independent Samples Pop Up Window for generating Kruskal-Wallis H Test

SPSS Output (Figure 15.18)

Kruskal-Wallis Test

Ranks

	Cities	N	Mean Rank
Peak ozone levels	Nairobi	10	17.75
	Kisumu	10	26.55
	Mombasa	10	19.70
	Eldoret	10	18.00
	Total	40	

Test Statistics[a, b]

	Peak Ozone Levels
Kruskal-Wallis H	3.764
df	3
Asymp. sig.	.288

[a] Kruskal Wallis test.
[b] Grouping variable: cities.

FIGURE 15.18   Kruskal-Wallis H Test

In the SPSS output, select the Kruskal-Wallis H test using the Chi-square statistic from the Test Statistics table, degree of freedom and p-value (Asymp. Sig.)

$$H = \chi^2(3) = 3.764$$

Step 4: Determine the p-value,
   From the SPSS output, select the p-value = 0.288 (Asymptotic Sig. (2-sided test)
Step 5: Decision on H_0 using p-value:
   For p-value: we compare p-value = 0.288 with $\alpha = 0.05$
   To make decision: *If p-value < α, reject H_0; otherwise, do not reject H_0.*
   Since *p-value > α, we fail reject H_0*
Step 6: Make conclusions at a given level of confidence based on the decision in Step 5.
   At 95% level of confidence, the population medians across all the four cities are equal.

Practice Exercise

1. Consider the following data showing the cost of attending private and public colleges in thousands of dollars.

Private Colleges

62.8	53.2	55.1	43.2	44.0	40.6	55.7	47.7	60.5	52.0	43.5	52.3
61.5	48.3	52.1	50.6	46.2	53.9	49.3	56.3	60.9	48.4		

Public Colleges

30.4	32.5	38.2	25.6	34.1	38.6	32.8	35.8	28.7	35.7	24.4	31.8
32.2	36.4	28.9	31.4	26.3	33.1	38.3	28.9				

Is the cost of attending public colleges significantly different from the cost of attending private colleges? At 97% confidence level, what will be your conclusion? Use p-value approach.

2. For the data set *brakes.sav* test if the median braking performance for machine 5 is significantly different from 8. Use 5% level of significance.

3. Consider the following data collected from sample of professionals to measure the job satisfaction for pharmacist, teachers, accountants, auditors and engineers.

Pharmacists	Teachers	Accountants	Auditors	Engineers
43	55	54	43	64
44	69	65	73	56
72	70	79	72	84
42	76	69	60	86
54	50	79	64	72
60	48	64	66	69
45	53	59	41	78
48	42	78	55	55
74	40	61	44	
41	62			

Test for any significant differences in the job satisfaction among the different professions. At $\alpha = 0.05$, what conclusion do you draw from your results?

4. Consider the following data on the distance covered in kilometers by ten individuals participating in two races.

Individual	Race 1	Race 2
1	44	55
2	42	78
3	74	80
4	42	86
5	53	60
6	50	59
7	45	62
8	48	52
9	64	55
10	38	50

Do these data suggest a significant difference in the distance covered in the two races? What is the p-value? Using $\alpha = 0.04$ level of significance, what is your conclusion?

16

Analysis of Time Series Data

16.1 Introduction to Time Series Analysis and Forecasting

Time series is an ordered sequence of values of a variable at equally spaced time intervals. Time series may be measured continuously or discretely. *Time series data* are *data gathered collected sequentially over time on a given characteristic*. For example, the numbers unemployed for each quarter of four successive years. The series should contain enough observations for proper parameter estimation. Time-series forecasting techniques attempt to account for changes over time by examining patterns, cycles or trends, or using information about previous time periods to predict the outcome for a future time period. Time series analysis is the evaluation and extraction of the components of a model that divides a particular series into understandable and explainable portions so that: trends can be identified, extraneous factors can be identified and to make forecasts. There are two main approaches used to analyze time series; *in the time domain* or *in the frequency domain*. Time series can be presented in graphs or plots i.e. time sequence plot, time plot or graphical plot of series.

Application of Time Series Forecasting:

- Investment analysts use time series forecasting to predict stock price by getting the closing price of the stock on each given day.
- Retail companies use forecasting to predict sales and units sold for different products.
- Procurement specialists use time series to determine future demand of products by looking at the current and previous product utilization.
- Weather prediction is another application that can be done using time series forecasting.
- Financial institutions use previous customer bank transactions to determine chance of a client qualifying to take up bank products.
- Education institutions apply time series techniques to determine future student enrollment trends.
- It is used by government departments to predict a state's population, at any particular region, or the nation as a whole.

16.1.1 Time Series Components

There are four components of time-series data. These are the *trend, cyclical effects, seasonality and irregularity/error component.*

- *Trend* is the long-term general direction of data; Trends can be deterministic or stochastic; they may be linear or curvilinear. Curvilinear trends can be polynomial, exponential or dampened. Trends can also be short run or long run.
- *Cycles* are patterns of highs and lows through which data move over time periods usually of more than a year.
- *Seasonality* are shorter cycles, which usually occur in time periods of less than one year. Seasonal effects are measured over monthly period or quarterly, or in as smaller time frame such as a weekly or even a daily. Seasonal effects may be additive or multiplicative.
- *Irregular fluctuations* are rapid changes in the data, which occur in even shorter time frames than seasonal effects. These fluctuations are the unforeseen, unpredictable, and unexplained factors by the model. This is also the error component.

DOI: 10.1201/9781003292654-20

16.1.2 Understanding Time Series

In time series each data point (Y_t) at time t can be expressed as either as an addition or multiplication of the four components: Seasonality (S_t), Trend (T_t), Cyclical (C_t) and Irregularity (I_t).

The additive time series mode is given by,

$$Y_t = S_t + T_t + C_t + I_t$$

The multiplicative time series mode is given by,

$$Y_t = S_t \times T_t \times C_t \times I_t$$

16.1.3 Stationary Time Series

Time series may be stationary or non-stationary. Time-series data that has no trend, cyclical or seasonal effects is said to be *stationary*.

A time series is stationary if:

i. The mean value of time-series is constant over time, implying that there is no trend component.

ii. The variance does not increase over time.

iii. There is minimal seasonality effect.

Non-stationary series have random walk, drift, trend or changing variance.

16.1.4 Creating Lags of Time Series

A lag of time series is created when the time base is shifted by a given number of periods. Lags of a time series are often used as explanatory variables to model the actual time series itself. This is because the state of the few back time series periods may still has an influence on the series current state.

16.1.5 The Measurement of Forecasting Error

Once you have forecasted time series values using a given technique/fitted a time series model, it is good to determine the forecasting error for a given technique method using forecast fit measures. They are also known as goodness-of-fit indicators. Time series forecasting methods include Autoregressive Integrated Moving Average (ARIMA) Model, Seasonal Autoregressive Integrated Moving Average (SARIMA) Model, Vector Autoregression (VAR) method models and Long Short Term Memory network (LSTM). There are several techniques that can be used to measure overall forecasting error: *mean error (ME), mean absolute error (MAE), mean square error (MSE), mean percentage error (MPE) and mean absolute percentage error (MAPE)*. These goodness-of-fit measures are available in SPSS.

16.1.5.1 Error

The *error of an individual forecast* is *the difference between the actual value and the forecast of that value*. It is computed as:

$$e_t = x_t - F_t$$

where

e_t = the error of the forecast

x_t = the actual value

F_t = the forecast value

Using the error of the forecast, we can generate MAE and MSE as shown ahead.

16.1.5.2 Mean Absolute Error (MAE)

The MAD is the mean of the absolute values of the errors and is computed as

$$MAE = \frac{|e_t|}{T}$$

where e_t are the values of the errors, t = time period, T = total number of observations

16.1.5.3 Mean Square Error (MSE)

The MSE is the sum of the squared errors divided by the degrees of freedom and is computed as:

$$MSE = \frac{\sum e_t^2}{T-k}$$

where e_t are the values of the errors, t = time period, T = total number of observations and k = number of parameters in the model.

16.1.6 Reading Time Series Data

The first thing that to do to analyze time series data will be to read it into SPSS, and to plot the time series to observe the trends in the data set.

Example: Consider the following two-year monthly sales data from a certain company. The time period is from January 2020 to December 2021.

Time Point	Sales in Thousands
1	12.23
2	11.84
3	12.25
4	11.10
5	10.97
6	8.75
7	7.75
8	10.50
9	6.71
10	7.60
11	12.46
12	8.47
13	12.27
14	12.57
15	8.87
16	11.15
17	11.86
18	11.07
19	10.38
20	8.71
21	12.07
22	12.74
23	9.82
24	11.51

We go to SPSS and create variable in the Variable View window then key data in the Data View window (Figures 16.1 and 16.2).

Variable View

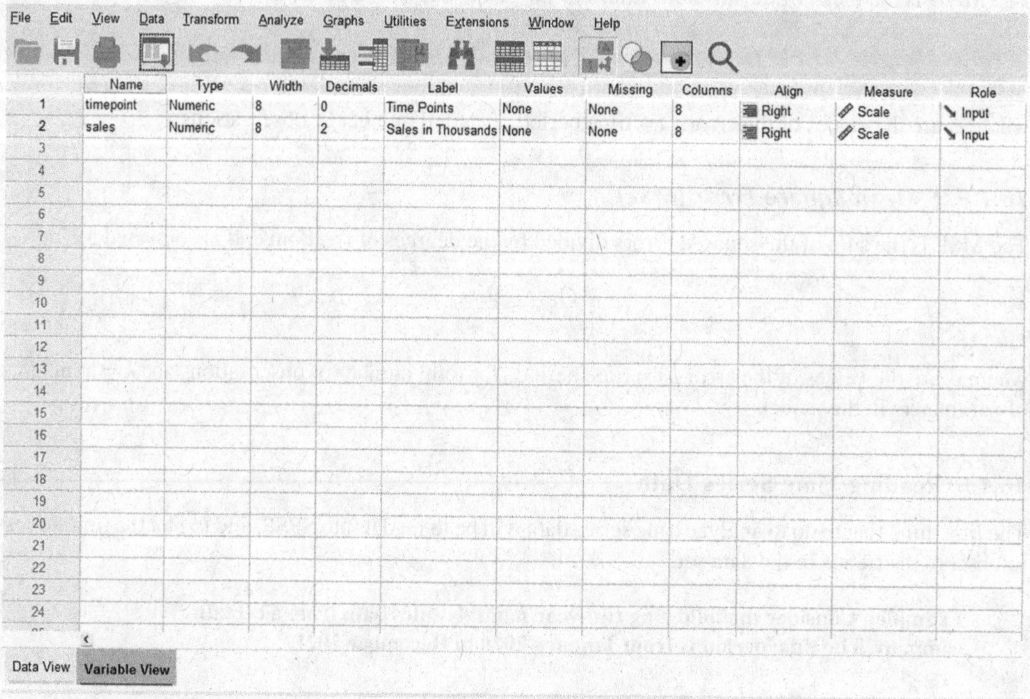

	Name	Type	Width	Decimals	Label	Values	Missing	Columns	Align	Measure	Role
1	timepoint	Numeric	8	0	Time Points	None	None	8	Right	Scale	Input
2	sales	Numeric	8	2	Sales in Thousands	None	None	8	Right	Scale	Input
3											
4											
5											
6											
7											
8											
9											
10											
11											
12											
13											
14											
15											
16											
17											
18											
19											
20											
21											
22											
23											
24											

Data View **Variable View**

FIGURE 16.1   SPSS Variable View Window for Time Points and Sales in Thousands

Data View

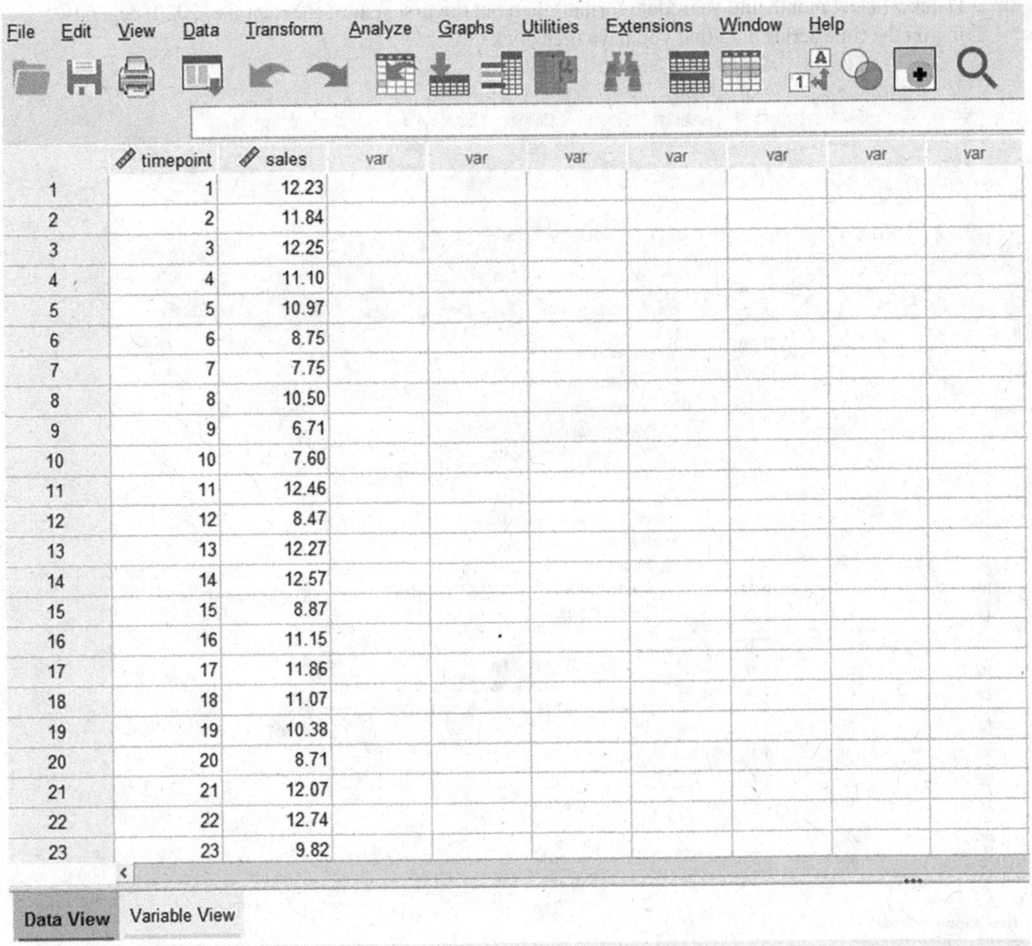

	timepoint	sales	var	var	var	var	var	var	var
1	1	12.23							
2	2	11.84							
3	3	12.25							
4	4	11.10							
5	5	10.97							
6	6	8.75							
7	7	7.75							
8	8	10.50							
9	9	6.71							
10	10	7.60							
11	11	12.46							
12	12	8.47							
13	13	12.27							
14	14	12.57							
15	15	8.87							
16	16	11.15							
17	17	11.86							
18	18	11.07							
19	19	10.38							
20	20	8.71							
21	21	12.07							
22	22	12.74							
23	23	9.82							

Data View Variable View

FIGURE 16.2 SPSS Data View Window for Time Points and Sales in Thousands

Convert the data into time series data and the type of time series data (Figures 16.3–16.6).

In SPSS you go to: Data > Define Date and Time > There will be a pop up window for Define Dates > Select month and year (date format) then put the first year of the data (i.e. 2020) depending on the time series data that you have then click OK

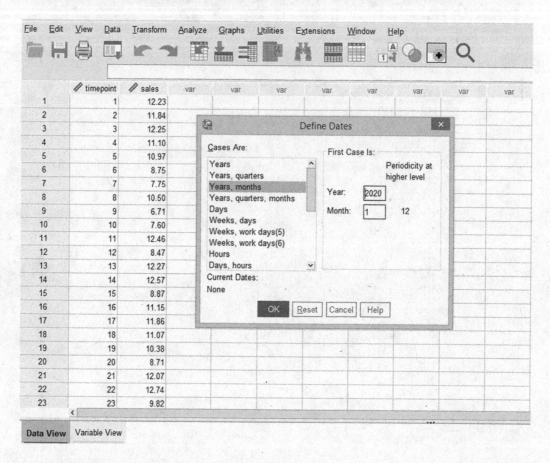

FIGURE 16.3 SPSS Pop Up Window for Defining Dates

Go to Data View window, the data set will be in time series format with months and dates.

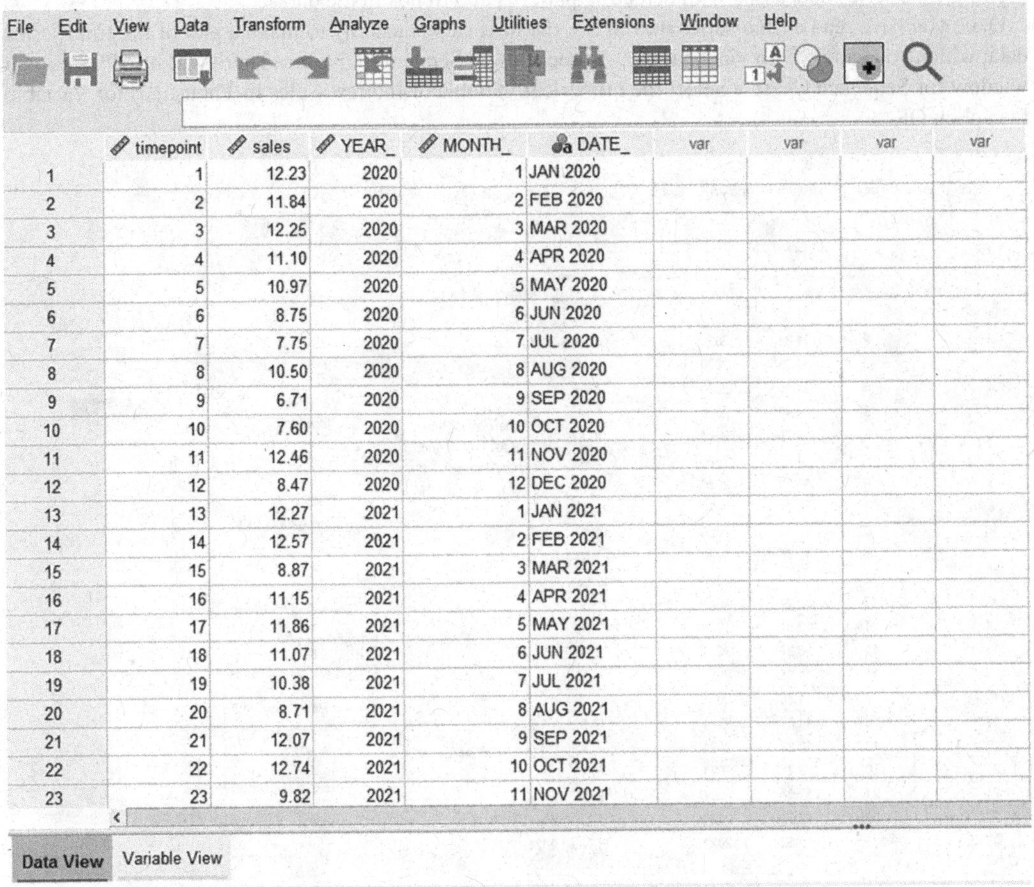

	timepoint	sales	YEAR_	MONTH_	DATE_	var	var	var	var
1	1	12.23	2020	1	JAN 2020				
2	2	11.84	2020	2	FEB 2020				
3	3	12.25	2020	3	MAR 2020				
4	4	11.10	2020	4	APR 2020				
5	5	10.97	2020	5	MAY 2020				
6	6	8.75	2020	6	JUN 2020				
7	7	7.75	2020	7	JUL 2020				
8	8	10.50	2020	8	AUG 2020				
9	9	6.71	2020	9	SEP 2020				
10	10	7.60	2020	10	OCT 2020				
11	11	12.46	2020	11	NOV 2020				
12	12	8.47	2020	12	DEC 2020				
13	13	12.27	2021	1	JAN 2021				
14	14	12.57	2021	2	FEB 2021				
15	15	8.87	2021	3	MAR 2021				
16	16	11.15	2021	4	APR 2021				
17	17	11.86	2021	5	MAY 2021				
18	18	11.07	2021	6	JUN 2021				
19	19	10.38	2021	7	JUL 2021				
20	20	8.71	2021	8	AUG 2021				
21	21	12.07	2021	9	SEP 2021				
22	22	12.74	2021	10	OCT 2021				
23	23	9.82	2021	11	NOV 2021				

Data View   Variable View

FIGURE 16.4 SPSS Data View Window displaying Time point, Sales in Thousands and Date

16.1.7 Plotting Time Series

Plotting time series data assist with detecting patterns and trends in the dataset.

Once you have read a time series into SPSS, the next step is usually to make a plot of the time series data, which you can do with via Analyze > Forecasting > Sequence Charts > There will be SPSS pop up window for Sequence Chart > Select the numerical variable of Interest (sales in Thousand) for Variable then click OK

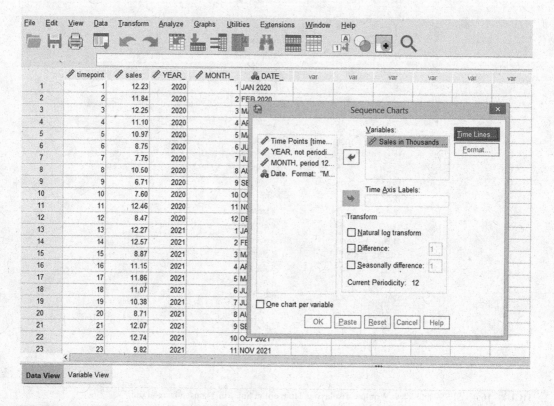

FIGURE 16.5   SPSS Pop Up Window for generating Sequence Charts

SPSS Output showing Time Series Plot

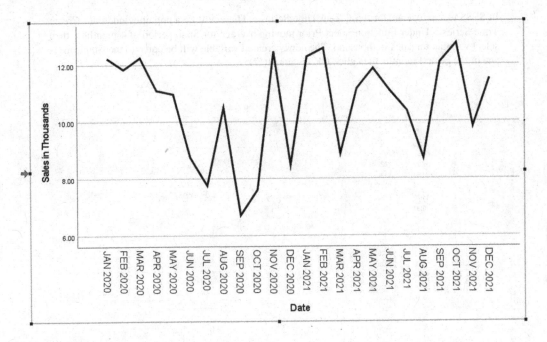

FIGURE 16.6   Time Series Plot for Sales in Thousands

We can see that there are monthly cycles plus increasing and decreasing trends over time.

16.2 Decomposing Time Series

Decomposing a time series means separating/splitting time series data into trend component, cyclic component, an irregular component and seasonal component if it is a seasonal time series.

16.2.1 Decomposing Nonseasonal Data

A nonseasonal time series consists of a trend component and an irregular component. Decomposing the time series involves trying to separate the time series into these components by estimating the trend component and the irregular component.

The *simple moving average of the time series* can be used to estimate the trend component of a nonseasonal time series described by the additive model. Simple moving average is one of the averaging techniques for "smoothing out" the irregular fluctuations in the time series data.

The *simple moving average of the time series* can be generated in SPSS. In SPSS we go to Transform > Create Time Series > There will be a pop up window for Create Time Series > Under Function select Prior moving average for Span period of interest depending on the sequence of the data > then select the time series data variable. The newly created variable will be equal to the simple average of the prior span period that was selected then click OK. Go to Data View window and you will see the moving average data depending on the chosen time span.

Example: Using the sales data, generate a four monthly moving average.
Plot the original data with the four monthly moving average.

In SPSS we go to Transform > Create Time Series > There will be a pop up window for Create Time Series > Under Function select Prior moving average for Span period of 4 months > then select variable for sales in thousand. The newly created variable will be equal to the simple average of the prior 4 months then click OK (Figure 16.7).

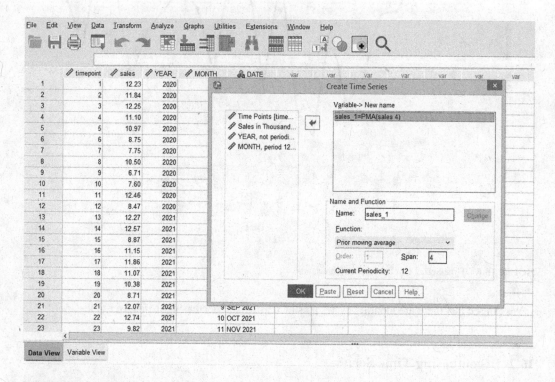

FIGURE 16.7   SPSS Create Time Series Pop Up Window for generating Moving Averages

Go to Data View window and you will see the four-month moving average data (Figure 16.8).

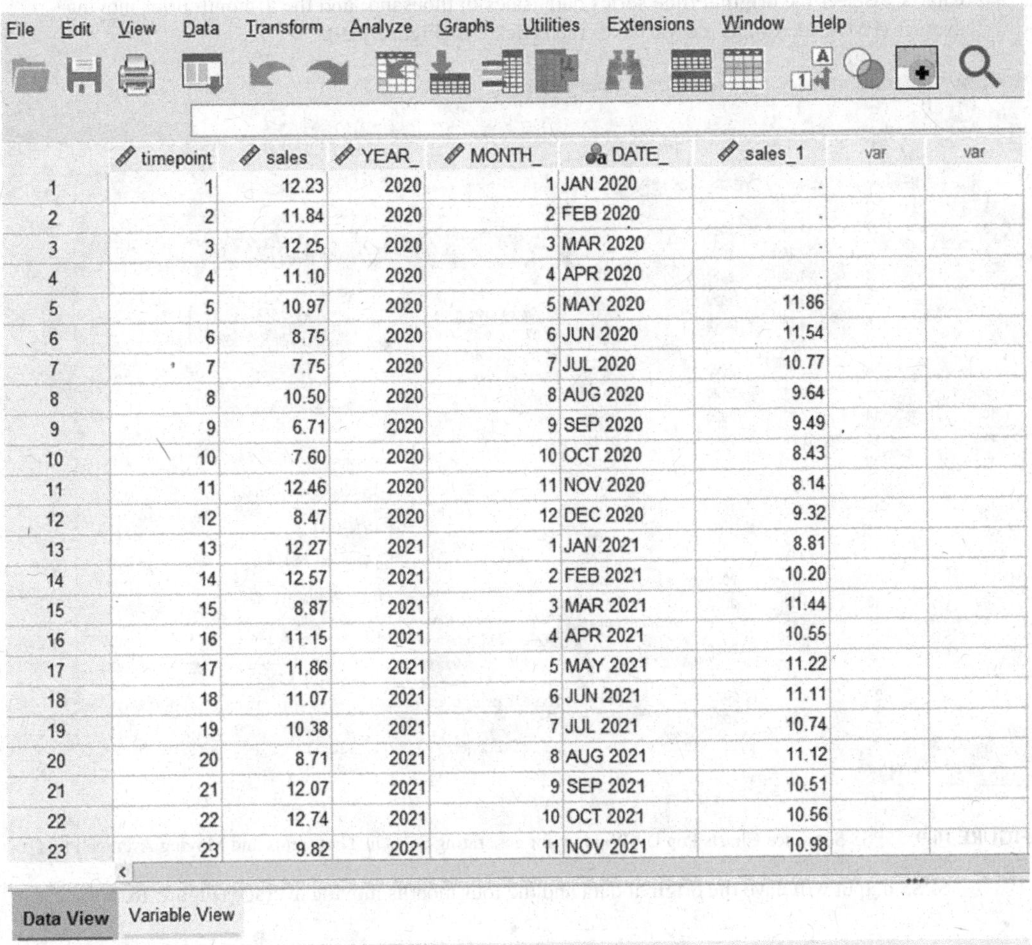

	timepoint	sales	YEAR_	MONTH_	DATE_	sales_1	var	var
1	1	12.23	2020	1	JAN 2020	.		
2	2	11.84	2020	2	FEB 2020	.		
3	3	12.25	2020	3	MAR 2020	.		
4	4	11.10	2020	4	APR 2020	.		
5	5	10.97	2020	5	MAY 2020	11.86		
6	6	8.75	2020	6	JUN 2020	11.54		
7	7	7.75	2020	7	JUL 2020	10.77		
8	8	10.50	2020	8	AUG 2020	9.64		
9	9	6.71	2020	9	SEP 2020	9.49		
10	10	7.60	2020	10	OCT 2020	8.43		
11	11	12.46	2020	11	NOV 2020	8.14		
12	12	8.47	2020	12	DEC 2020	9.32		
13	13	12.27	2021	1	JAN 2021	8.81		
14	14	12.57	2021	2	FEB 2021	10.20		
15	15	8.87	2021	3	MAR 2021	11.44		
16	16	11.15	2021	4	APR 2021	10.55		
17	17	11.86	2021	5	MAY 2021	11.22		
18	18	11.07	2021	6	JUN 2021	11.11		
19	19	10.38	2021	7	JUL 2021	10.74		
20	20	8.71	2021	8	AUG 2021	11.12		
21	21	12.07	2021	9	SEP 2021	10.51		
22	22	12.74	2021	10	OCT 2021	10.56		
23	23	9.82	2021	11	NOV 2021	10.98		

Data View   Variable View

FIGURE 16.8   SPSS Data View Window displaying Time Point, Sales in Thousands, Date and Moving Averages

Graph the original time series data and the four-month prior moving average.

Analyze > Forecasting > Sequence Charts > There will be a pop up window for Sequence Charts > Select the original time series data (sales in thousand) and the 4-month prior moving average (PMA(sales,4)) as the Variables then click OK (Figure 16.9).

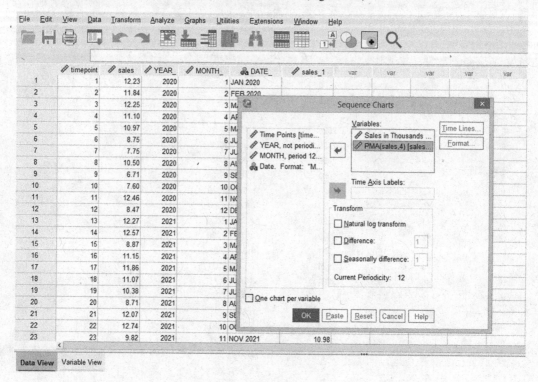

- **FIGURE 16.9**   SPSS Sequence Charts Pop Up Window for generating Sales in Thousands and Moving Average Plots

SPSS output will have the original data and the four months moving average (Figure 16.10).

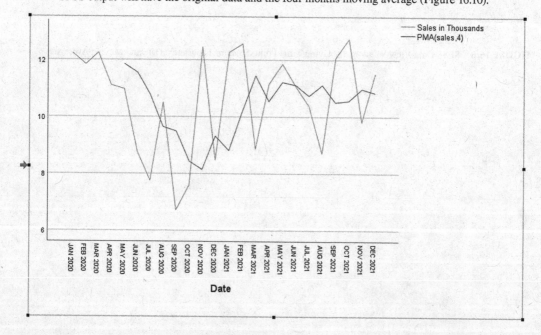

FIGURE 16.10   Sales in Thousands and Moving Average Plots

The Moving Average line follows the pattern of the observed data, but does not reach the extreme peaks and valleys of the original data. By averaging four values at a time, we "smooth out" the original curve.

We have to compute the mean of the preceding four months to forecast a future period beyond the range of the observed data.

Other averaging techniques that can be used to smooth a series and describe a series spanning a number of time periods include: a simple average, a centered moving average, a double moving average and weighted moving average.

16.2.2 Decomposing Seasonal Data

A seasonal time series consists of a trend component, a seasonal component, a cyclic component and an irregular component. Decomposing the time series means separating the time series into these four components: that is, estimating these four components.

The seasonal decomposition procedure has two different approaches for modeling the seasonal factors: multiplicative or additive.

- Multiplicative. The seasonal component is a factor by which the seasonally adjusted series is multiplied to yield the original series. In effect, seasonal components that are proportional to the overall level of the series. Observations without seasonal variation have a seasonal component of 1.
- Additive. The seasonal adjustments are added to the seasonally adjusted series to obtain the observed values. This adjustment attempts to remove the seasonal effect from a series in order to look at other characteristics of interest that may be "masked" by the seasonal component. In effect, seasonal components that do not depend on the overall level of the series. Observations without seasonal variation have a seasonal component of 0.

To perform seasonal decomposition in SPSS we go to Analyze > Forecasting > Seasonal Decomposition > There will be a pop up window for Seasonal Decomposition > Select the time series data for Variable(s) > Tick Model Type as Multiplicative or Additive > OK > There will be a pop up message to add four variables to the data file click OK

The SPSS Output will give additional data sets for seasonal adjusted series, seasonal factors, trend-cycle and error that can be viewed in the Data View window.

Example: Perform seasonal decomposition for the sales data.

First, we create quarterly time series data using the path:

Data > Define date and time > There will be a pop up window for Define Dates > Select Years, quarters under Cases Are: then put 2010 for Years and 1 for Quarter then click OK.

In the SPSS Data View window there will be the quarterly time series data.

Once you have the quarterly data, you can now perform seasonal decomposition using the path:

Analyze > Forecasting > Seasonal Decomposition > There will be a pop up window for Seasonal Decomposition > Select the sales data for Variable(s) > Tick Model Type as Multiplicative and for Moving Average Weight select All point equal > OK > There will be a pop up message to add four variables to the data file click OK (Figure 16.11).

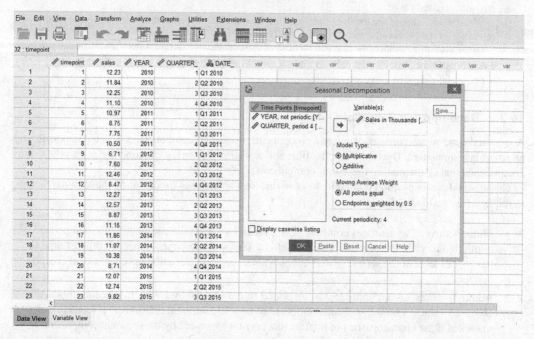

FIGURE 16.11   SPSS Seasonal Decomposition Pop Up Window for generating Time Series Components

The SPSS Output will give the sales data, seasonal, trend and error. Go to Data View window and you will see the sales data and the four components (seasonal adjusted series, seasonal factors, trend-cycle and irregular (Figure 16.12).

	timepoint	sales	YEAR_	QUARTER_	DATE_	ERR_1	SAS_1	SAF_1	STC_1	var
2	2	11.84	2010	2	Q2 2010	.95292	11.37453	1.04092	11.93644	
3	3	12.25	2010	3	Q3 2010	1.10121	13.09684	.93534	11.89311	
4	4	11.10	2010	4	Q4 2010	1.04094	11.74528	.94506	11.28331	
5	5	10.97	2011	1	Q1 2011	.99278	10.16986	1.07868	10.24386	
6	6	8.75	2011	2	Q2 2011	.89021	8.40601	1.04092	9.44277	
7	7	7.75	2011	3	Q3 2011	.92889	8.28576	.93534	8.92006	
8	8	10.50	2011	4	Q4 2011	1.28113	11.11040	.94506	8.67234	
9	9	6.71	2012	1	Q1 2012	.72621	6.22058	1.07868	8.56579	
10	10	7.60	2012	2	Q2 2012	.81064	7.30122	1.04092	9.00670	
11	11	12.46	2012	3	Q3 2012	1.33085	13.32136	.93534	10.00966	
12	12	8.47	2012	4	Q4 2012	.84324	8.96239	.94506	10.62856	
13	13	12.27	2013	1	Q1 2013	1.03403	11.37504	1.07868	11.00068	
14	14	12.57	2013	2	Q2 2013	1.10109	12.07583	1.04092	10.96717	
15	15	8.87	2013	3	Q3 2013	.86589	9.48318	.93534	10.95195	
16	16	11.15	2013	4	Q4 2013	1.07190	11.79818	.94506	11.00683	
17	17	11.86	2014	1	Q1 2014	1.00531	10.99495	1.07868	10.93684	
18	18	11.07	2014	2	Q2 2014	.98568	10.63480	1.04092	10.78933	
19	19	10.38	2014	3	Q3 2014	1.04937	11.09757	.93534	10.57551	
20	20	8.71	2014	4	Q4 2014	.87223	9.21634	.94506	10.56637	
21	21	12.07	2015	1	Q1 2015	1.02682	11.18963	1.07868	10.89737	
22	22	12.74	2015	2	Q2 2015	1.08535	12.23915	1.04092	11.27665	
23	23	9.82	2015	3	Q3 2015	.90204	10.49886	.93534	11.63904	
24	24	11.51	2015	4	Q4 2015	1.03036	12.17911	.94506	11.82023	

FIGURE 16.12   SPSS Data View Window displaying Time Point, Sales in Thousands, Date and Time Series Components

The generated data are then plotted using sequence charts.

In SPSS, we go to Analyze > Forecasting > Sequence Charts > There will be SPSS pop up window for Sequence Chart > Select the numerical variable of Interest one at a time (STC_1, then SAF_1, then ERR_1, then SAS_1) for Variable then click OK. SPSS Output Window will display four different plots for the time series components.

16.3 Forecasts Using Exponential Smoothing

Exponential smoothing can be used to make short-term forecasts for time series data. Exponential smoothing gives a declining weight to observations. The more current observation, the more importance it will have on the forecast. Parameters can also be added. Holt method is when you add a trend parameter while Holt-Winters is when seasonality is added.

16.3.1 Simple Exponential Smoothing

Simple exponential smoothing can be used to make short-term forecasts when we have a time series that is described using an additive model with constant level and no seasonality. The simple exponential smoothing method provides a way of estimating the level at the current time point. Smoothing is controlled by the parameter alpha; for the estimate of the level at the current time point. The value of alpha lies between 0 and 1. Values of alpha that are close to 0 mean that little weight is placed on the most recent observations when making forecasts of future values.

Simple exponential smoothing procedure is available in SPSS through the path:

Analyze > Forecasting > Create Traditional Models > There will be a pop up window for Time Series Modeler > Select time series data as the Dependent Variable > Under Method select Exponential Smoothing > click Criteria and there will be a pop up window for Time Series Modeler: Exponential Smoothing Criteria there will be Simple for Nonseasonal and Seasonal Simple.

To forecast: Under Options select First case after end of estimation period through a specified data > then put the year and months of forecast, i.e. Year 2022 and month 12 > click Continue then OK. The SPSS output will give the time series model and in the Data View window you will be able to see the forecasted values.

16.3.2 Holt's Linear Exponential Smoothing

If you have a time series that can be described using an additive model with increasing or decreasing trend and no seasonality, then Holt's linear exponential smoothing can be used to make short-term forecasts. Holt's exponential smoothing estimates the level and slope at the current time point. Smoothing is controlled by two parameters, alpha, for the estimate of the level at the current time point, and beta for the estimate of the slope b of the trend component at the current time point. As with simple exponential smoothing, the parameters alpha and beta have values between 0 and 1, and values that are close to 0 mean that little weight is placed on the most recent observations when making forecasts of future values.

Holt's exponential smoothing procedure is available in SPSS through the path:

Analyze > Forecasting > Create Traditional Models > There will be a pop up window for Time Series Modeler > Select time series data as the Dependent Variable > Under Method select Exponential Smoothing > click Criteria and there will be a pop up window for Time Series Modeler: Exponential Smoothing Criteria there will be Simple for Nonseasonal and Seasonal Simple.

To forecast: Under Options select First case after end of estimation period through a specified data > then put the year and months of forecast, i.e. Year 2022 and month 12 > click Continue then OK. The SPSS output will give the time series model and in the Data View window you will be able to see the forecasted values.

16.3.3 Holt-Winters Exponential Smoothing

If you have a time series that can be described using an additive model with increasing or decreasing trend and seasonality, then Holt-Winters exponential smoothing can be used to make short-term forecasts. Holt-Winters exponential smoothing estimates the level, slope and seasonal component at the current time point. Smoothing is controlled by three parameters: alpha, beta and gamma, for the estimates of the level, slope b of the trend component and the seasonal component, respectively, at the current time point. The parameters alpha, beta and gamma all have values between 0 and 1, and values that are close to 0 mean that relatively little weight is placed on the most recent observations when making forecasts of future values.

In SPSS we go to Analyze > Forecasting > Create Traditional Models > There will be a pop up window for Time Series Modeler > Select time series data as the Dependent Variable > Under Method select Exponential Smoothing > click Criteria and there will be a pop up window for Time Series Modeler: Exponential Smoothing Criteria and select Winter's Multiplicative > To forecast: Under Options select First case after end of estimation period through a specified data > the put the year and months of forecast, i.e. Year 2022 and month 12 > click Continue then OK. The SPSS output will give the time series model and in the Data View window you will be able to see the forecasted values

> **Example: Forecast the sales data for the next 1 year using the exponential smoothing technique.**
>
> In SPSS we go to Analyze > Forecasting > Create Traditional Models > There will be a pop up window for Time Series Modeler > Select Sales in thousands as the Dependent Variable > Under Method select Exponential Smoothing > click Criteria and there will be a pop up window for Time Series Modeler: Exponential Smoothing Criteria and select Winter's Multiplicative > (Figure 16.13)

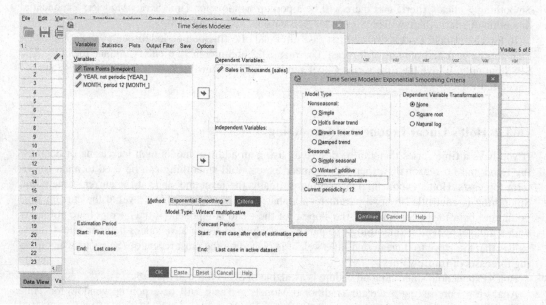

FIGURE 16.13 SPSS Time Series Modeler Pop Up Window for performing Exponential Smoothing

To forecast: Under Options select First case after end of estimation period through a specified data > the put the year and months of forecast, i.e. Year 2022 and month 12 > then click OK. (Figure 16.14)

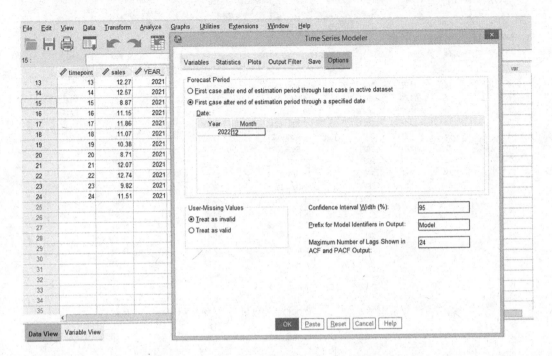

FIGURE 16.14   SPSS Time Series Modeler Pop Up Window for performing Forecasts

SPSS output (Figure 16.15).

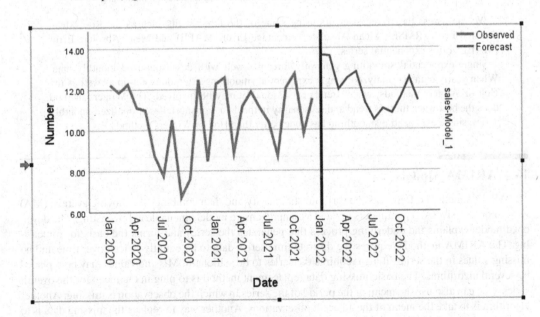

FIGURE 16.15   Time Series Forecast using Exponential Smoothing Method

From the Data View window we will have the predicted values in additional column (Figure 16.16).

File Edit View Data Transform Analyze Graphs Utilities Extensions Window Help

1 : Predicted_sales_... 12.25234786778292

	timepoint	sales	YEAR_	MONTH_	DATE_	Predicted_sales_Model_1	LCL_sales_Model_1	UCL_sales_Model_1	NResidual_sales_Model_1	var
16	16	11.15	2021	4	APR 2021	11.81	7.53	16.09	-.66	
17	17	11.86	2021	5	MAY 2021	11.72	7.44	16.01	.14	
18	18	11.07	2021	6	JUN 2021	9.45	5.16	13.73	1.62	
19	19	10.38	2021	7	JUL 2021	8.39	4.10	12.67	1.99	
20	20	8.71	2021	8	AUG 2021	11.07	6.78	15.35	-2.36	
21	21	12.07	2021	9	SEP 2021	7.37	3.09	11.65	4.70	
22	22	12.74	2021	10	OCT 2021	8.28	4.00	12.57	4.46	
23	23	9.82	2021	11	NOV 2021	13.19	8.91	17.47	-3.37	
24	24	11.51	2021	12	DEC 2021	9.10	4.82	13.39	2.41	
25	.	.	2022	1	JAN 2022	13.34	9.05	17.62	.	
26	.	.	2022	2	FEB 2022	13.39	9.10	17.67	.	
27	.	.	2022	3	MAR 2022	11.06	6.77	15.34	.	
28	.	.	2022	4	APR 2022	12.14	7.85	16.42	.	
29	.	.	2022	5	MAY 2022	12.57	8.29	16.86	.	
30	.	.	2022	6	JUN 2022	11.11	6.83	15.40	.	
31	.	.	2022	7	JUL 2022	10.23	5.95	14.51	.	
32	.	.	2022	8	AUG 2022	10.29	6.00	14.57	.	
33	.	.	2022	9	SEP 2022	10.90	6.62	15.19	.	
34	.	.	2022	10	OCT 2022	11.72	7.43	16.00	.	
35	.	.	2022	11	NOV 2022	11.89	7.61	16.18	.	
36	.	.	2022	12	DEC 2022	11.26	6.98	15.54	.	

Data View | Variable View

FIGURE 16.16 SPSS Data View Window displaying Predicted values using Exponential Smoothing

We can assess the forecast errors using some of the forecast fits such as the Root Mean Squared Error (RMSE), Mean Absolute Percentage Error (MAPE) and Mean Absolute Error (MAE) for this exponential series.

Single exponential smoothing generally forecasts well with deseasonalized monthly data. When yearly data are analyzed, single exponential smoothing often does not do as well as the Holt or Winters methods, where trends or seasonality may be involved. The Winters method does the best when there is trend and seasonality in the data. For short, deseasonalized, monthly series, single exponential smoothing has done better than Holt or Winters methods.

16.4 ARIMA Models

The ARIMA method differences the series to stationarity and then combines the moving average (MA) with autoregressive(AR) parameters to yield a comprehensive model amenable to forecasting. The developed model explains the underlying process that generates the series and forms the basis for forecasting. The ARIMA methods requires the discrete time series data to be equally spaced over time and no missing values in the series. It also requires the series to be stationary. Missing values may be replaced by several algorithms. The basic missing data replacement method is to plug in the mean for the overall series. One can also use the mean of the period of the series in which the observation is missing. Another algorithm is to take the mean of the adjacent observations. Another way to replace the missing data is to take the median of nearby points. Linear interpolation may be employed to impute a missing value, as can linear trend at a point.

In SPSS, we can replace the missing values using the path:

Transform > Replace Missing Values > There will be SPSS pop up window for Replace Missing Values > Click the drop down for Method and select the preferred method for replacing the missing values (i.e. Series Mean, Mean of nearby points, Median of nearby points, Linear interpolation and Linear trend at point) > Select the time series data of interests (e.g. sales) and put in the New Variable(s) window then click OK. Go to SPSS Data View window and you find the missing values replaced.

ARIMA is a combination of the Autoregressive (AR) and Moving Average (MR) model. The AR model forecast corresponds to a linear combination of past values of the variable. The moving average model forecast corresponds to a linear combination of past forecast errors. The "I" represents the data values that are replaced by the difference between their values and the previous values.

ARIMA models contain three parameters (p, d, q):

- AR(p): AR part of the model. Means that we use p past observations from the time series as predictors. This can be estimated from the PACF plot.
- Differencing (d): Used to transform the time series into a stationary one by taking the differences between successive observations at appropriate lags d.
- MA(q): uses q past forecast errors as predictors. This can be estimated from the ACF plot.

If the data set is seasonal, then you will have SARIMA which adds a seasonal part to the model.

16.4.1 Formulation of ARIMA Models

Exponential smoothing methods are useful for making forecasts, and make no assumptions about the correlations between successive values of the time series. ARIMA models include an explicit statistical model for the irregular component of a time series that allows for nonzero autocorrelations in the irregular component. The first step in time series analysis is to plot the time series data using sequence charts to examine if there are outliers, missing data, or elements of nonstationarity. If there are elements of nonstationarity then first differencing is used to transform the series into a condition of stationarity.

The process of developing ARIMA models includes examination of the characteristic patterns of the correlograms. Correlograms (ACF and PACF plots) for stationary processes exhibit characteristic patterns.

Once the process has been transformed into stationarity, we can proceed with the analysis.

The series is then examined for autoregressive or moving average components. We consider the autoregressive process first then the moving average process. The autoregressive process is a function of previous observations in the series while the moving average process is a function of the series innovations.

16.4.2 Differencing a Time Series

ARIMA models are meant for stationary time series. Therefore, if you start off with a nonstationary time series, you will first need to 'difference' the time series until you obtain a stationary time series. If you have to difference the time series d times to obtain a stationary series, then you have an ARIMA(p,d,q) model, where d is the order of differencing used. We first plot the sequence charts to examine if there are outliers, missing data, or elements of nonstationarity. If there are elements of nonstationarity then first differencing is used to transform the series into a condition of stationarity. We can also examine the characteristic pattern for the correlograms (ACF and PACF plots). If the ACF attenuates very slowly, that is evidence of nonstationarity.

To make the time series data stationary: In SPSS we go to Analyze > Forecasting > Sequence Charts > There will be a pop up window for Sequence Charts > select the time series data as Variable(s) then under Transform select Difference of 1 click OK. Look at the SPSS output to see for stationarity

There are two statistical tests for nonstationarity: Dicky–Fuller test and Augmented Dickey–Fuller test. The Dickey–Fuller test is used to test for difference stationarity and trend stationarity. The augmented Dickey–Fuller test is conducted to test if there is autocorrelation in the series.

16.4.3 Selecting a Candidate ARIMA Model

If your time series is stationary, or if you have transformed it to a stationary time series by differencing d times, the next step is to select the appropriate ARIMA model, which means finding the values of most appropriate values of p and q for an ARIMA(p,d,q) model. To do this, you usually need to examine the correlogram and partial correlogram (ACF and PACF plots) of the stationary time series.

In SPSS we go to Analyze > Forecasting > Autocorrelation > there will be a pop up window for Autocorrelation > Select sales as the Variable and Display Autocorrelation and Partial Autocorrelation then click OK.

Once the series has been rendered stationary, the ACF and PACF are examined to determine the type and order of the model. The autoregressive process is identified by the characteristic patterns of its ACF and PACF. The ACF has a gradual attenuation and the PACF possesses the same number of spikes as the order of the model.

The MA process is identified by the number of significant spikes in the ACF that are followed by no subsequent significant spikes.

16.4.4 Forecasting Using an ARIMA Model

Once you have selected the best candidate ARIMA(p,d,q) model for your time series data, you can estimate the parameters of that ARIMA model, and use that as a predictive model for making forecasts for future values of your time series.

The Time Series Modeler procedure estimates exponential smoothing, univariate ARIMA and multivariate ARIMA (or transfer function models) models for time series, and produces forecasts.

In SPSS we go to Analyze > Forecasting > Create Traditional Models > There will be a pop up window for Time Series Modeler > Select the time series data as Dependent Variables > Under Method select ARIMA > click Criteria > There will be a pop up window for Time Series Modeler: ARIMA Criteria > Click Model and put the values for p, d and q for nonseasonal ARIMA Orders then Continue > click Statistics select the goodness-of-fit measures under Fit Measures (i.e. Stationary R Square, R Square, Root mean square error, Mean absolute percentage error, Mean absolute error, Maximum absolute percentage error, Maximum absolute error and Normal BIC) > click Options and under Forecast Period select First case after end of estimation period through a specified data then put the forecast that you want (e.g. Year 2022 and Month 12) click then OK. The SPSS output will give the model statistics and the graphical display of the forecast. SPSS Data View Window will display the forecasted values.

Example: Forecast the sales data for the next 1 year using the ARIMA model.

Key the data into SPSS and tell convert your data into Time Series by specifying the time.

Preliminary evidence to assess stationarity is to plot the time series data and observe the trends.

Go to Analyze > Forecasting > Sequence Chart > There will be a pop up window for Sequence Charts > select the Sales as the Variable(s) then click OK.

Examine the SPSS output if there are seasonal patterns in the series. This will imply that the series does not appear to be stationary if it is fluctuating up and down over the trend.

Generate and view the ACF and PACF charts to check for stationarity.

In SPSS go to Analyze > Forecasting > Autocorrelation > there will be a pop up window for Autocorrelation > Select sales as the Variable. Under Display tick Autocorrelation and Partial Autocorrelation then click OK.

The SPSS Output will have the ACF and PACF plots.

Examine the patterns of the ACF chart. If the ACF chart has values above the top and bottom border lines, then it shows that there is a significant correlation and also confirms that the data is not stationary. For ARIMA model, the data has to be stationary. We need to remove nonstationarity and trends.

We can difference to make the data stationary.

Go to Analyze > Forecasting > Sequence Charts > There will be a pop up window for Sequence Charts > select the time series data as Variable(s) then under Transform select Difference of 1 click OK.

The SPSS Output will have the plot of the differenced data.

Examine the trend for the differenced data. If data is oscillating around zero, then it shows that the data is now stationary.

Generate and assess the ACF and PACF with Difference at 1.

Analyze > Forecasting > Autocorrelations > Select sales as the Variables > Under Display select Autocorrelations and Partial autocorrelations > Under Transform Click Difference and put 1 then click OK.

The SPSS output will have the ACF plot. Look for the point/ lag for which the ACF correlation is insignificant (values below the line). This will imply stationarity.

To determine the possible values for p and q for candidate ARIMA model: go to ACF plot and PACF plot. The significant lag point in the ACF will be the moving average value.

Assess the residual plots for ACF and PACF for any significant correlations using the estimated values for p, d, and q.

In SPSS go to Analyze > Forecasting > Create Traditional Models > There will be a pop up window for Time Series Modeler > Select the time series data as Dependent Variables > Under Method select ARIMA > click Criteria > There will be a pop up window for Time Series Modeler: ARIMA Criteria > Click Model and put the values for p (3), d (1) and q(0) for nonseasonal ARIMA Orders then Continue > click Statistics select Parameter estimates, Residual ACF, Residual PACF, Display forecast , Normalized BIC > click Plots select Residual ACF, Residual PACF > click Options and put the forecast that you want (Year 2022 and Month 12) click then OK.

Go to SPSS Output and select the PACF and ACF residual plots.

Assess the PACF and ACF Residual plots for no significant autocorrelations. If there is no significant autocorrelation, go and fit the ARIMA model.

Once you have determined the values for p (3), d (1) and q (0) for the time series data, you go ahead and estimate the parameters of that ARIMA model, and use that as a predictive model for making forecasts for future values of your time series.

In SPSS, you go to Analyze > Forecasting > Create Traditional Models > There will be a pop up window for Time Series Modeler > Select Sales in thousands as the Dependent Variable > Under Method select ARIMA > click Criteria and there will be a pop up window for Time Series Modeler: ARIMA Criteria and select Model the Nonseasonal and put 3 for p, 1 for d and 0 for q > click Statistics select the goodness-of-fit measures under Fit Measures (i.e. Stationary R Square, R Square, Root mean square error, Mean absolute percentage error, Mean absolute error, Maximum absolute percentage error, Maximum absolute error and Normal BIC) > click Options and under Forecast Period select First case after end of estimation period through a specified data then put the forecast that you want (e.g. Year 2022 and Month 12) click then OK (Figure 16.17).

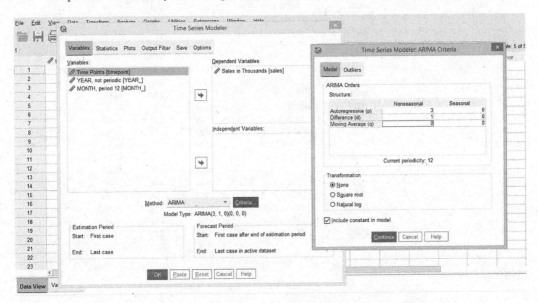

FIGURE 16.17 SPSS Time Series Modeler Pop Up Window for generating ARIMA Models

SPSS output (Figure 16.18) shows the graphical display of the forecast (i.e. values after Jan 2022) and the observed values (i.e. values before Jan 2022).

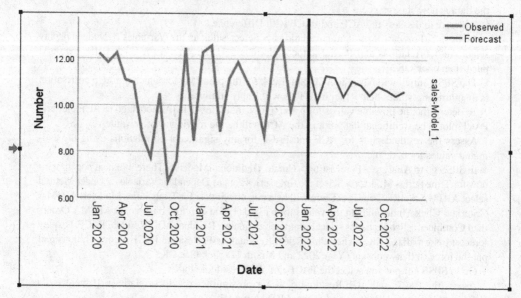

FIGURE 16.18   Time Series Forecast using ARIMA Method

From the Data View window we will have the predicted values in additional column (Figure 16.19).

	timepoint	sales	YEAR_	MONTH_	DATE_	Predicted _sales_M odel_1	LCL_sale s_Model_ 1	UCL_sale s_Model_ 1	NResidua l_sales_ Model_1	var
17	17	11.86	2021	5	MAY 2021	11.25	7.42	15.08	.61	
18	18	11.07	2021	6	JUN 2021	9.64	5.81	13.47	1.43	
19	19	10.38	2021	7	JUL 2021	11.70	7.87	15.53	-1.32	
20	20	8.71	2021	8	AUG 2021	11.15	7.32	14.98	-2.44	
21	21	12.07	2021	9	SEP 2021	9.65	5.82	13.48	2.42	
22	22	12.74	2021	10	OCT 2021	10.55	6.72	14.38	2.19	
23	23	9.82	2021	11	NOV 2021	10.61	6.78	14.44	-.79	
24	24	11.51	2021	12	DEC 2021	11.92	8.09	15.75	-.41	
25			2022	1	JAN 2022	11.73	7.90	15.56		
26			2022	2	FEB 2022	10.19	6.02	14.37		
27			2022	3	MAR 2022	11.29	6.88	15.71		
28			2022	4	APR 2022	11.22	5.72	16.73		
29			2022	5	MAY 2022	10.40	4.60	16.19		
30			2022	6	JUN 2022	11.06	4.97	17.16		
31			2022	7	JUL 2022	10.90	4.17	17.62		
32			2022	8	AUG 2022	10.46	3.46	17.46		
33			2022	9	SEP 2022	10.84	3.54	18.15		
34			2022	10	OCT 2022	10.67	2.92	18.41		
35			2022	11	NOV 2022	10.43	2.42	18.44		
36			2022	12	DEC 2022	10.64	2.33	18.94		
37										

Data View   Variable View

FIGURE 16.19   SPSS Data View Window displaying Predicted values using ARIMA

We can assess the forecast errors using some of the forecast fits such as the Root Mean Squared Error (RMSE), Mean Absolute Percentage Error (MAPE) and Mean Absolute Error (MAE) for this exponential series.

16.5 Autocorrelation and Partial Autocorrelation

16.5.1 Autocorrelation Function

The autocorrelation function, ACF(k), is a standardization of the autocovariance function. The autocovariance function (ACV) shows the covariance in a series between one observation and another observation in the same series k lags away.

The standardization is performed by dividing the autocovariance (with a distance of k lags between observations) by a quantity equal to the variance—that is, the product of the standard deviation at lag 0 and the standard deviation at lag 0. This is equivalent to computing the Pearson product moment correlation of the series by dividing the covariance of a series and its lagged form by the product of the standard deviation of the series times itself.

Given this definition of the autocorrelation function, different characteristic patterns emerge for various autoregressive and moving average autoregressive processes. It is important to examine the characteristic differences between the ACFs of those two processes. The characteristic pattern of the autoregressive process AR(p) has gradual attenuation of the magnitude of the autocorrelation. A moving average process exhibits a different characteristic autocorrelation function pattern. The characteristic pattern consists of sharp spikes up to and including the lag, indicating the order of the MA(q) process under consideration.

Once we know the magnitude of the standard error of the ACF, we can estimate the confidence limits formed by standard errors. ACFs with magnitudes beyond the confidence limits are those worthy of attention. The significance of the autocorrelation coefficient can also be determined by either the Box-Pierce portmanteau Q statistic or the modified Ljung-Box Q statistic.

16.5.2 Partial Autocorrelation Function

Partial autocorrelation function, used in conjunction with the autocorrelation function to distinguish a first-order from a higher order autoregressive process. It works in the same way as a partial correlation. This function, when working at k lags, controls for the confounding autocorrelations in the intermediate lags. The effect is to partial out those autocorrelations, leaving only the autocorrelation between the current and kth observation.

For an autoregressive process, the partial autocorrelation function exhibits diminishing spikes through the lag of the process, after which those spikes disappear. In an AR(1) model, there will be one spike in the PACF. If the autocorrelation is positive, the partial autocorrelation function will exhibit positive spikes. If the autocorrelation is negative, then the PACF for the AR(1) model will exhibit negative spikes. Because the model is only that of an AR(1) process, there will be no partial spikes at higher lags. Similarly, in an AR(2) model, there will be two PACF spikes with the same sign as those of the autocorrelation. No PACF spikes will appear at higher lags. Therefore, the PACF very clearly indicates the order of the autoregressive process. The PACF is not as useful in identifying the order of the moving average process as it is in identifying the order of the autoregressive process. The estimated standard errors of the partial autocorrelation are the same as those of the autocorrelation.

Other correlation functions that have been found to be useful in identifying univariate time series models are the Inverse Autocorrelation Function (IACF) and the Extended Sample Autocorrelation Function (EACF or ESACF).

To generate Autocorrelation and Partial Autocorrelation plots in SPSS, we go to In SPSS we go to: Analyze > Forecasting > Autocorrelation > There will be a pop up window for Autocorrelation > Select the time series data as the Variables > Under Display select Autocorrelations and Partial Autocorrelations then click OK

Example: Generate the autocorrelation and partial autocorrelation plots for the sales data.

In SPSS we go to: Analyze > Forecasting > Autocorrelation > There will be a pop up window for Autocorrelation > Select sales in thousand as the Variables > Under Display select Autocorrelations and Partial Autocorrelations then click OK (Figure 16.20)

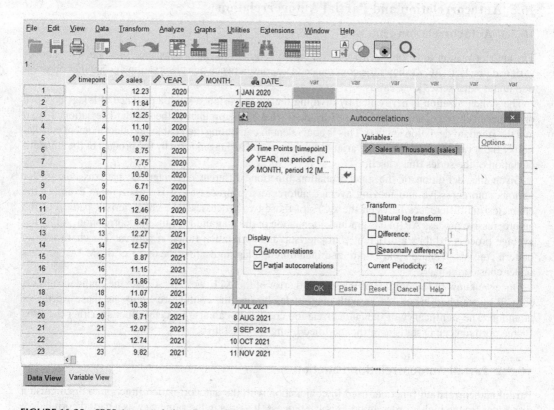

FIGURE 16.20   SPSS Autocorrelations Pop Up Window for generating Autocorrelations and Partial Autocorrelations

SPSS Output (Figure 16.21).

			Box-Ljung Statistic		
Lag	Autocorrelation	Std. Error[a]	Value	df	Sig.[b]
1	.098	.192	0.262	1	.609
2	.040	.188	0.308	2	.857
3	.346	.183	3.874	3	.275
4	−.280	.179	6.313	4	.177
5	−.069	.174	6.469	5	.263
6	−.172	.170	7.490	6	.278
7	−.352	.165	12.048	7	.099
8	−.007	.160	12.050	8	.149
9	−.140	.155	12.865	9	.169
10	.055	.150	13.001	10	.224
11	.102	.144	13.503	11	.262
12	−.168	.139	14.975	12	.243
13	.055	.133	15.149	13	.298
14	−.008	.127	15.154	14	.368
15	−.132	.120	16.359	15	.359
16	.002	.113	16.360	16	.428

Autocorrelations

Series: Sales in Thousands

[a] The underlying process assumed is independence (white noise).
[b] Based on the asymptotic chi-square approximation.

FIGURE 16.21 Time Series Autocorrelations Values

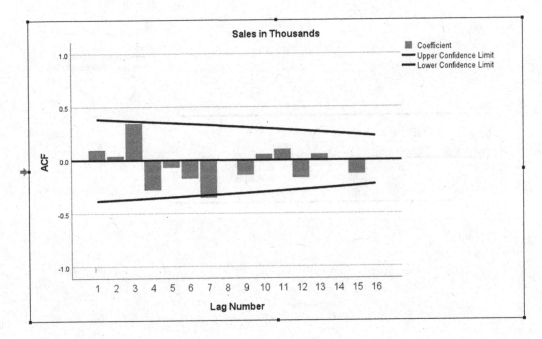

FIGURE 16.22 Time Series Plot of Autocorrelation Function Values

The ACF plot can be used to identify the possible structure of the time series data. The lag (time span between observations) is shown along the horizontal line and the autocorrelation is on the vertical. The upper and the lower lines are the bounds for statistical significance. The ACF plot in Figure 16.22 shows that there is a significant autocorrelation in the residuals at lag 7 followed by non-significant values. The pattern is useful for determining the MA process.

Partial Autocorrelations		
Series: Sales in Thousands		
Lag	Partial Autocorrelation	Std. Error
1	.098	.204
2	.031	.204
3	.343	.204
4	−.392	.204
5	.023	.204
6	−.374	.204
7	−.011	.204
8	−.076	.204
9	.058	.204
10	.146	.204
11	−.118	.204
12	−.224	.204
13	−.221	.204
14	−.024	.204
15	−.017	.204
16	.006	.204

FIGURE 16.23   Time Series Partial Autocorrelation Values

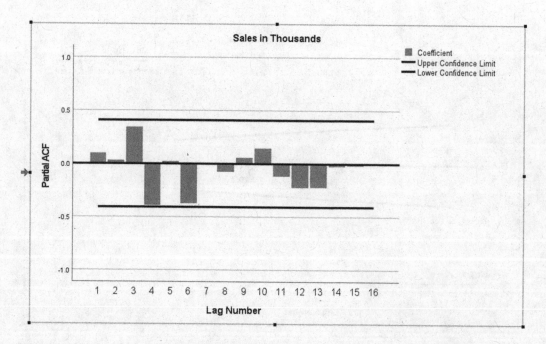

FIGURE 16.24   Time Series Plot of Partial Autocorrelation Function Values

The partial autocorrelations in figure 16.24 for all the values are not significant suggesting a possible AR(0) model for this data.

Practice Exercise

1. Consider the following quarterly sales data for 5 years.

	Sales of a Company			
Years	Qtr 1	Qtr 2	Qtr 3	Qtr 4
2010	19	31	62	9
2011	20	32	65	17
2012	24	36	78	14
2013	25	42	85	24
2014	22	53	45	36

i. Plot the time series data
ii. Perform seasonal decomposition
iii. Forecast for the next 1 year using:
   a. Three quarterly moving average
   b. Different exponential smoothing techniques
   c. ARIMA Models
iv. Estimate the forecasting error for the different forecasting techniques

17

Statistical Quality Control

17.1 Introduction

Statistical Process Control (SPC) is a method of quality control which employs statistical methods to measure, monitor and control a process. It uses a scientific visual method to monitor, control and improve the process by eliminating points that are out of control (where there are variation in a process). SPC has three parts: *Statistics, Process* and *Control. Statistics* is the process of collecting, analyzing, presenting and interpreting data with an aim of drawing information from the data. *Process* converts input resources into the desired output (goods or services) with a combination of available resources. *Control* is the system, policies and procedures that have been put in place to ensure that the overall output meets the requirement. SPC helps to increase productivity, improving overall quality, streamlining the process and providing data to support decision making among others.

17.2 Objectives of Statistical Process Control

SPC aims to optimize continuous improvement by using statistical tools to analyze data, make inferences about process behavior, and then make appropriate decisions. SPC assumes that all processes are subject to variation. *Variation* is the measure of spread of the data around the central tendency. Variation can be classified as either *random cause variation* or *assignable cause variation*. The random cause of variation in the process is due to chance and not due to any assignable factor. It is the variation that is inherent in the process. Assignable cause of variation is when the variation in a process is not due to chance but due to other factors that can be identified and eliminated.

How to Perform SPC

1. Identify the processes: Identify the key process that impacts the output of the products or the processes that are very critical to the customer.
2. Determine measurable attributes of the process: Identify the attributes that need to be measured during the production.
3. Determine the measurement method procedure and the measuring instrument.
4. Determine the subgroup size and the sampling plan.
5. Collect the data as per the sample size and plot the appropriate SPC chart based on data type (Continuous or Discrete) and also subgroup size.
6. Describe natural variation of attributes by calculating the control limits, i.e. the upper control limit (UCL) and lower control limit (LCL).
7. Monitor process variation by interpreting the control chart and checking whether any point is out of control and the pattern.

DOI: 10.1201/9781003292654-21

17.3 SPC Control Charts Using SPSS

A Control chart is one of the main techniques of SPC. The control chart is a graphical display of quality characteristics that have been measured or computed from a sample versus the sample number or time. Control charts contain a center line represents the average value of the quality characteristics and two other horizontal lines known as UCL and LCL.

Selection of an appropriate control chart is very important in control charts mapping. If not properly done then it will end up with inaccurate control limits for the data. The selection of control chart depends on the data type: whether Continuous or Discrete

17.3.1 Continuous/Variable Control Charts

Continuous/variable control charts measure the output on a continuous scale. They include X-bar chart, R chart, Run Charts, X-MR Charts, X-bar – S Charts and EWMA Chart.

17.3.1.1 X-Bar and R Control Charts

An X-bar and R chart is used to monitor the process performance of a continuous data and the data to be collected in subgroups at a set time periods. It is actually a two plots to monitor the process mean and the process variation over the time – usually between three and five pieces per subgroup.

X-bar chart: displays the mean or average change in process over time from subgroup values. The control limits on the X-Bar brings the sample's mean and center into consideration.

R chart: displays the range of the process over the time from subgroups values. This monitors the spread of the process over the time.

Constructing X-bar and R Charts
Compute the average of each subgroup, i.e. $\bar{X}$, then compute grand average of all $\bar{X}$ value, this will be center line for X-bar chart.

Compute the range of each subgroup, i.e. range, then measure grand averages of all range values, i.e. $\bar{R}$ and this will be the center line for R chart.

Plot the X-bar and R values for each subgroup in time series.

Determine the Control Limits

For X-bar chart:

$$UCL_{\bar{X}} = \bar{\bar{X}} + 3\frac{\hat{\sigma}}{\sqrt{n}} = \bar{\bar{X}} + A_2\bar{R}$$

$$LCL_{\bar{X}} = \bar{\bar{X}} - 3\frac{\hat{\sigma}}{\sqrt{n}} = \bar{\bar{X}} - A_2\bar{R}$$

For R chart

$$UCL_{\bar{R}} = D_4\bar{R}$$
$$LCL_{\bar{R}} = D_3\bar{R}$$

where

- X is the individual value (data)
- n is the sample size
- $\bar{X}$ is the average of reading in a sample
- R is the Range, in other words the difference between largest and smallest value in each sample
- D_4 is the appropriate constant from the table of control chart constants
- A_2 is a factor (based on subgroup size)
- $\bar{R}$ is the average of all the ranges
- UCL is Upper control limit
- LCL is Lower control limit

In SPSS we go to We go Analyze > Quality Control > Control Charts > there will be a pop up window for Control Charts > Under Variable Charts select X-bar, R, s. Under Data Organization select Cases are subgroups then click Define > there will be a pop window for X-bar, R, s: cases are subgroups > Click Options > select number of sigmas = 3 and Minimum subgroup sample size (number of subgroups) > Click Control Rules then Select all control rules then click Continue then OK > the SPSS Output will give X-bar chart and R chart

Example

Generate X-bar and R control Chart for the following data.

	Measured Values			
Sample	x_1	x_2	x_3	x_4
1	50	48	24	43
2	32	28	51	31
3	36	16	26	26
4	21	46	23	44
5	29	24	44	49
6	26	20	22	23
7	24	52	24	28
8	19	22	26	12
9	8	28	27	24
10	32	12	18	50

We go to SPSS and create variable in the Variable View window then key data in the Data View window

Variable View (Figure 17.1)

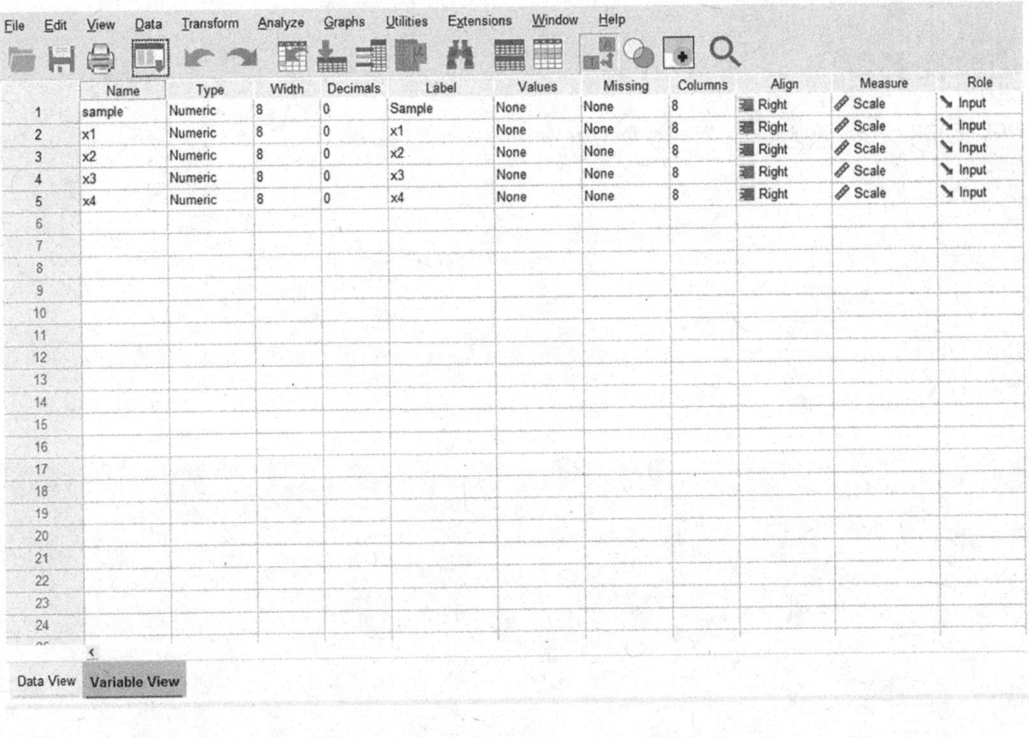

	Name	Type	Width	Decimals	Label	Values	Missing	Columns	Align	Measure	Role
1	sample	Numeric	8	0	Sample	None	None	8	Right	Scale	Input
2	x1	Numeric	8	0	x1	None	None	8	Right	Scale	Input
3	x2	Numeric	8	0	x2	None	None	8	Right	Scale	Input
4	x3	Numeric	8	0	x3	None	None	8	Right	Scale	Input
5	x4	Numeric	8	0	x4	None	None	8	Right	Scale	Input

FIGURE 17.1 SPSS Variable View Window for Four Variables

Key the data in Data View (Figure 17.2)

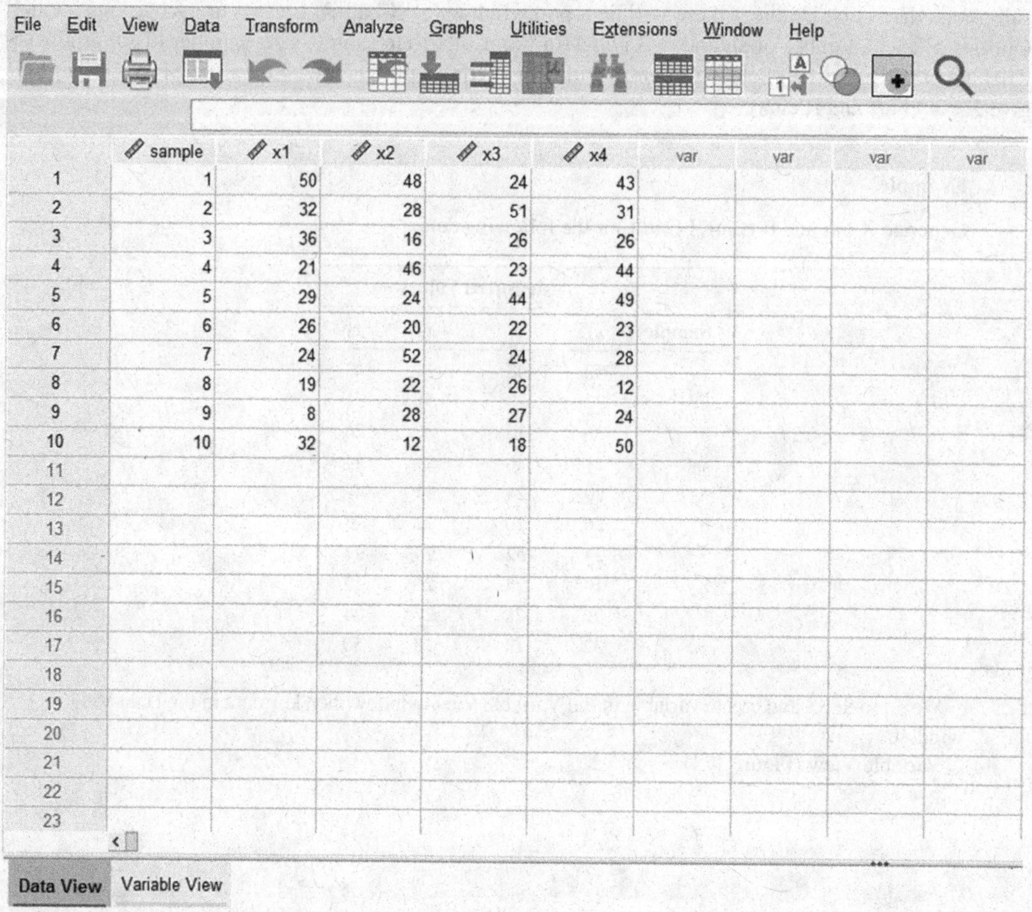

File	Edit	View	Data	Transform	Analyze	Graphs	Utilities	Extensions	Window	Help

	sample	x1	x2	x3	x4	var	var	var	var
1	1	50	48	24	43				
2	2	32	28	51	31				
3	3	36	16	26	26				
4	4	21	46	23	44				
5	5	29	24	44	49				
6	6	26	20	22	23				
7	7	24	52	24	28				
8	8	19	22	26	12				
9	9	8	28	27	24				
10	10	32	12	18	50				
11									
12									
13									
14									
15									
16									
17									
18									
19									
20									
21									
22									
23									

Data View   Variable View

FIGURE 17.2   SPSS Variable View Window for Four Variables

We go Analyze > Quality Control > Control Charts > There will be a pop up window for Control Charts > Under Variable Charts select X-bar, R, s. Under Data Organization select Cases are subgroups then click Define (Figure 17.3)

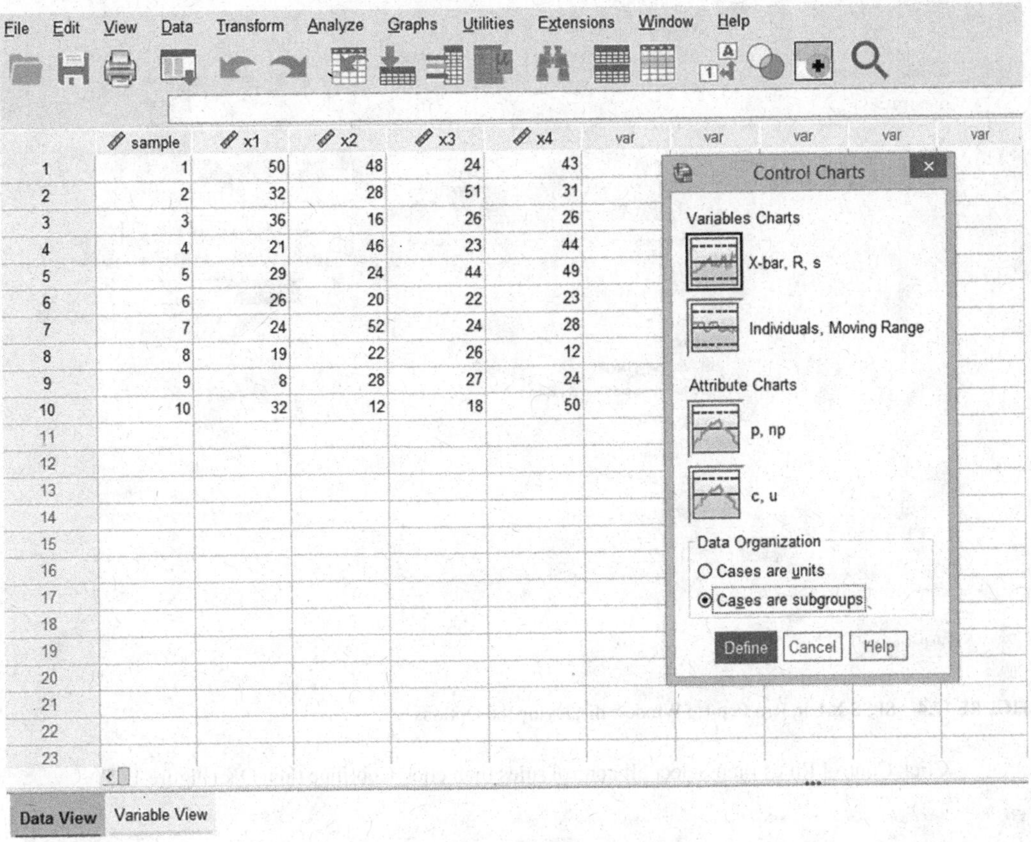

	sample	x1	x2	x3	x4	var	var	var	var	var
1	1	50	48	24	43					
2	2	32	28	51	31					
3	3	36	16	26	26					
4	4	21	46	23	44					
5	5	29	24	44	49					
6	6	26	20	22	23					
7	7	24	52	24	28					
8	8	19	22	26	12					
9	9	8	28	27	24					
10	10	32	12	18	50					

File   Edit   View   Data   Transform   Analyze   Graphs   Utilities   Extensions   Window   Help

Control Charts

Variables Charts

X-bar, R, s

Individuals, Moving Range

Attribute Charts

p, np

c, u

Data Organization
○ Cases are units
● Cases are subgroups

Define   Cancel   Help

Data View   Variable View

FIGURE 17.3   SPSS Control Charts Pop Up Window for selecting Variables Charts

There will be a pop window for X-bar, R, s: cases are subgroups > Under Samples select x1, x2, x3, and x4 variables > Click Options > select number of sigmas = 3 and Minimum subgroup sample size = 4 > click Continue then OK (Figure 17.4)

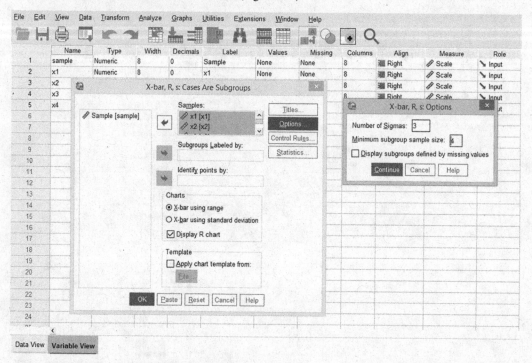

FIGURE 17.4   SPSS X-bar, R, s Pop Up Window displaying the Options

Click Control Rules then Select all control rules then click Continue then OK (Figure 17.5)

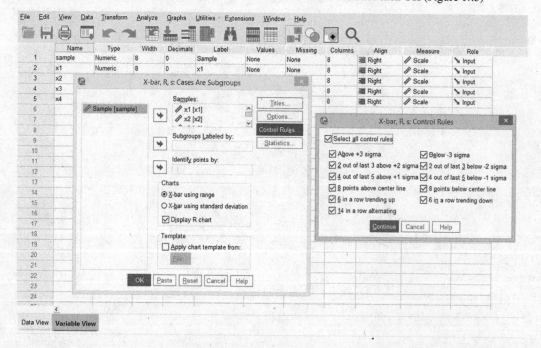

FIGURE 17.5   SPSS X-bar, R, s Pop Up Window displaying the Control Rules

The SPSS Output will give X-bar chart and R chart

X-bar chart

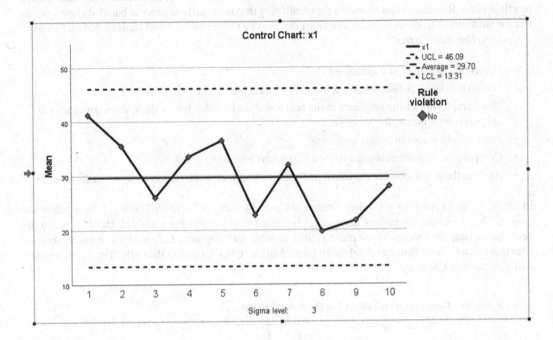

FIGURE 17.6 X-bar Control Chart

Range Chart

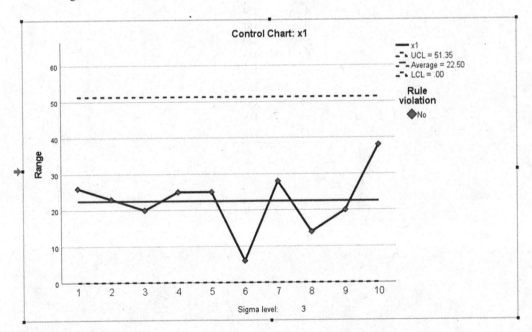

FIGURE 17.7 R Control Chart

Interpretation: The above charts shows that none of the data points were out of control process.

17.3.1.2 Run Charts

A run chart displays observed data as they evolve over time. It is a basic graph that displays data values in a time order. Run charts can be useful for identifying trends or shifts in process but also allows one to test for randomness in the process. A run chart can also be referred to as trend chart or time series plot.

Constructing Run Charts

- Determine the data to be measured
- Obtain the data – collect a minimum of 10–15 data points in a time sequence.
- Plot a graph with a time sequence in the horizontal x-axis (like, hours, days, weeks) and a vertical y-axis with measuring variable.
- Plot the data values in a time sequence
- Compute the mean/median and draw a horizontal line in the graph
- Analyze the graph, observe the trends and patterns to detect special cause variation in the process

In SPSS, we go to Analyze > Quality Control > Control Charts > There will be a pop up window for Control Charts > Under Variables Charts select Individuals, Moving Range and Click Define > Select the outcome variable for Process Measurement, time variable for Subgroups Labeled by and Individuals for Charts > Click Control Rules and Select all control rules> Click Continue then OK. The SPSS window will give the Run Chart

Example: Generate a run chart for the following data.

Week	Pay
1	126
2	130
3	143
4	154
5	157
6	140
7	143
8	131
9	127
10	151
11	148
12	146

Create the variables in the Variable View Window (Figure 17.8)

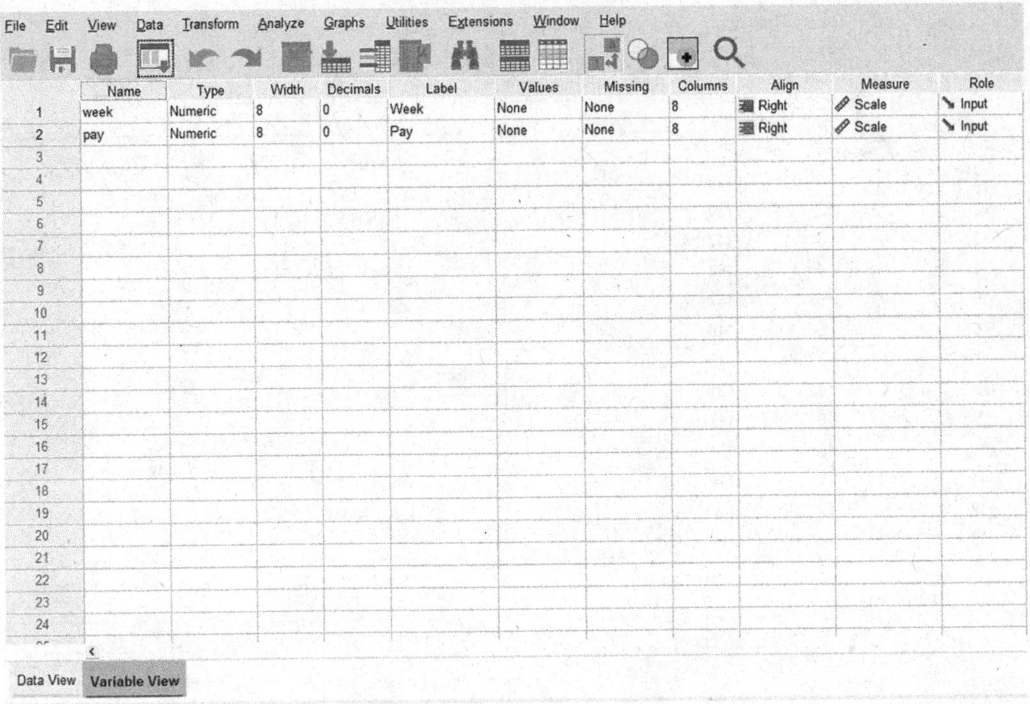

	Name	Type	Width	Decimals	Label	Values	Missing	Columns	Align	Measure	Role
1	week	Numeric	8	0	Week	None	None	8	Right	Scale	Input
2	pay	Numeric	8	0	Pay	None	None	8	Right	Scale	Input

FIGURE 17.8   SPSS Variable View Window for Week and Pay

Key Data in Data View (Figure 17.9)

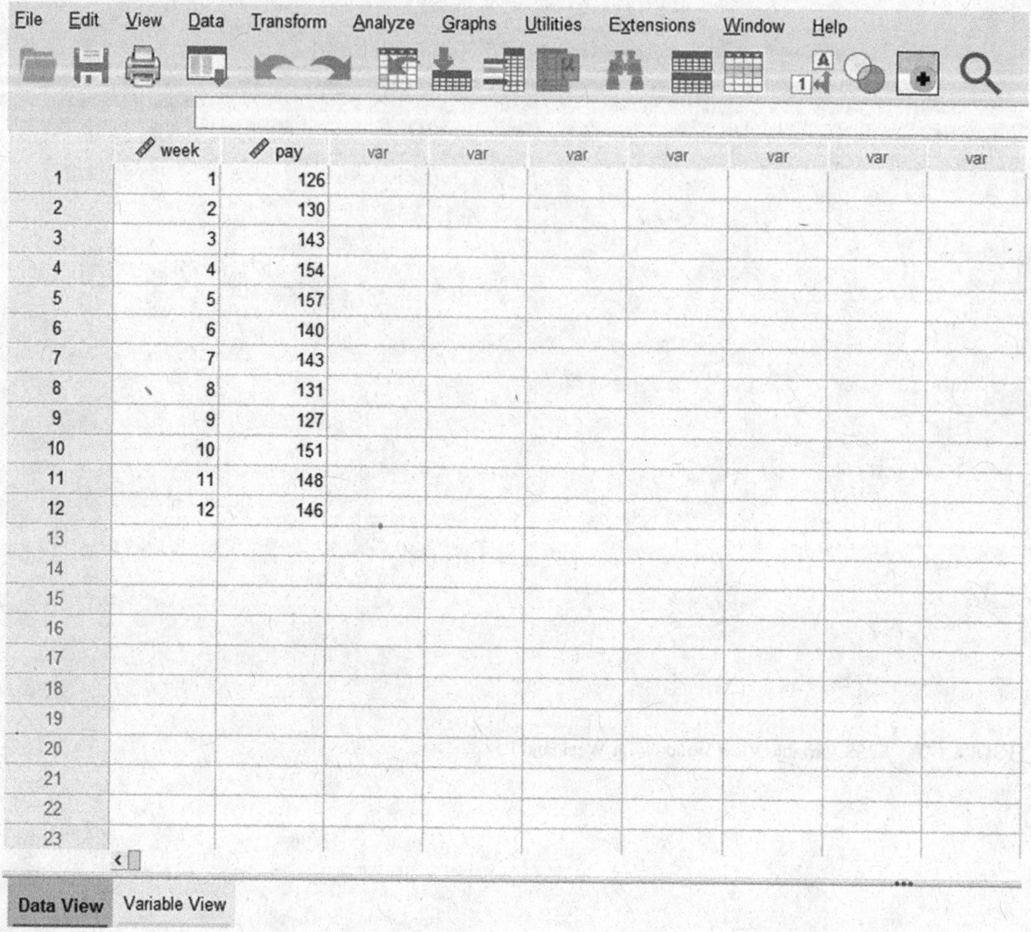

	week	pay	var	var	var	var	var	var	var
1	1	126							
2	2	130							
3	3	143							
4	4	154							
5	5	157							
6	6	140							
7	7	143							
8	8	131							
9	9	127							
10	10	151							
11	11	148							
12	12	146							
13									
14									
15									
16									
17									
18									
19									
20									
21									
22									
23									

Data View   Variable View

FIGURE 17.9   SPSS Data View Window for Week and Pay

In SPSS, we go to Analyze > Quality Control > Control Charts > There will be a pop up window for Control Charts > Under Variables Charts select Individuals, Moving Range > Under Data Organization select Cases are units then Click Define (Figure 17.10)

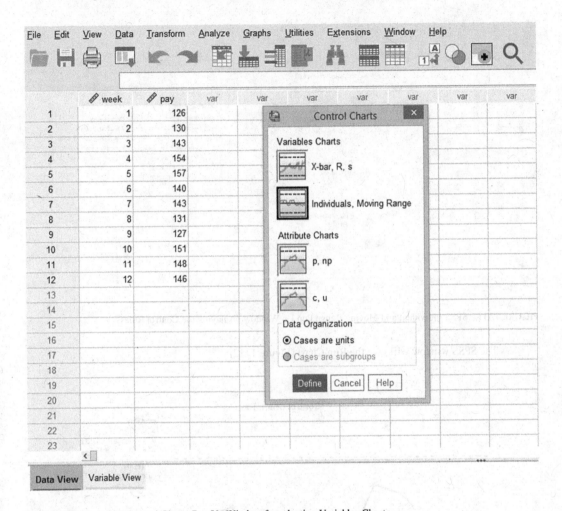

FIGURE 17.10 SPSS Control Charts Pop Up Window for selecting Variables Charts

There will be a pop up window for Individuals and Moving Range > Select Pay for Process Measurement, Week for Subgroups Labeled by and Individuals for Charts > Click Control Rules and Select all control rules > Click Continue then OK (Figure 17.11).

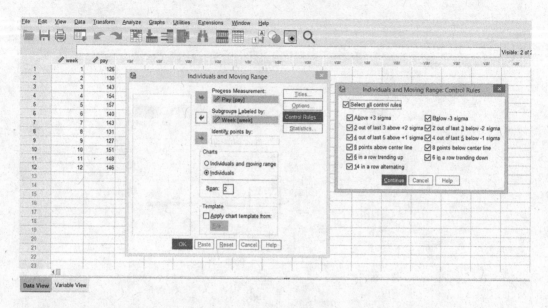

FIGURE 17.11   SPSS Individual and Moving Range Pop Up Window displaying the Control Rules

The SPSS window will give the Run Chart (Figure 17.12)

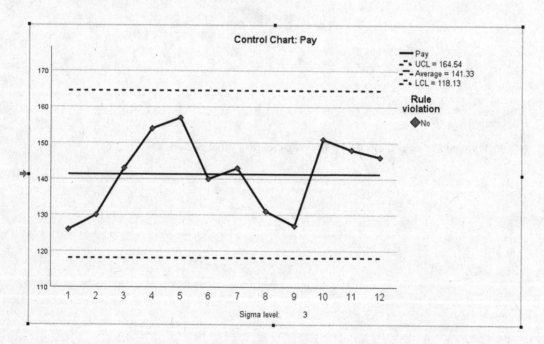

FIGURE 17.12   Run Control Chart

Interpretation: The above charts shows that none of the data points were out of control process.

17.3.1.3 X–MR Charts (I–MR, Individual Moving Range)

An Individual moving range (I-MR) chart is used when data are continuous and not collected in subgroups, i.e. when we collect the single observation at a time. An I-MR chart provides process variation over time in graphical method. The I-MR chart also known as X-MR chart.

Constructing X-MR Charts

- List all of your measurements
- Calculate the moving range by absolute difference between each measure by subtracting one from the other in sequential order.
- From the above example, if you have measures of 4, 6, 3 and 5, you will then get the following differences:
  - $(4–6) = 2$
  - $6–3 = 3$
  - $(3–5) = 2$
- Calculate the mean of the samples.
  - In our example the mean is $4 + 6 + 3 + 5 = 18$. $18/4 = 4.5$
- Calculate the mean of the individual moving ranges. This will act as the control limit – plot this horizontally on the graph.
  - $2 + 3 + 2 = 7$. $7/3 = 2.333$
- Calculate the UCL & LCL.
  - UCL = Sample mean $+ 3 \times$ MR mean/d2
  - LCL = Sample mean $- 3 \times$ MR mean/d2
  - d2 is obtained from the table of control chart constants
  - The 3 refers to 3 standard deviations.
  - UCL in our example would $= 4.5 + (3 \times 2.333/d2)$
  - LCL in our example would $= 4.5 – (3 \times 2.333/d2)$
- Plot UCL and LCL

In SPSS, we go to Analyze > Quality Control > Control Charts > There will be a pop up window for Control Charts > Under Variables Charts select Individuals, Moving Range and Click Define > Select the outcome variable for Process Measurement and time variable for Subgroups Labeled by > Click Control Rules and Select all control rules> Click Continue then OK. The SPSS window will give the Run Chart

Example: Generate the I-MR chart for the following data.

Week	Pay
1	126
2	130
3	143
4	154
5	157
6	140
7	143
8	131
9	127
10	151
11	148
12	146

Create the variables in the Variable View Window

	Name	Type	Width	Decimals	Label	Values	Missing	Columns	Align	Measure	Role
1	week	Numeric	8	0	Week	None	None	8	Right	Scale	Input
2	pay	Numeric	8	0	Pay	None	None	8	Right	Scale	Input
3											
4											
5											
6											
7											
8											
9											
10											
11											
12											
13											
14											
15											
16											
17											
18											
19											
20											
21											
22											
23											
24											

Data View   Variable View

SPSS Variable View Window for Week and Pay

Key Data in Data View

	week	pay	var	var	var	var	var	var	var
1	1	126							
2	2	130							
3	3	143							
4	4	154							
5	5	157							
6	6	140							
7	7	143							
8	8	131							
9	9	127							
10	10	151							
11	11	148							
12	12	146							
13									
14									
15									
16									
17									
18									
19									
20									
21									
22									
23									

Data View Variable View

SPSS Data View Window for Week and Pay

In SPSS, we go to Analyze > Quality Control > Control Charts > There will be a pop up window for Control Charts > Under Variables Charts select Individuals, Moving Range and Click Define (Figure 17.13)

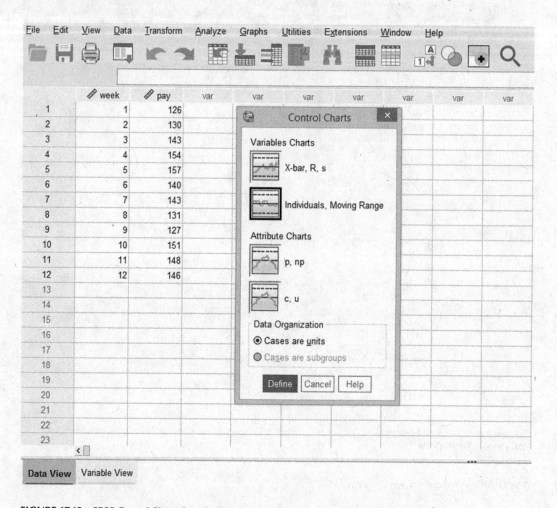

FIGURE 17.13   SPSS Control Charts Pop Up Window for selecting Variables Charts

Select Pay for Process Measurement, Week for Subgroups Labeled by and for Charts select
Individuals and moving range > Click Control Rules and Select all control rules> Click Continue
then OK (Figure 17.14).

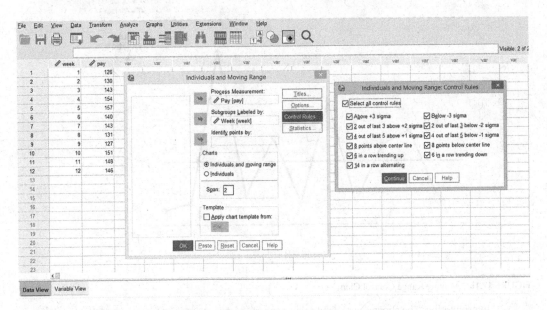

FIGURE 17.14 SPSS Individual and Moving Range Pop Up Window displaying the Control Rules

The SPSS window will give the I-MR chart

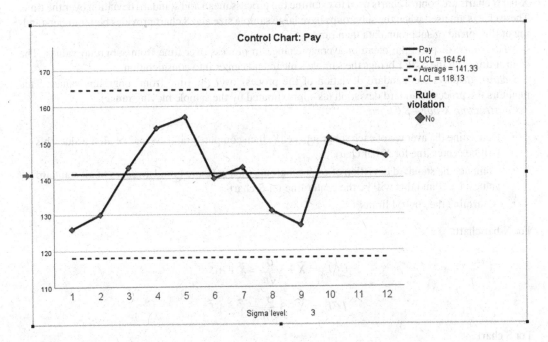

FIGURE 17.15 Individual Control Chart

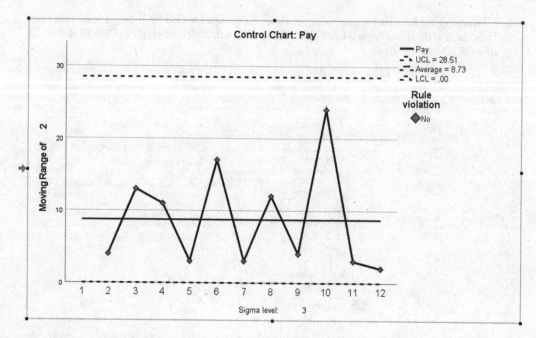

FIGURE 17.16   Moving Range Control Chart

Interpretation: The above charts shows that none of the data points were out of control process.

17.3.1.4 X-Bar – S Charts

X-bar S charts are control charts used to examine the process mean and standard deviation over the time. These charts are used when the subgroups have large sample size and S chart provides better understanding of the spread of subgroup data than range.

X-bar chart: displays the mean or average change in process over time from subgroup values. The control limits on the X-Bar brings the sample's mean and center into consideration.

S-chart: displays the standard deviation of the process over the time from subgroups values. This monitors the process standard deviation (as approximated by the sample moving range).

Constructing X-bar – S Charts

- Determine the average of each subgroup, i.e. $\bar{X}$, then compute grand average of all $\bar{X}$ value, this will be center line for X-bar chart
- Compute the standard deviation of each subgroup, then measure grand averages of all standard values, i.e. $\bar{S}$ and this will be the center line for S chart
- Determine the control limits

For X-bar chart:

$$UCL_{\bar{X}} = \bar{\bar{X}} + 3\frac{\hat{\sigma}}{\sqrt{n}} = \bar{\bar{X}} + A_3\bar{S}$$

$$LCL_{\bar{X}} = \bar{\bar{X}} - 3\frac{\hat{\sigma}}{\sqrt{n}} = \bar{\bar{X}} - A_3\bar{S}$$

For S chart

$$UCL_{\bar{S}} = B_4\bar{S}$$

$$LCL_{\bar{S}} = B_4\bar{S}$$

where
- X is the individual value (data)
- n is the sample size
- $\bar{X}$ is the average of reading in a sample
- S is the standard deviation
- $\bar{S}$ is the average of all the standard deviation
- B_4 is the appropriate constant from the table of control chart constants
- A_3 is a factor (based on subgroup size) from the table of control chart constants
- UCL is Upper control limit
- LCL is Lower control limit

In SPSS we go to We go Analyze > Quality Control > Control Charts > there will be a pop up window for Control Charts > Under Variable Charts select X-bar, R, s. Under Data Organization select Cases are subgroups then click Define > there will be a pop window for X-bar, R, s: cases are subgroups > Click Options > select number of sigmas = 3 and Minimum subgroup sample size (number of subgroups) > Under Charts select X-bar using standard deviation > Click Control Rules then Select all control rules then click Continue then OK > the SPSS Output will give X-bar chart and R chart

Example

Generate X-bar and S control Chart for the following data.

	Measured Values			
Sample	x_1	x_2	x_3	x_4
1	50	48	24	43
2	32	28	51	31
3	36	16	26	26
4	21	46	23	44
5	29	24	44	49
6	26	20	22	23
7	24	52	24	28
8	19	22	26	12
9	8	28	27	24
10	32	12	18	50

We create variables in Variable View

	Name	Type	Width	Decimals	Label	Values	Missing	Columns	Align	Measure	Role
1	sample	Numeric	8	0	Sample	None	None	8	Right	Scale	Input
2	x1	Numeric	8	0	x1	None	None	8	Right	Scale	Input
3	x2	Numeric	8	0	x2	None	None	8	Right	Scale	Input
4	x3	Numeric	8	0	x3	None	None	8	Right	Scale	Input
5	x4	Numeric	8	0	x4	None	None	8	Right	Scale	Input

SPSS Variable View Window for Four Variables

Key the data in Data View

	sample	x1	x2	x3	x4	var	var	var	var
1	1	50	48	24	43				
2	2	32	28	51	31				
3	3	36	16	26	26				
4	4	21	46	23	44				
5	5	29	24	44	49				
6	6	26	20	22	23				
7	7	24	52	24	28				
8	8	19	22	26	12				
9	9	8	28	27	24				
10	10	32	12	18	50				
11									
12									
13									
14									
15									
16									
17									
18									
19									
20									
21									
22									
23									

Data View   Variable View

SPSS Variable View Window for Four Variables

We go Analyze > Quality Control > Control Charts > There will be a pop up window for Control Charts > Under Variable Charts select X-bar, R, s. Under Data Organization select Cases are subgroups then click Define

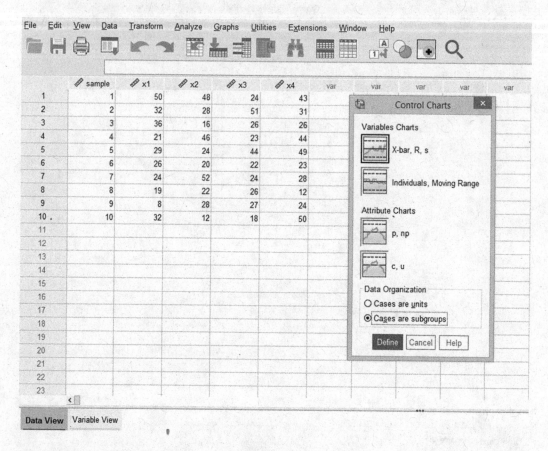

SPSS Control Charts Pop Up Window for selecting Variables Charts

There will be a pop window for X-bar, R, s: cases are subgroups > Under Samples select x1, x2, x3, and x4 variables > Click Options > select number of sigmas = 3 and Minimum subgroup sample size = 4 > Under Charts select X-bar using standard deviation > Click Control Rules then Select all control rules then click Continue then OK (Figure 17.17)

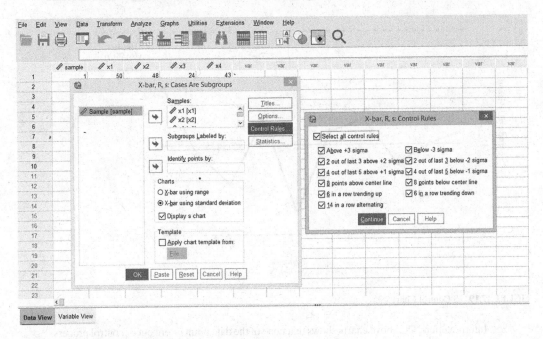

FIGURE 17.17   SPSS X-bar, R, s Pop Up Window displaying the Control Rules

The SPSS Output will give X-bar chart and s chart
X-bar chart

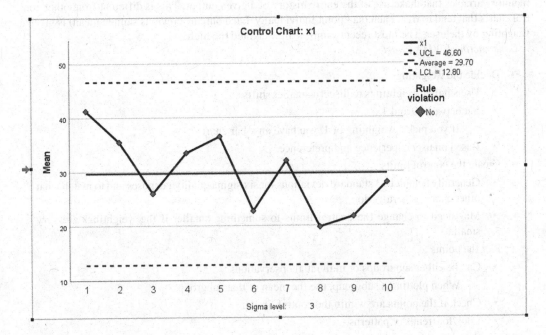

FIGURE 17.18   X-bar Control Chart

s Chart

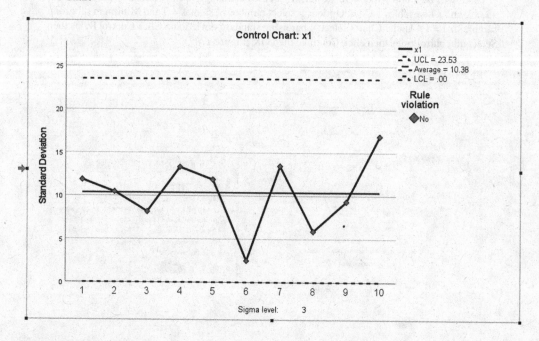

FIGURE 17.19   S Control Chart

Interpretation: The above charts shows that none of the data points were out of control process.

17.3.1.5 EWMA Chart

The Exponentially Weighted Moving Average (EWMA) chart is used in statistical process control to monitor variables that make use of the entire history of a given output. This is different from other control charts that tend to treat each data point individually. Each output (previous sample mean) is given a weighting by the user. The most recent samples are weighted the highest.

Constructing EWMA Chart

- Decide the weightings
  - Use smaller weightings to discern smaller shifts.
  - Set between 0 and 1.
    - If you pick a weighting of 1, you have an x-bar chart.
  - Based on user experience and preference.
- Create the control limits
  - Generally default to 3 standard deviations for six sigma quality purposes and to match what other charts generally do.
  - May need to change the control limits to something smaller if the weightings are very small.
- Plot the points
  - Can be either subgroups or individual observations.
    - When plotting a subgroup, use the mean of that subgroup.
  - Check if the points are within the control limits.
  - Look for trends or patterns.

EWMA chart is not built in SPSS.

17.3.2 Discrete/Attributes Control Charts

Discrete/attributes control charts are used when the output is a decision or counting on characteristics of a product. It is based on the visual inspection on the characteristics like good or bad, fail or pass, accept or reject.

17.3.2.1 p Charts

p chart is used when the data are the fraction defective of some set of process output. It can also be expressed as percentage of defective. The points plotted on p-chart are the fraction of nonconfirming units or defective pieces found in n samples. p chart is also known as the control chart for proportions and it uses binomial distribution to measure the proportion of defectives or nonconfirming units in a sample.

Constructing p Charts

- Determine the subgroup size. The subgroup size must be large enough for the p chart; otherwise, control limits may not be accurate when estimated from the data.
- Calculate each subgroups nonconformities rate = np/n
- Compute $\bar{p}$ = total number of defectives/total number of samples = $\Sigma np/\Sigma n$
- Calculate UCL and low control limit (LCL). If LCL is negative, then use 0. Since the sample sizes are unequal, the control limits vary from sample interval to sample interval.

$$\bar{n} = \frac{\Sigma n}{k}$$

$$\bar{p} = \frac{\Sigma np}{\Sigma n}$$

$$UCL_p = \bar{p} + 3\sqrt{\frac{\bar{p}(1-\bar{p})}{n}}$$

$$LCL_p = \bar{p} - 3\sqrt{\frac{\bar{p}(1-\bar{p})}{n}}$$

where
np = number of defectives in the sample
k = number of lots
n = sample size

- Plot the graph with proportion on the y-axis, lots on the x-axis: Draw centerline, UCL and LCL.
- Interpret the data to determine whether the process is in control.

In SPSS, we go to Analyze > Quality Control > Control Charts > There will be a pop up window for Control Charts > Select p, np under Attribute Charts > Click Cases are subgroups then Define > There will be a pop up window for p, np: Cases Are Subgroups > Put number of defective under Number Nonconforming, Lot under Subgroups Labelled by: sample size under Variables and select p under Chart > Click Control Rules then Select all control rules > Click Continue then OK > There will be SPSS output for p chart

Example: Generate a p chart for the following data.

Lot	Sample Size	Number Defective in the Sample
1	1350	26
2	1050	30
3	1000	13
4	800	34
5	850	17
6	1250	20
7	1200	23
8	1150	11
9	950	27
10	1560	31
11	1230	18
12	1100	26

Create the variables in the Variable View Window (Figure 17.20)

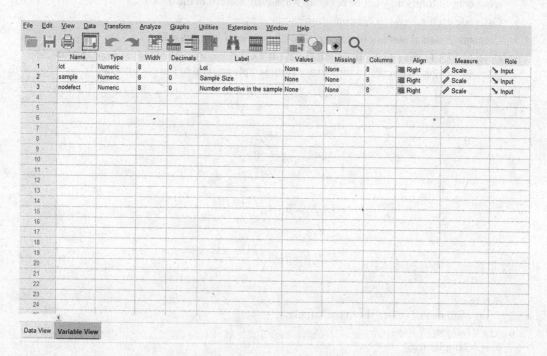

FIGURE 17.20   SPSS Variable View for Lot, Sample Size and Number Defective in the Sample

Key Data in Data View (Figure 17.21)

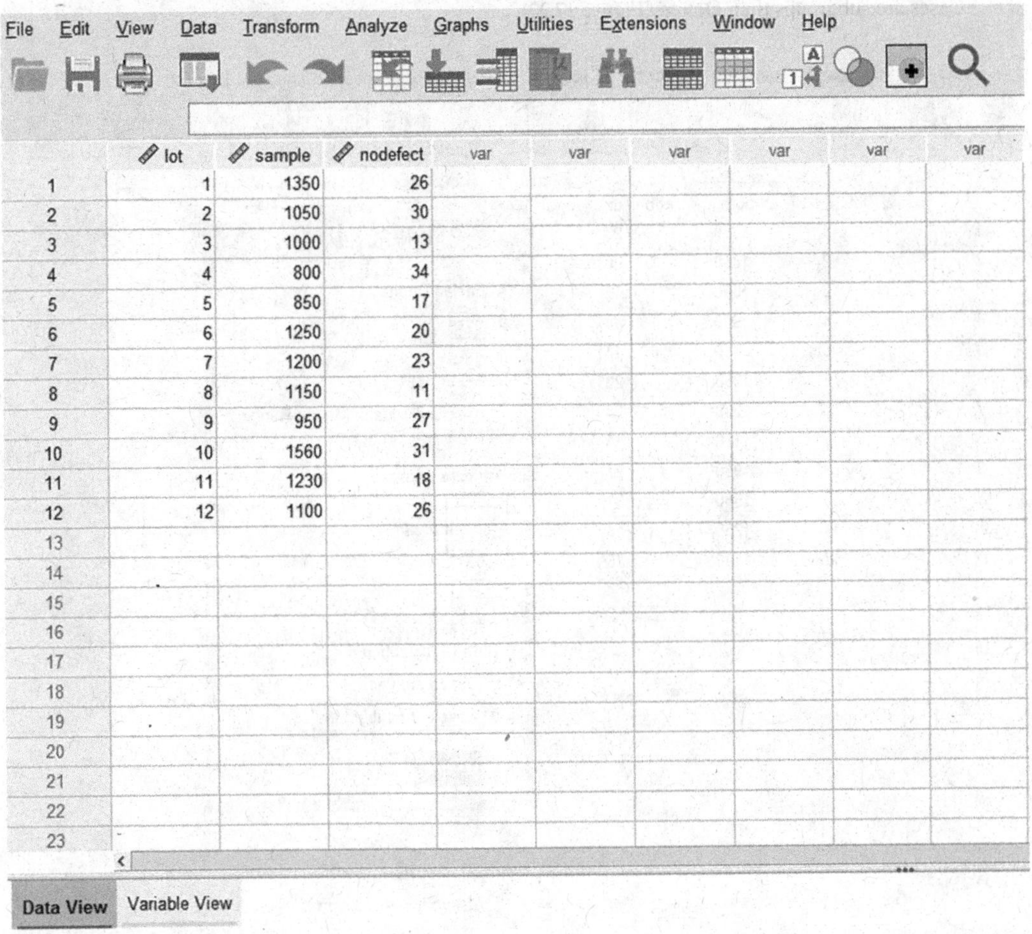

	lot	sample	nodefect	var	var	var	var	var	var
1	1	1350	26						
2	2	1050	30						
3	3	1000	13						
4	4	800	34						
5	5	850	17						
6	6	1250	20						
7	7	1200	23						
8	8	1150	11						
9	9	950	27						
10	10	1560	31						
11	11	1230	18						
12	12	1100	26						
13									
14									
15									
16									
17									
18									
19									
20									
21									
22									
23									

Data View Variable View

FIGURE 17.21 SPSS Data View for Lot, Sample Size and Number Defective in the Sample

In SPSS, we go to Analyze > Quality Control > Control Charts > There will be a pop up window for Control Charts > Select p, np under Attribute Charts > Under Data Organization click Cases are subgroups then Define (Figure 17.22)

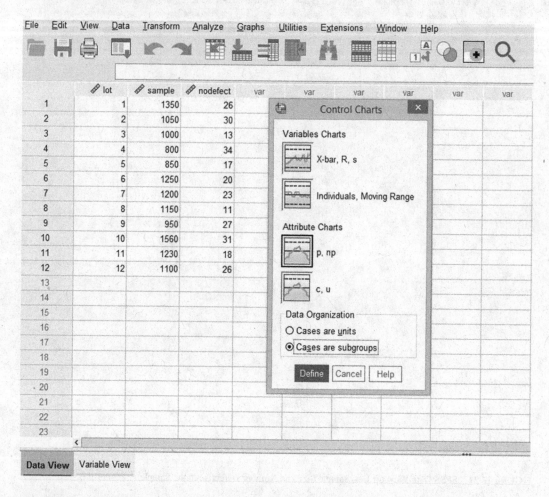

FIGURE 17.22   SPSS Control Charts Pop Up Window for selecting Attribute Charts

There will be a pop up window for p, np: Cases Are Subgroups > Put number of defective under Number Nonconforming, Lot under Subgroups Labelled by: sample size under Variables and select p under Chart > Click Control Rules then Select all control rules > Click Continue then OK (Figure 17.23)

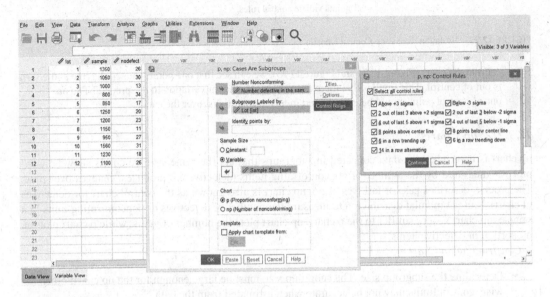

FIGURE 17.23 SPSS p, np Pop Up Window displaying the Control Rules

There will be SPSS output for p chart (Figure 17.24).

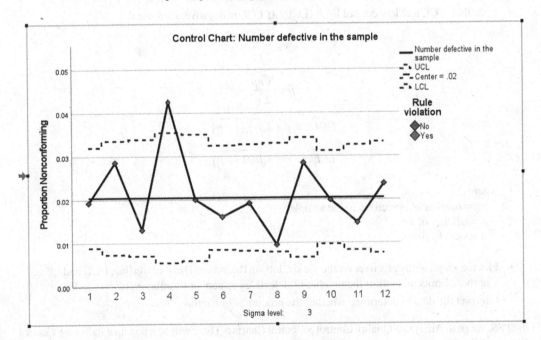

FIGURE 17.24 p Control Chart

Rule Violations	
Lot	Violations for Points
4	Greater than +3 sigma
1 points violate control rules.	

FIGURE 17.25   Violation Rule for p Control Chart

From the chart (Figure 17.24 and Figure 17.25), we can see that one point violated the rule and is out of control. This implies that there are some non-random variation in the process. The one out-of-control point should be investigated in an attempt to discover the cause.

17.3.2.2 np Charts

np chart is used when the data collected in subgroups that are the same size. np charts show how the process, measured by the number of nonconforming items (defectives) it produces, changes over time. The process describes pass or fail, yes or no. np chart is also known as the control chart for defectives (d-chart). It uses binomial distribution to measure the number of defectives or nonconfirming units in a sample. np chart is very similar to the p chart. np chart plots the number of items, while p chart plot the proportion of defective items.

Constructing np Chart

- Determine the subgroup size. The subgroup size must be large enough for the np chart; otherwise, control limits may not be accurate when estimated from the data
- Count the number of defectives in each sample
- Compute $\bar{p}$ = total number of defectives/total number of samples = $\Sigma np/\Sigma n$
- Calculate centerline $n\bar{p}$ = total number of defectives/no of lots = $\Sigma np/k$
- Calculate UCL and low control limit (LCL), If LCL is negative, then use 0

$$n\bar{p} = \frac{\Sigma np}{k}$$

$$\bar{p} = \frac{\Sigma np}{\Sigma n}$$

$$UCL_{np} = \bar{p} + 3\sqrt{n\bar{p}(1-\bar{p})}$$

$$LCL_{np} = \bar{p} - 3\sqrt{n\bar{p}(1-\bar{p})}$$

where
   np = number of defectives in the sample
   k = number of lots
   n = sample size

- Plot the graph with defectives on the y-axis, lots on the x-axis: Draw centerline, UCL and LCL. Use these limits to monitor the number of defectives or nonconforming going forward.
- Interpret the data to determine whether the process is in control.

In SPSS, we go to Analyze > Quality Control > Control Charts > There will be a pop up window for Control Charts > Select p, np under Attribute Charts > Click Cases are subgroups then Define > There will be a pop up window for p, np: Cases Are Subgroups > Put number of defective under Number nonconforming, Lot under Subgroups Labeled by, put sample size under Variables, and Chart select np > Click Control Rules then Select all control rules > Click Continue then OK > There will be SPSS output for np chart.

Example: Generate np chart for the following data.

Lot	Sample Size	Number Defective in the Sample
1	1150	26
2	1150	30
3	1150	13
4	1150	34
5	1150	17
6	1150	20
7	1150	23
8	1150	11
9	1150	27
10	1150	31
11	1150	18
12	1150	26

Create the variables in the Variable View Window (Figure 17.26)

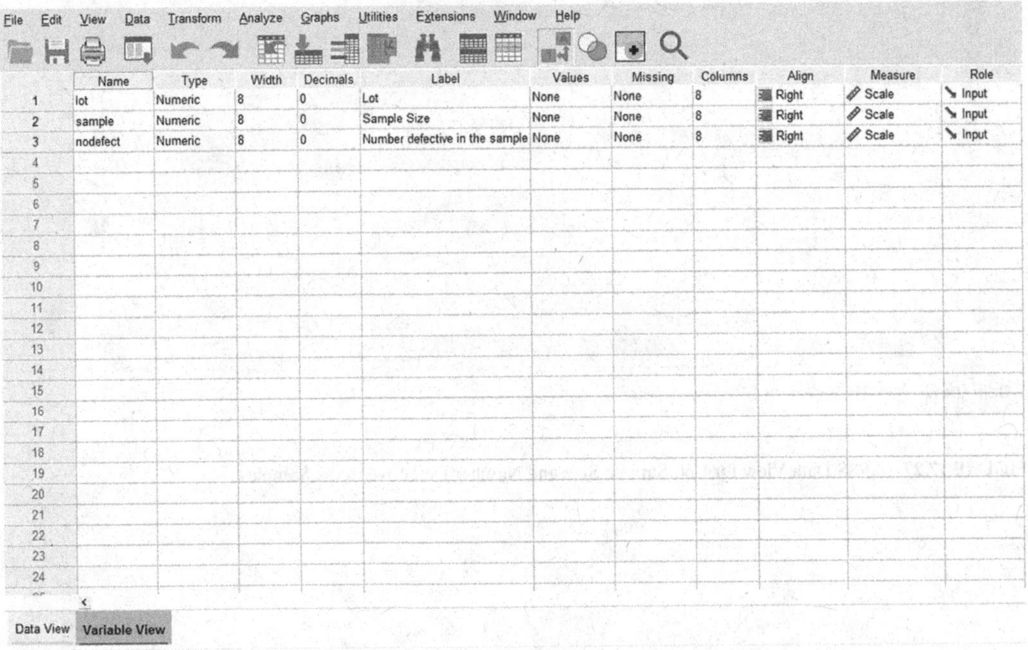
File Edit View Data Transform Analyze Graphs Utilities Extensions Window Help
Name Type Width Decimals Label Values Missing Columns Align Measure Role
1 lot Numeric 8 0 Lot None None 8 Right Scale Input
2 sample Numeric 8 0 Sample Size None None 8 Right Scale Input
3 nodefect Numeric 8 0 Number defective in the sample None None 8 Right Scale Input
Data View
Variable View

FIGURE 17.26 SPSS Variable View for Lot, Sample Size and Number Defective in the Sample

Key Data in Data View (Figure 17.27)

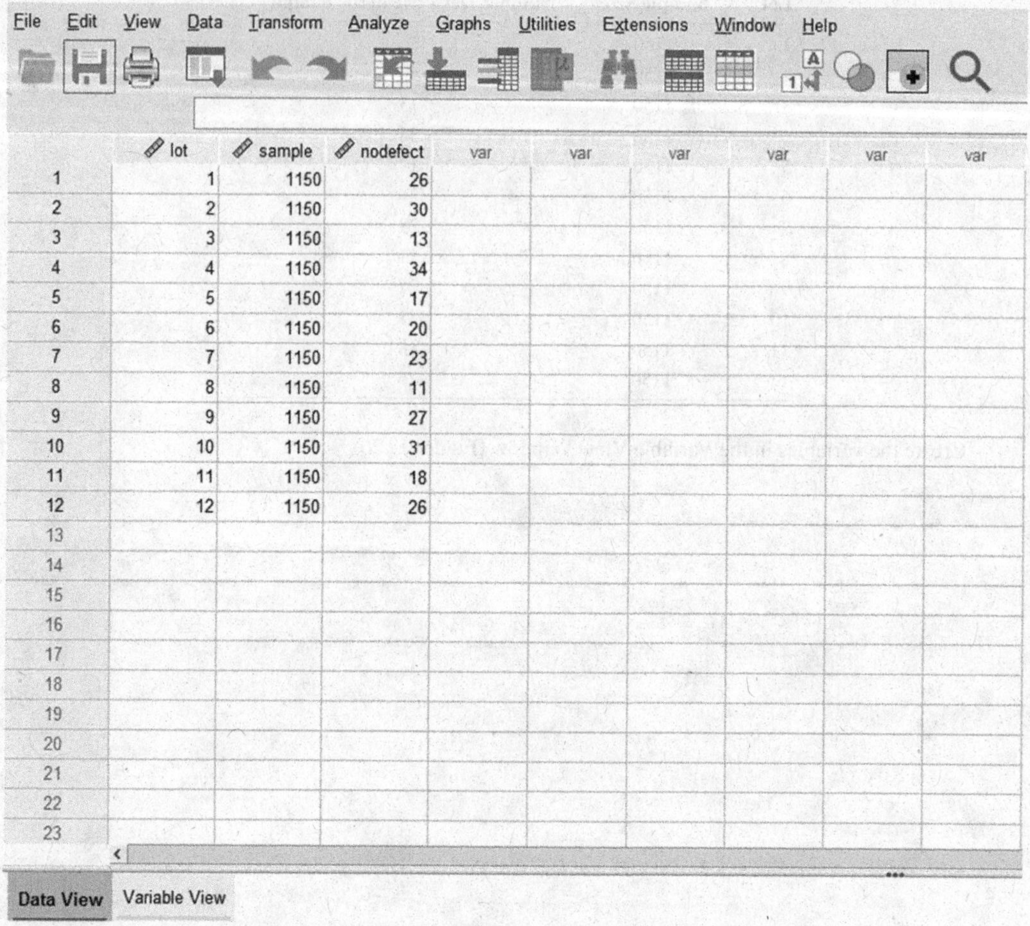

	lot	sample	nodefect	var	var	var	var	var	var
1	1	1150	26						
2	2	1150	30						
3	3	1150	13						
4	4	1150	34						
5	5	1150	17						
6	6	1150	20						
7	7	1150	23						
8	8	1150	11						
9	9	1150	27						
10	10	1150	31						
11	11	1150	18						
12	12	1150	26						
13									
14									
15									
16									
17									
18									
19									
20									
21									
22									
23									

Data View   Variable View

FIGURE 17.27   SPSS Data View for Lot, Sample Size and Number Defective in the Sample

In SPSS, we go to Analyze > Quality Control > Control Charts > There will be a pop up window for Control Charts > Select p, np under Attribute Charts > Click Cases are subgroups then Define (Figure 17.28)

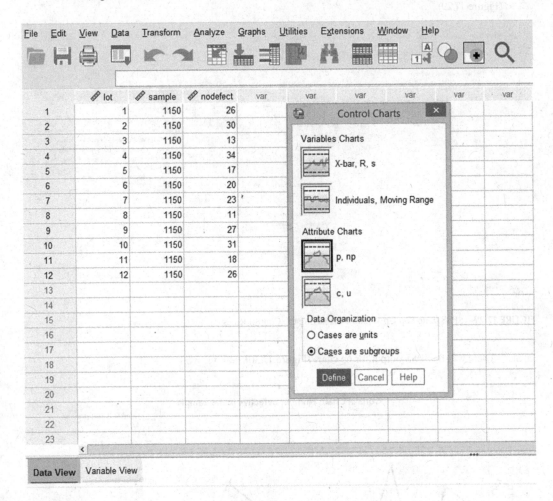

FIGURE 17.28   SPSS Control Charts Pop Up Window for selecting Attribute Charts

There will be a pop up window for p, np: Cases Are Subgroups > Put number of defective under Number nonconforming, Lot under Subgroups Labeled by, put sample size under Variables, and Chart select np > Click Control Rules then Select all control rules > Click Continue then OK (Figure 17.29)

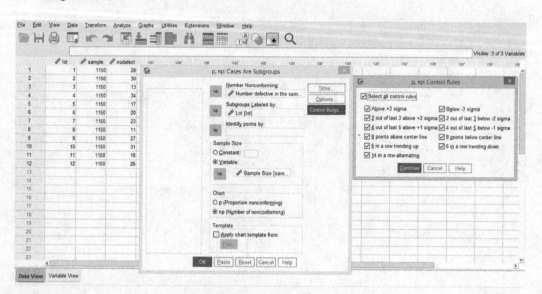

FIGURE 17.29   SPSS p, np Pop Up Window displaying the Control Rules

There will be SPSS output for np chart (Figure 17.30).

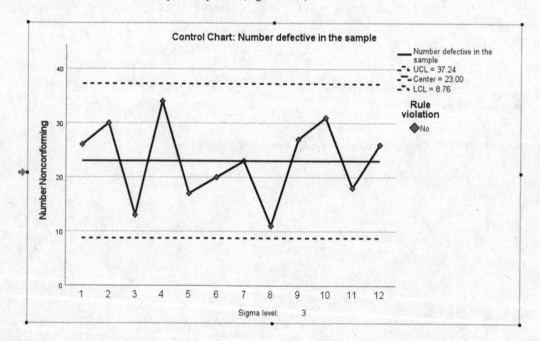

FIGURE 17.30   np Control Chart

Interpretation: The above chart (Figure 17.30) shows that none of the data points were out of control process.

17.3.2.3 c Charts

c chart is used when the data are concerned with the number of defects in a product. c chart is also known as the control chart for defects (counting of the number of defects). It is generally used to monitor the number of defects in constant size units. c chart tracks the total number of defects in each unit and it assumes the underlying data approximate the Poisson distribution.

Constructing c Charts

- Determine the subgroup size. The subgroup size must be large enough for the c chart; otherwise, control limits may not be accurate when estimated from the data.
- Count the number of defects in each sample
- Compute centreline $\bar{c}$ = total number of defects/number of samples = $\Sigma c/k$
- Calculate UCL and low control limit (LCL). If LCL is negative, then consider it as 0

$$\bar{c} = \frac{\Sigma c}{k}$$

$$UCL_c = \bar{c} + 3\sqrt{\bar{c}}$$

$$LCL_c = \bar{c} - 3\sqrt{\bar{c}})$$

where
  c = number of defects
  k = number of samples

- Plot the graph with number of defects on the y-axis, lots on the x-axis: Draw centerline, UCL and LCL. Use these limits to monitor the number of defects going forward.
- Finally, interpret the data to determine whether the process is in control.

In SPSS, we go to Analyze > Quality Control > Control Charts > There will be a pop up window for Control Charts > Select c, u under Attribute Charts > Click Cases are subgroups then Define > There will be a pop up window for c, u: Cases Are Subgroups > Put number of defective under Number nonconforming, sample size under Subgroups Labeled by, Lot under Identify points by, for Sample size put constant to be the sample size for defective and Chart select p > Click Control Rules then Select all control rules > Click Continue then OK > There will be SPSS output for p chart

Example: Generate a c chart for the following data

Lot	Sample Size	Number Defective in the Sample
1	650	26
2	650	30
3	650	13
4	650	34
5	650	17
6	650	20
7	650	23
8	650	11
9	650	27
10	650	31
11	650	18
12	650	26

Create the variables in the Variable View Window (Figure 17.31)

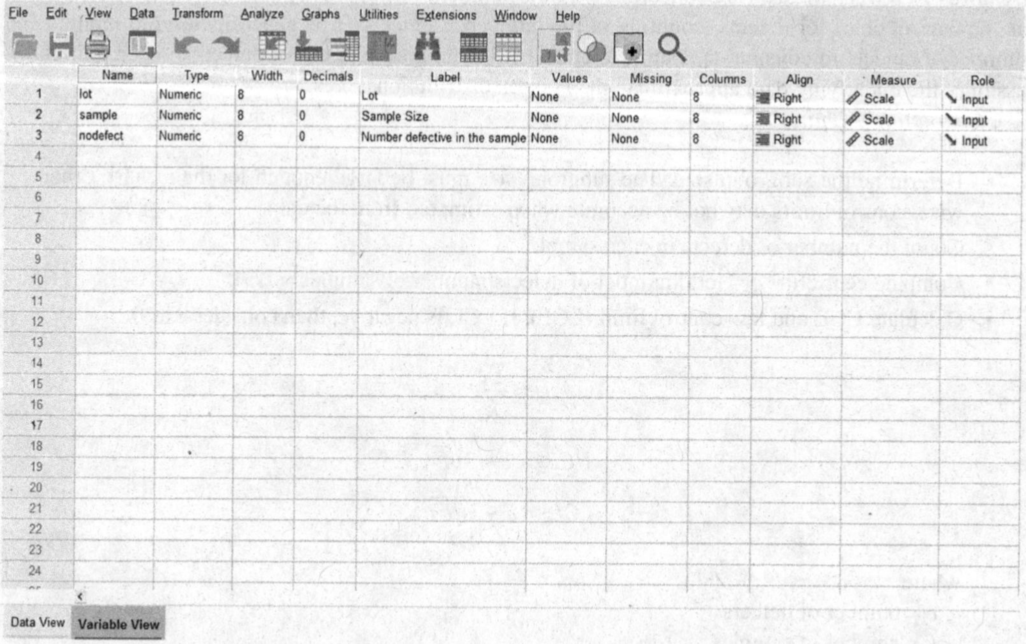

	Name	Type	Width	Decimals	Label	Values	Missing	Columns	Align	Measure	Role
1	lot	Numeric	8	0	Lot	None	None	8	Right	Scale	Input
2	sample	Numeric	8	0	Sample Size	None	None	8	Right	Scale	Input
3	nodefect	Numeric	8	0	Number defective in the sample	None	None	8	Right	Scale	Input

FIGURE 17.31   SPSS Variable View for Lot, Sample Size and Number Defective in the Sample

Key Data in Data View (Figure 17.32)

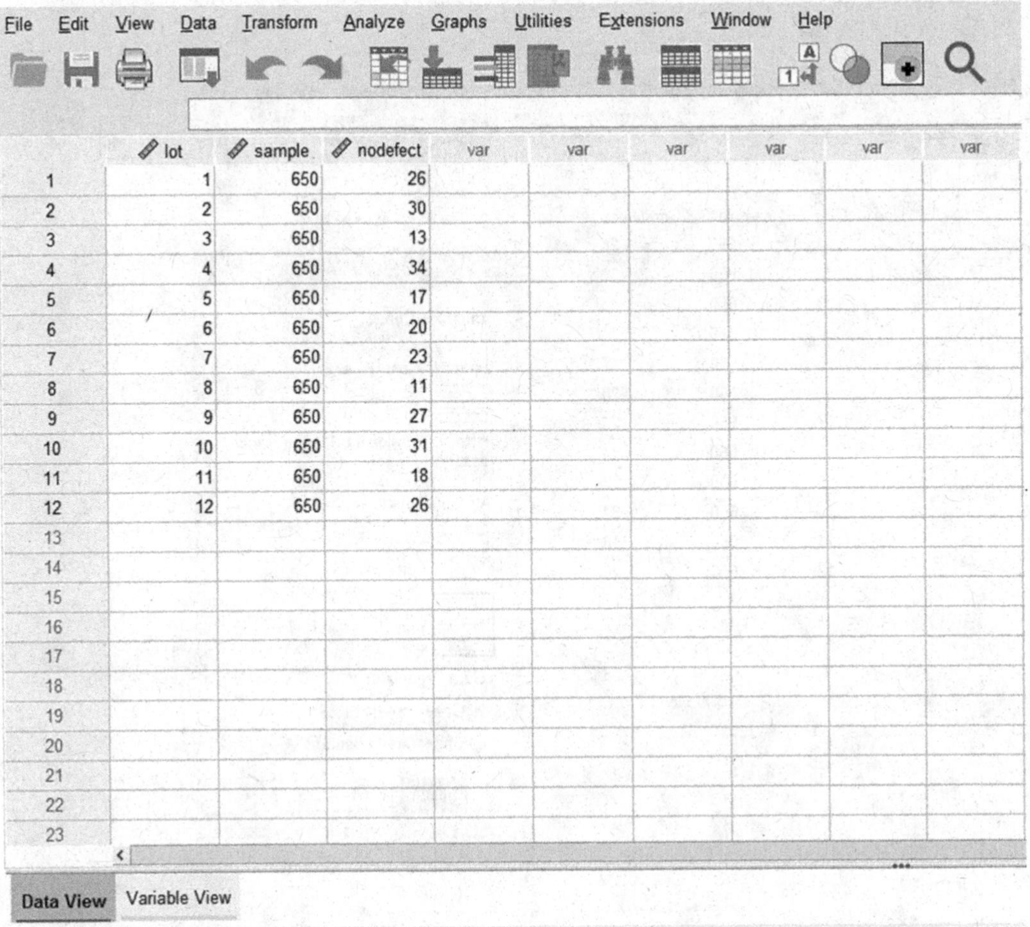

	lot	sample	nodefect	var	var	var	var	var	var
1	1	650	26						
2	2	650	30						
3	3	650	13						
4	4	650	34						
5	5	650	17						
6	6	650	20						
7	7	650	23						
8	8	650	11						
9	9	650	27						
10	10	650	31						
11	11	650	18						
12	12	650	26						
13									
14									
15									
16									
17									
18									
19									
20									
21									
22									
23									

File Edit View Data Transform Analyze Graphs Utilities Extensions Window Help

Data View   Variable View

FIGURE 17.32   SPSS Data View for Lot, Sample Size and Number Defective in the Sample

In SPSS, we go to Analyze > Quality Control > Control Charts > There will be a pop up window for Control Charts > Select c, u under Attribute Charts > Click Cases are units then Define (Figure 17.33)

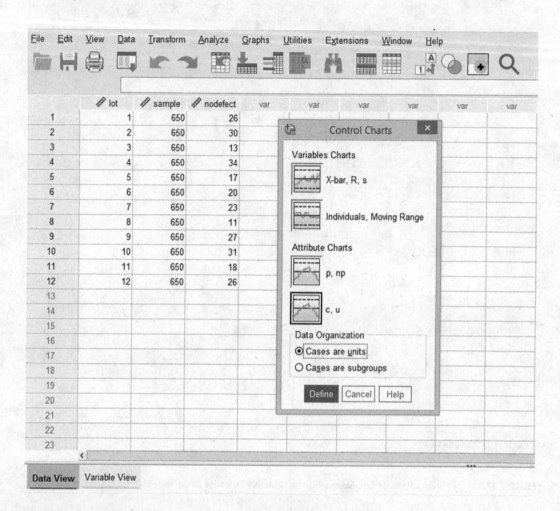

FIGURE 17.33 SPSS Control Charts Pop Up Window for selecting Attribute Charts

There will be a pop up window for c, u: Cases Are Units > Put number of defective under Characteristic, Lot under Subgroups Defined by, sample size under Identify points by, and select c under Chart > Click Control Rules then Select all control rules > Click Continue then OK (Figure 17.34)

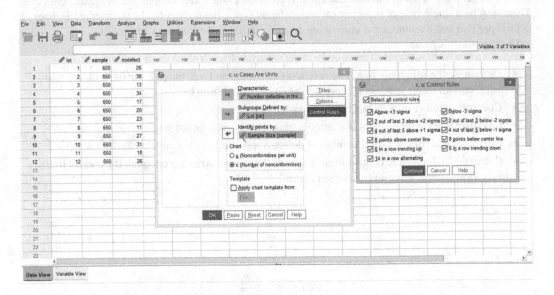

FIGURE 17.34   SPSS c, u Pop Up Window displaying the Control Rules

There will be SPSS output for c chart (Figure 17.35).

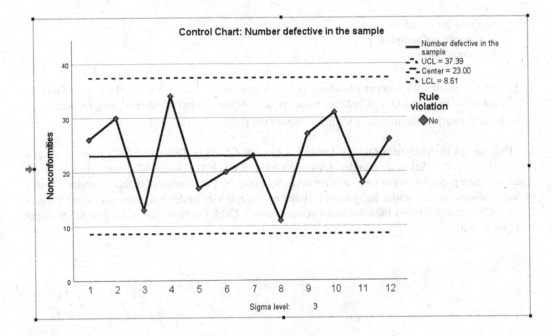

FIGURE 17.35   c Control Chart

Interpretation: The above chart (Figure 17.35) shows that none of the data points were out of control process.

17.3.2.4 u Chart

A u chart is an attribute control chart that displays how the frequency of defects, or nonconformities, is changing over time for a process or system. u chart is also known as the control chart for defects per unit chart. It is generally used to monitor the count type of data where the sample size is greater than one. There may be a single type of defect or several different types, but u chart tracks the average number of defects per unit and it assumes the underlying data approximate the Poisson distribution.

    Constructing u Charts

- Determine the subgroup size. The subgroup size must be large enough for the c chart; otherwise, control limits may not be accurate when estimated from the data.
- Count the number of defects in each inspected unit
- Calculate u value for each lot u = number of defects in each lot/lot size
- Compute centerline $\bar{u}$ = total number of defects/total number of samples = $\Sigma c/\Sigma n$
- Calculate UCL and low control limit (LCL). If the sample sizes are unequal, the control limits vary from sample interval to sample interval.

$$u = \frac{c}{n}$$

$$\bar{u} = \frac{\Sigma c}{\Sigma n}$$

$$UCL_u = \bar{u} + 3\frac{\sqrt{\bar{u}}}{n_i}$$

$$LCL_u = \bar{u} - 3\frac{\sqrt{\bar{u}}}{n_i}$$

    where
        c = number of defects
        k = number of samples
        n = sample size

- Plot the graph with number of defects per each unit on the y-axis, lots on the x-axis: Draw centerline, UCL and LCL. Use these limits to monitor the number of defects going forward.
- Finally, interpret the data to determine whether the process is in control.

In SPSS, we go to Analyze > Quality Control > Control Charts > There will be a pop up window for Control Charts > Select c, u under Attribute Charts > Click Cases are Subgroups then Define > There will be a pop up window for c, u: Cases Are Subgroups > Put number of defective under Number of Nonconformities, Lot under Subgroups Defined by, sample size under Variables and select u under Chart > Click Control Rules then Select all control rules > Click Continue then OK. The SPSS output will give u chart

Example: Generate a u chart for the following data.

Lot	Sample Size	Number Defective in the Sample
1	150	26
2	80	30
3	130	13
4	125	34
5	140	17
6	90	20
7	100	23
8	120	11
9	90	27
10	85	31
11	115	18
12	120	26

Create the variables in the Variable View Window (Figure 17.36)

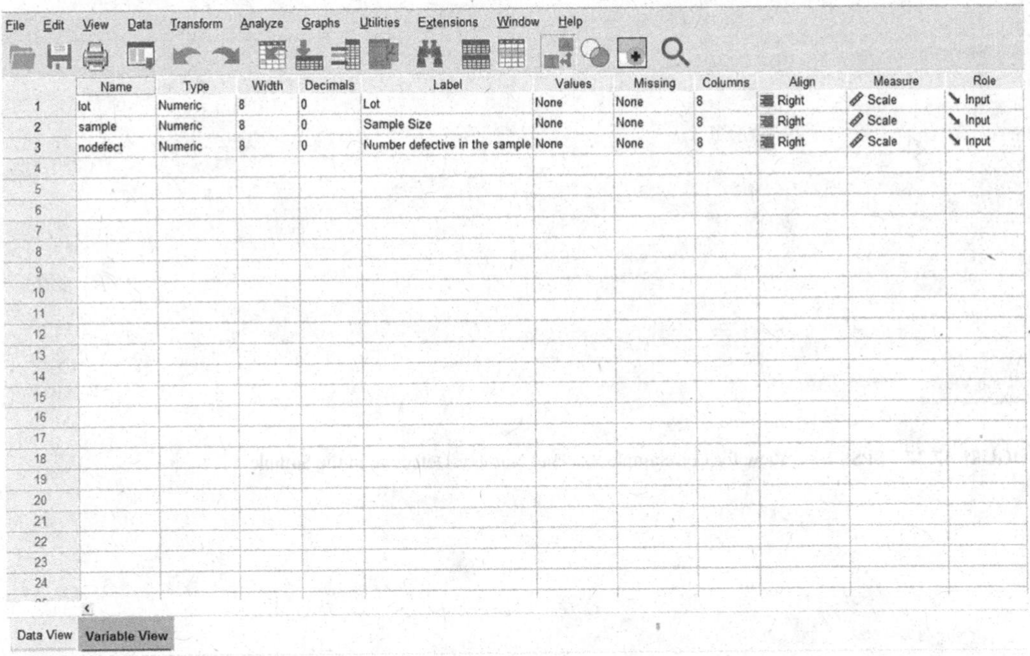
File Edit View Data Transform Analyze Graphs Utilities Extensions Window Help
Name Type Width Decimals Label Values Missing Columns Align Measure Role
1 lot Numeric 8 0 Lot None None 8 Right Scale Input
2 sample Numeric 8 0 Sample Size None None 8 Right Scale Input
3 nodefect Numeric 8 0 Number defective in the sample None None 8 Right Scale Input
Data View Variable View

FIGURE 17.36 SPSS Variable View for Lot, Sample Size and Number Defective in the Sample

Key Data in Data View (Figure 17.37)

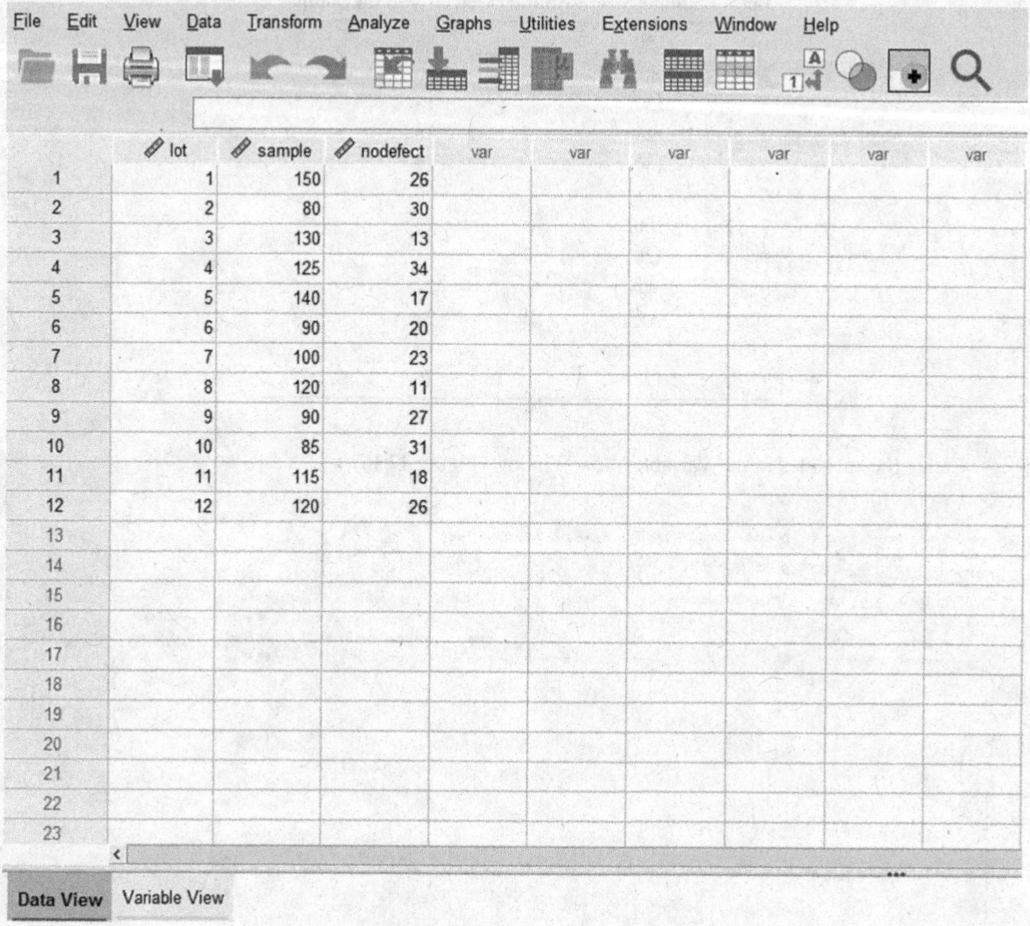

	lot	sample	nodefect	var	var	var	var	var	var
1	1	150	26						
2	2	80	30						
3	3	130	13						
4	4	125	34						
5	5	140	17						
6	6	90	20						
7	7	100	23						
8	8	120	11						
9	9	90	27						
10	10	85	31						
11	11	115	18						
12	12	120	26						
13									
14									
15									
16									
17									
18									
19									
20									
21									
22									
23									

Data View   Variable View

FIGURE 17.37   SPSS Data View for Lot, Sample Size and Number Defective in the Sample

In SPSS, we go to Analyze > Quality Control > Control Charts > There will be a pop up window for Control Charts > Select c, u under Attribute Charts > Click Cases are Subgroups then Define (Figure 17.38)

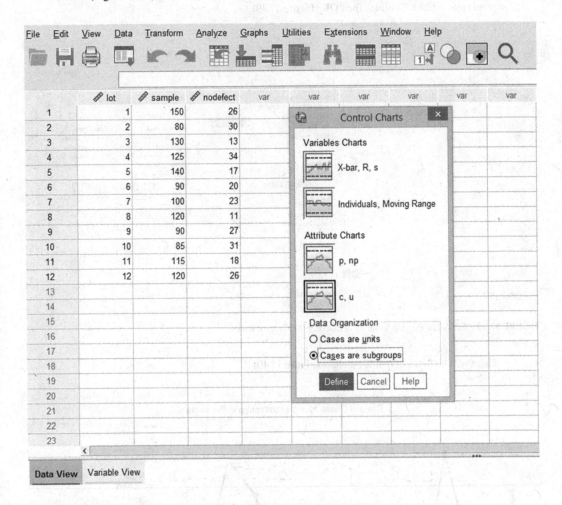

FIGURE 17.38   SPSS Control Charts Pop Up Window for selecting Attribute Charts

There will be a pop up window for c, u: Cases Are Subgroups > Put number of defective under Number of Nonconformities, Lot under Subgroups Defined by, sample size under Identify points by, sample size under Variables and select u under Chart > Click Control Rules then Select all control rules > Click Continue then OK (Figure 17.39)

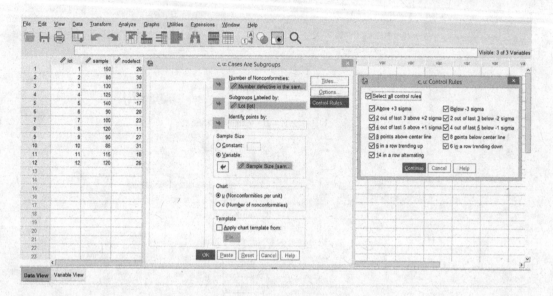

FIGURE 17.39   SPSS c, u Pop Up Window displaying the Control Rules

There will be SPSS output for u chart (Figure 17.40)

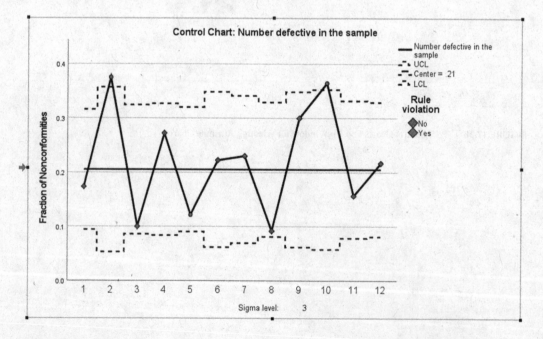

FIGURE 17.40   u Control Chart

Rule Violations	
Lot	**Violations for Points**
2	Greater than +3 sigma
5	2 points out of the last 3 below −2 sigma
10	Greater than +3 sigma
3 points violate control rules.	

FIGURE 17.41   Violation Rule for u Control Chart

We can see that there are three points that violated the control rules and are out of control (Figure 17.40 and Figure 17.41). The three out-of-control points should be investigated in an attempt to discover the cause.

Practice Exercise

1. Consider customer_subset.sav data. Using the appropriate variables and sample size, construct the following listed control charts using SPSS.

   a. $\bar{X}$ and R chart

   b. X and I-MR

   c. $\bar{X}$ and S chart

2. Consider customer_subset.sav data. Using the appropriate variables and sample size, construct the following listed control charts using SPSS.

   a. p chart

   b. np chart

   c. c chart

   d. u chart

Index

Note: Locators in *italics* represent figures and **bold** indicate tables in the text.

Printed in the United States
by Baker & Taylor Publisher Services